Stromlo

Dr Tom Frame is a Visiting Fellow at the ANU's Research School of Astronomy and Astrophysics. He is a professional historian and the author of the bestselling *Where Fate Calls: The HMAS* Voyager *Tragedy, HMAS* Sydney*: Loss and Controversy, Binding Ties: An Experience of Adoption and Reunion in Australia* and *A Church for a Nation: The Anglican Diocese of Canberra and Goulburn*. Born in 1962, Dr Frame grew up in Wollongong and served in the Royal Australian Navy for fourteen years before being ordained as an Anglican priest. He is presently Bishop to the Australian Defence Force and lives in Canberra with his wife, Helen, and their daughters Megan and Kelly.

Dr Don Faulkner is a Visiting Fellow at the ANU's Research School of Astronomy and Astrophysics. Born in Brisbane in 1937, he received his DSc from the University of Queensland and his PhD from the ANU. His main research interest is in the internal workings of stars, and how the physical processes taking place inside them determine how they are born, live their lives and die. Dr Faulkner spent virtually his whole professional career at Mount Stromlo Observatory and retired in 1998 as Associate Director for Education and Outreach. He was President of the Astronomical Society of Australia from 1989 to 1991. He is an enthusiastic exponent of astronomy, and his series of evening lectures, 'Astronomy for Fun', has been running for twenty years at Mount Stromlo. He lives in Canberra with his wife June.

Stromlo

An Australian Observatory

Tom Frame
and
Don Faulkner

Colour images by Mike Bessell

ALLEN&UNWIN

First published in 2003

Allen & Unwin
83 Alexander Street
Crows Nest NSW 2065
Australia
Phone: (61 2) 8425 0100
Fax: (61 2) 9906 2218
Email: info@allenandunwin.com
Web: www.allenandunwin.com

National Library of Australia
Cataloguing-in-Publication entry:

Frame T.R. (Thomas R.).
Stromlo: an Australian observatory.

Includes index.
ISBN 1 86508 659 2.

1. Mount Stromlo Observatory—History.
2. Astronomical observatories—Australian Capital Territory.
I. Faulkner, Donald. II. Title.

522.109947

Set in 11/14pt Fairfield LH Light by Midland Typesetters, Victoria
Printed by Griffin Press, South Australia
10 9 8 7 6 5 4 3 2 1

Foreword

Professor Jeremy R. Mould
Director, Mount Stromlo & Siding Spring Observatories
13 December 1993–30 January 2001

ASTRONOMY IS EXPLORING THE UNIVERSE. For observatories founded in the nineteenth century, much of that exploration was in the two dimensions of the positions of stars on the sky. For astronomers, exploring the three dimensions of our Milky Way galaxy was a theme introduced in the early part of the twentieth century. The 80 years of Mount Stromlo Observatory have been the 80 years since the discovery of the expanding Universe. Exploring the dimension of the time since the Big Bang is where astronomy's Captain Cooks can now be found. The coastline, estuaries and harbours they are discovering form the outline of the evolution of the Universe.

Human curiosity being what it is, all of us now know something of the extraordinary Universe we live in, thanks to untiring efforts to build better observatories with larger telescopes, and to apply physical insights to understanding how the Universe works. Exploring the Universe is now part of the global and national consciousness. The part that Mount Stromlo and its astronomers have played is the theme of this book. It is particularly timely, because the oral history of the Observatory's foundation and early years has all been accessible to the authors' research.

Eighty years is a long time for a national institution in a young federation such as the Commonwealth of Australia. Setting aside the national parliament and the military, few national institutions have had a history as exciting as that of the national observatory on Mount Stromlo. The underlying reasons for this include the very nature of astronomy, the time in the history of astronomy when this institution arrived, the discretionary character of astronomy in a national budget, the insatiable curiosity we all have about the outside world and the desire of people to have, and therefore to maintain, an institution like a national observatory.

Reading the history of Mount Stromlo Observatory is like riding a roller-coaster. The metaphor is too restrictive, however. Roller-coasters have tracks. Mount Stromlo has explored the ups and downs of solar physics, 'optical munitions', astrophysics, international projects, science education from kindergarten to doctorate, and cosmology. Its social history has taken it from bush outpost, through the development of the national capital, to dual roles as a university research centre and a public observatory.

There is something of the character of NASA's space program in this history. This is an appropriate reflection after the destruction of the space shuttle *Columbia* and the second 'great fire of Stromlo' in the first two months of 2003. In both cases the instinctive reaction of the people who are called on to support our national institutions, which is all of us, is to redesign and rebuild, so that the mission can continue.

Message from the Director

Professor Penny D. Sackett
Director, Research School of Astronomy and Astrophysics
22 July 2002 –

As you read these words, a new Mount Stromlo Observatory is in its infancy. Like the native gum trees on its slopes, Mount Stromlo emerged from the ashes of the 2003 firestorm that engulfed it even before the first rains of relief began to fall. Embedded just below the surface of our institution's history are the seeds of its renaissance.

To understand the factors that have shaped Mount Stromlo and its place in world astronomy and Australian heritage over the past 80 years is, in large part, to understand what will mould its next eight decades. The elements of the Mount Stromlo story are those common to many human sagas: excellence and determination, invention and discovery, challenge and setback, self-reliance and teamwork, foresight and good fortune, sinew and spirit.

Read on.

Contents

Abbreviations and acronyms

2dF	two-degree field (instrument at the AAT)
AAO	Anglo-Australian Observatory
AAS	Australian Academy of Science
AAT	Anglo-Australian Telescope
AATB	Anglo-Australian Telescope Board
AC	alternating current
ACI	Australian Consolidated Industries
ACT	Australian Capital Territory
AGN	active galaxy nucleus
AID	Army Inventions Directorate
ALP	Australian Labor Party
alt-az	altitude-azimuth
ANITA	Australian National Institute for Theoretical Astrophysics
ANU	Australian National University
ANZAAS	Australian & New Zealand Association for the Advancement of Science
AO	adaptive optics
ARC	Australian Research Council
ASA	Astronomical Society of Australia
ASO	Australian Space Office
AST	Australian Synthesis Telescope (never built)
ATC	Astrophysical Theory Centre (of the ANU)

BSc	Bachelor of Science
Caltech	California Institute of Technology
CASPIR	Cryogenic Array Spectrograph-Imager
CCD	charge-coupled device
CFPA	Center for Particle Astrophysics (University of California)
CMB	cosmic microwave background (radiation)
CMG	Companion of the Order of St Michael and St George
COBE	(NASA's) Cosmic Background Explorer (satellite)
CRDF	Cathode Ray Direction Finder
CSIR	Council for Scientific and Industrial Research (forerunner to CSIRO)
CSIRO	Commonwealth Scientific and Industrial Research Organisation
CSO	Commonwealth Solar Observatory
DC	direct current
DEC	Digital Equipment Corporation
DIMM	Differential Image Motion Monitor
DITAC	Department of Industry, Technology and Commerce
DSc	Doctor of Science
ELS	Eggen, Lynden-Bell & Sandage (see Chapter 8, note 4)
EROS	*Expérience de recherche d'objets sombres*
ESA	European Space Agency
ESO	European Southern Observatory
EST	Eastern Standard Time
FCT	Federal Capital Territory
FRGS	Fellow of the Royal Geographical Society
FRS	Fellow of the Royal Society (of London)
FUSE	Far Ultraviolet Spectroscopic Explorer
GMT	Great Melbourne Telescope
GR	(Einstein's theory of) General Relativity
HF	high frequency
HFDF	High Frequency Direction Finding
HST	Hubble Space Telescope
IAS	Institute of Advanced Studies (of the ANU)
IAU	International Astronomical Union
IBM	International Business Machines
IPS	Ionospheric Prediction Service
ISI	Institute for Scientific Information
IUE	International Ultraviolet Explorer

JACARA	Joint Australian Centre for Astrophysical Research in Antarctica
JPC	Joint Policy Committee (of the AAT project)
KAO	Kuiper Airborne Observatory
Lab	laboratory
LLNL	Lawrence Livermore National Laboratory
LMC	Large Magellanic Cloud
LOT	large optical telescope
MACHO	massive astronomical compact halo object
MSc	Master of Science
MSO	Mount Stromlo Observatory
MSOAR	Annual Report of the Observatory on Mount Stromlo
MSSSO	Mount Stromlo and Siding Spring Observatories (of the ANU)
NASA	National Aeronautics and Space Administration
NCA	National Committee for Astronomy (of the AAS)
NIFS	Near-infrared Integral Field Spectrograph
NSDR	Narrow Sector Directional Recorder
OBE	Officer of the Order of the British Empire
OGLE	Optical Gravitational Lensing Experiment
OMP	Optical Munitions Panel
OSU	Observatory Services Unit (of the ANU)
PCA	photon-counting array
PhD	Doctor of Philosophy
PN(e)	planetary nebula(e)
PZT	Photographic Zenith Tube
QSO	quasi-stellar object
Quasar	quasi-stellar radio source
RAAF	Royal Australian Air Force
RAN	Royal Australian Navy
RFC	Royal Flying Corps
RN	Royal Navy
RRB	Radio Research Board
RSAA	Research School of Astronomy and Astrophysics (of the ANU)
RSPhysS	Research School of Physical Sciences (of the ANU)
SID	sudden ionospheric disturbances
SMC	Small Magellanic Cloud
SN (Ia)	supernova (of Type Ia)
SNR(s)	supernova remnant(s)
SSO	Siding Spring Observatory

TEST	Two- (or Three-) Element Synthesis Telescope (at Parkes)
UNESCO	United Nations Educational, Scientific and Cultural Organization
URSI	International Union of Radio Science
VLT	Very Large Telescope (of ESO)
WIMP	weakly-interacting massive particle
WFI	Wide-Field Imager (for SSO telescopes)
YWCA	Young Women's Christian Association

Standard international abbreviations for units of measurement are also used.

Abbreviations for Periodicals

Standard international abbreviations are used for the titles of periodicals. Those below are non-standard further contractions commonly used by astronomers (usually without terminating periods).

A&A	*Astronomy & Astrophysics*
A&AS	*Astronomy & Astrophysics, Supplement Series*
Acta Astron	*Acta Astronomica*
AJ	*The Astronomical Journal*
AJSci	*Australian Journal of Science*
AJPhys	*Australian Journal of Physics*
ApJ	*The Astrophysical Journal*
ApJS	*The Astrophysical Journal, Supplement Series*
ApSS	*Astrophysics & Space Science*
ARAA	*Annual Review of Astronomy and Astrophysics*
Astron & Geophys	*Astronomy & Geophysics*
Astrophys. Lett	*Astrophysical Letters & Communications*
Aust J Sci Res	*Australian Journal of Scientific Research*
Gerl. Beitr. Z. Geophys	*Gerlands Beitrage zur Geophysik*
J. Hist. Astron	*Journal of the History of Astronomy*
Memoirs R. Astron. Soc	*Memoirs of the Royal Astronomical Society*
MN	*Monthly Notices of the Royal Astronomical Society*
Ned. Tijdschr. Natuurkunde	*Nederland Tijdschrift Natuurkunde*
Obs	*Observatory Magazine*
PASP	*Publications of the Astronomical Society of the Pacific*

Phy. Rev	*Physical Review*
ProcASA	*Proceedings of the Astronomical Society of Australia*
Proc. Nat. Acad. Sci	*Proceedings of the National Academy of Science*
Proc. Phys Soc	*Proceedings of the Physical Society*
Publ. Astron. Soc. Aust	*Publications of the Astronomical Society of Australia*
Publ. Astr. Soc. Japan	*Publications of the Astronomical Society of Japan*
QJRAS	*Quarterly Journal of the Royal Astronomical Society*
RAAS	*Records of the Australian Academy of Science*
Rep. Prog. Physics	*Reports on Progress in Physics*
S&T	*Sky & Telescope*
SMH	*Sydney Morning Herald*
Terr. Mag Atm. Electr	*Terrestrial Magnetism and Atmospheric Electricity*

Preface

In an interview for the Australian National University's newspaper, the then Director of Mount Stromlo Observatory, Professor Jeremy Mould, spoke confidently of astronomy entering a 'golden age'.

> Like two other sciences whose discoveries fill the media, molecular biology and information technology, right now astronomy's secrets all seem to be tumbling out at once. We aim to tell the story of the evolution of the Universe from the Big Bang, through to the formation of our Galaxy, the Milky Way, to the condensation of our solar system and others like it. Much of the understanding has come just in the last few years and the pace of discovery will continue as long as the technology for probing the Universe continues to advance.[1]

He went on to describe the place of the Mount Stromlo and Siding Spring Observatories and the extent to which they allowed Australia to participate in research with a bearing on our understanding of the origins and destiny of the Universe. In a nation that publicly reveres its sporting heroes but often overlooks its leading scientists, there is a great deal of ignorance about Australia's role as a leading international player in the exciting discipline of astronomy. This ignorance was one reason for a meeting at Mount Stromlo on 3 June 1997 to consider the publication of a history of the Observatory to coincide with the 75th anniversary of its establishment in 1999. That timetable could not be met, but the need to explain and proclaim the Observatory's many achievements remained. Dr Tom Frame, a former naval officer with a continuing interest in celestial navigation and a professional historian, was commissioned by the Australian National University to produce a history of the Observatory in late 1999. Dr Don Faulkner, the Observatory's former Associate Director for Education and Outreach, was asked to act as scientific adviser. With Dr Frame's unexpected appointment as Anglican Bishop to the Australian Defence Force in early 2001, before the research was completed and the manuscript finalised, Dr Faulkner accepted an invitation to become co-author. It was, with hindsight, a fortunate development: both authors believe this is a much better work because of their cooperation. The narrative is divided into two parts. The first covers the years 1905 to 1956 and concentrates on the establishment of the Observatory and the acquisition of the facilities, equipment and staff necessary to undertake world-class astronomical research. The second part focuses on the Observatory's scientific and technological achievements and their significance for the direction and progress of contemporary astronomy. In brief, Dr Frame was largely responsible for the introduction and the first six chapters while Dr Faulkner produced the last five chapters and the epilogue. The views expressed in this history are, of course, those of the authors, who were given complete freedom in matters of content and style subject to any reasonable requests from the Observatory Director. Nothing that we have written should be taken to represent the views of the Observatory or its staff, the Australian National University or the Commonwealth government.

The manuscript had just been completed when a devastating firestorm engulfed the western flank of Canberra on 18 January 2003. A little more than 50 years after an earlier bushfire had swept over the mountain, flames again destroyed the roof and contents of the Administration Building, five telescopes and eight staff residences. Although an in-principle decision has been made to rebuild on Mount Stromlo, the shape and substance of the 'new' observatory were still being discussed as this book entered production. Other than some minor changes to this preface and the addition of an epilogue, the narrative has not been changed.

As astronomy will continue on Mount Stromlo, this book is intended primarily for those who will visit the 'new' observatory and the facilities at Siding Spring in years to come. The number of people touring the observatories was in excess of 65,000 each year before the January 2003 fire. There is no doubt they will return to Mount Stromlo when the proposed reconstruction is complete. We have also had in mind general readers of Australian history with more than a passing interest in astronomy. This work is not, of course, intended for professional or academic astronomers, although we hope they will find much of interest within the narrative. They are encouraged to read the more technical historical analyses listed in the bibliography. Although our aims were very broad, we were naturally limited by strict editorial guidelines. In brief, we have attempted to describe Mount Stromlo's foundation as the Commonwealth Solar Observatory in 1924 and its subsequent evolution into Australia's premier optical astronomical institution. In addition to accounting for the Observatory's incorporation into the Australian National University (ANU) during 1957, the subsequent establishment of the ANU Research School of Astronomy and Astrophysics, and the more recent development of public education programs designed to encourage interest in astronomy, this history will explain the principal scientific achievements of the Observatory and highlight their continuing influence on theories of human origins and explanations of the cosmic context in which we live. This is not the final word on the Observatory and its history. There is still much to be written on the many and diverse people who have worked at the Observatory, the communities they have formed, the role of technicians and support staff, and the particular contributions of Mount Stromlo's scientific work to astronomy as a discipline. We trust that the inevitable brevities and omissions in this general work will prompt others to complete what is a fascinating and remarkable story of scientific, academic and technical achievement on the fringe of Australia's national capital.

Tom Frame
Don Faulkner
Mount Stromlo Observatory, Canberra
February 2003

Acknowledgements

THE AUTHORS WISH TO THANK THE FOLLOWING individuals whose assistance with this project was greatly appreciated: Professor Jeremy Mould, former Director of Mount Stromlo and Siding Spring Observatories, who approved the project and believed in its importance for the Observatory, the University and the people of Canberra; Mr Ian Bowring, of Allen & Unwin, for his enthusiasm for the project and management of the publication; Professor John Norris, Interim Director of the Research School of Astronomy and Astrophysics, who encouraged the authors in their work and read earlier drafts of the manuscript; Professor Penny Sackett, currently Director of the Research School of Astronomy and Astrophysics, for her opening message and support for the project; Professor Ken Freeman, who read drafts of Chapters 7–11, and Professor Mike Bessell, who very kindly provided the colour images including the cover illustration; Professor Don Mathewson, former Director of Mount Stromlo, for his comments on Chapter 9; Dr Sigrid McCausland for assisting the authors with locating and accessing material from the ANU archives; Mrs Fiona Aplin for her assistance on numerous occasions in accessing departmental files; Ms Jeanette Regan, Librarian at Mount Stromlo, for her kind help; Miss Carola Parke, Mrs Gay Kennedy, Ms Theresa Gallagher and Ms Wendy Slater for their assistance in obtaining the graduate information for Appendix 3; Professor Ben Gascoigne, who spent the majority of his professional life as an astronomer at Mount Stromlo and who has written a number of articles and papers on the history of the Observatory, who allowed the authors to make use of his own research and who also read earlier drafts of the manuscript; Professor Walter Stibbs, a former staff member whose early career was spent at Mount Stromlo, who provided valuable material and read an earlier draft of the first six chapters; Mrs Rose Allen, widow of Professor Cla Allen; Miss Joan Duffield, daughter of Professor Geoffrey Duffield; Mr Michael Nelmes, Canberra Astronomical Society, who provided much valuable material on the telescopes and their operation; Dr John Kennewell, Learmouth Solar Observatory, IPS Radio and Space Services, for material related to Cla Allen's work; Mr Frank Cook, a wartime staff member, who offered his recollections of life on Mount Stromlo; and Mrs Louise Taylor, a graduate student at the University of Canberra, who conducted research into community life at Mount Stromlo. Mr Robert Hyslop ISO, a fellow of the Australian Society of Indexers, very kindly prepared the Index. The authors also wish to thank Ms Gabrielle Hyslop and the staff at the National Archives of Australia in Canberra for their assistance with obtaining illustrative material. Without the goodwill and generosity of these individuals, this history would have been poorer.

The authors are grateful for a publication subsidy from the Australian National University Publications Committee.

Introduction

HUMAN BEINGS HAVE ALWAYS BEEN CURIOUS about the Universe of which they are but a miniscule part. They have gazed at the heavens in order to make sense of the Earth. At times there has been an extraordinary hunger for information about the Sun, stars, planets and the Earth, and for abiding or absolute truths that can make sense of human life and impart some point to individual lives. Some do not want to know that human life ultimately could be pointless, while others remain committed to searching for fundamental explanations to cope with existence in a world they might even disdain. In its broadest conception, astronomy is the study of the Universe. Over the last two millennia it has prompted an intellectual revolution, produced innumerable social benefits and posed compelling spiritual questions. It is a science with as many philosophical as practical dimensions, as a brief survey of its progress reveals.

The Greek philosopher Aristotle produced *On the Heavens* in 340BC. He thought that the Earth was a stable object around which the Sun, Moon and other planets revolved. Ptolemy agreed with Aristotle, and in his work *The Almagaest* assumed that each planet moved in a circle called an 'epicycle' whose centre was in turn carried around the Earth in a second circular orbit they called a 'deferent'. A Greek astronomer, Aristarchus of Samos, thought that the Earth might revolve around the Sun but his theory did not gain currency after it was dismissed by no less an authority than Aristotle. Judaism, Islam and Christianity all adopted the depiction of the Earth at the centre of all things as being consistent with a scriptural outlook on Creation.

In the 250 years between the birth in Poland of Nikolaj Kopernik—known to the world as Copernicus—in 1473 and the death of Sir Isaac Newton in 1727, there was a very remarkable change in the way that people thought about the world around them. The development is sometimes called 'The Copernican Revolution'; it occurred after Copernicus published *De Revolutionibus* (initially anonymously to avoid the charge of heresy) in 1543, when he was an elderly canon. This revolution was not fought with swords or guns, but with scientific equipment—particularly the telescope. This instrument seemed to be able to divide truth from falsehood, to identify and disperse outdated thinking and superstitious belief, and it reigned supreme. When Newton's *Philosophiae Naturalis Principia Mathematica* appeared in 1687, a new worldview had been planted. Newton was an Englishman and feared neither political pressure nor religious sanction. In any event, the world was going to change because the astronomers revealed what the Universe really was rather than what some preferred it to be.

Many philosophers of science consider that those 250 years actually witnessed the birth of science. Some would contest this, but it certainly marked the birth of the scientific method. In addition to Copernicus and Newton, three other major players participated in the unfolding revolution: Tycho Brahe, Johannes Kepler and Galileo Galilei. Each of the five was absorbed in trying to describe and explain the motions of the naked-eye planets—or the 'wandering stars', as they were called in the ancient world. It was an entirely astronomical enterprise. The planets are quite remote from the Earth, and their motions can have no direct bearing on our lives. Interestingly, however, this was not apparent at the outset of the revolution. The early players made no distinction between astronomy and astrology. Kepler, for instance, was required to cast an annual horoscope for the province as one of his formal university duties! There is no clearer testament to the change wrought by the Copernican Revolution in the way human beings think about the world of nature than that which occurred in the general attitude to astrology. By the time the revolution had run its course, the idea that the planets could influence human destiny was seen as manifestly ridiculous. When it was shown that the Earth was not the centre of the Universe, and that there were no fixed points in the Universe, it became clear that the relationship of the planets to the stars was merely a matter of perspective, and thus the prescriptions of astrologers were entirely arbitrary.

Each of the five key players made his own special contribution, not only to solving the immediate problem but also to the strategy for tackling all scientific problems—the scientific method. Copernicus had the original conceptual insight; Brahe was 'the father of observational astronomy', who established the importance of painstaking, careful measurement; Kepler introduced mathematics into the equation, encapsulating Brahe's thousands of observations into three simple mathematical 'laws'; Galileo (and his telescope) brought a no-nonsense dimension to the whole enterprise—he believed only in what he saw; and it was Newton's genius that brought the whole edifice together. He asked, and then answered, the most important question of all: '*Why* do the planets move as they do?'. His laws of gravitation and mechanics not only accounted for everything that the other four had seen and described, they also set in place the framework of the discipline of mechanics. They can be used to predict and explain the workings of any mechanical system, from the operation of a clock's pendulum to the propulsion required to launch a rocket into space.

It is now plain that, during the sixteenth and seventeenth centuries, Copernicus, Galileo, Brahe, Kepler and Newton changed the human view of the Universe. Their discoveries fuelled rather than quenched humankind's curiosity. The next revolution would start early in the twentieth century, with new observing and interpreting techniques, and astronomy is now enjoying a golden age, with practitioners—professional and amateur—united in their insatiable desire to know more about the Universe. They want to know about the nature of the Universe and the material from which the Universe is made. This desire to know has produced an enormous body of knowledge about the physical aspects of the Universe, from the huge clusters of galaxies down to the infinitesimal world of the atomic nuclei that are subjected to unimaginable forces deep in the interiors of stars.

Astronomers consider themselves fortunate to work in a science that attracts such widespread interest. Newspapers seem happy to report the discovery of each successive 'most distant quasar', or massive black hole, or any new conclusion concerning the ultimate fate of the Universe itself. An event visible to the naked eye—a comet or a nearby supernova, like the 1987 outburst in the Large Magellanic Cloud—arouses particular interest. Public lectures on astronomical topics attract large and curious audiences; if the speaker is willing to entertain questions (as most astronomers are), the ensuing discussion can easily last an hour or more. Not infrequently, towards the end of such a session, someone will ask the question that's been at the back of everyone else's mind: 'The study of the heavens is all very fascinating; but what practical use is it?'.

This is a question much loved by astronomers. In fact, some astronomers would admit there have been times when, sensing that no one is going to ask such a question, they have been guilty of gently prodding the audience until they do. Unless that question is asked, there is a danger that people will go out into the night under the misconception that, since the entities studied by astronomers are so remote from our Earth, and since they seem to have no tangible interaction with our existence, the study of astronomy should be regarded chiefly as an entertainment. Some may concede that the enterprise has a certain philosophical value—most people would, after all, like to know 'how it all started and what our end will be'. Nevertheless, given the choice of seeing their taxes directed to an observatory rather than a medical institute, or even an engineering school, most people would simply declare 'no contest'. 'The study of astronomy,' they would say, 'will never affect my life in even the slightest degree.' Astronomers would argue vehemently that nothing could be further from the truth.

It would not be an overstatement to claim that the Copernican Revolution—an entirely astronomical enterprise—provided about half of the raw material for the Industrial Revolution that began in the latter half of the eighteenth century. The rest was provided by the discipline of thermodynamics. So the question about whether there is any practical point or purpose to astronomy has its reply in the many and varied outcomes of the Industrial Revolution which, of course, affect every living person. We also study astronomy because of its practical benefits for things like time keeping, navigation and tides.

The start of the twentieth century witnessed an equally powerful development in science, sometimes referred to as the 'new physics'. The concepts of 'quantum mechanics' and 'general relativity' went to the core of astronomy and provided new insights into the intricate inner workings of our physical Universe. Einstein's development of 'general relativity' was especially pertinent for astronomy. The first hints that something was amiss with the Newtonian view of the world were, once again, astronomical in nature. Light did not seem to travel the way it ought. The 1887 Michelson–Morley experiment—to measure the motion of the Earth by detecting a difference in the speed of light travelling along, and then at right angles to, the Earth's direction of travel—proved completely unsuccessful. It became apparent to Einstein that Newton's laws simply ceased to apply for things travelling at velocities approaching that of light. Indeed, the velocity of light itself occupies a special place in the physical framework of the Universe—hence the significance of Einstein's Theory of General Relativity and all its related consequences for the discipline of astronomy. Its most profound consequence to date is bound up in the world's best known physical equation: $E = Mc^2$ (where E is energy, M is mass and c is the speed of light). Mass can be converted into energy. This was a prediction that became an awesome reality when the

United States Air Force dropped atomic bombs on Hiroshima and Nagasaki in August 1945. Once again, no one can claim that the study of astronomy is completely remote from developments in our society unless they are also willing to claim that nuclear weapons and the nuclear power industry have had no effect on our lives over the last 60 years.

The Universe is, by definition, the largest and most comprehensive physical laboratory that humanity could imagine. While it is true that humanity cannot arrange experiments on the Universe itself, it is equally true that those arranged by nature are of a magnitude that could never be performed in any terrestrial environment. Clearly this is a major advantage. Through the study of astronomy, humanity has become aware of the most fundamental laws at work in the physical world. Once properly understood, there is universal acknowledgement that these laws profoundly affect our lives. Immensely fundamental insights such as those offered by Newton or Einstein seldom occur, but grasping their consequences takes many decades. No astronomer goes to the observatory with the expectation that the day's discoveries will see commercial application within the month. The truly awesome developments of science are never like that: they take time to mature. But mature they certainly have over recent centuries, and mature they inevitably will in the centuries that lie ahead. The patient study of the Universe has already had, and will continue to have, the most far-reaching effects on our society. Whereas the first modern astronomical conclusions were resisted out of ignorance and fear, our society now affirms that understanding dispels ignorance, and with it prejudice and conceit, suspicion and superstition.

When the pioneer astronomers claimed that the Earth was simply a small planet orbiting around the Sun, their conclusions had a bearing on anthropology and the environment, philosophy and theology. They affected humanity's self-view: we were in fact not the centre of everything. And what if the Sun, the centre of our Universe, is a totally unimportant member of the galaxy? Is the existence of the Universe a mere sideshow to something bigger? Does the Universe have certain given features, embedded laws or customs that might disclose a purpose, intention or goal, thus imparting some meaning to our lives? Is there a pattern to the Universe that gives away the existence of a designer, architect or sustainer? Does the Universe disclose a God or Gods? Are we alone? If so, what do we make—if anything—of our uniqueness? Is the Universe benign or malevolent? Is it a transitory entity on its way to something else? When did it begin and does it have an end towards which it is inexorably moving?

These questions have been uttered in different forms for centuries. Inasmuch as the study of the stars was thought to have the potential to reveal some crucial existential questions, astronomy is an ancient science. It has attempted to explain not only the composition of the Universe and how it operates, but whether it has a starting point and what its destiny might be.

But we live in a pragmatic age where truthfulness is measured by practical outcomes. At the same time, postmodernism rejects the notion that there is objective truth. This creates a difficulty for science, which aims to establish some truth through the investigation and description of what can be observed publicly. Science does not rely upon a special vantage point that is available to some and not to others, or insights that are not available to all but only to the members of a specific social group. While scientists might be able to make some objective statements about an object or to deduce a physical truth, the philosophical apprehension and the social construction of these statements or truths are matters that go beyond the discipline of science and are not subject to its usual rubrics. This is why cosmology straddles the disciplines of physics and metaphysics.

Cosmology began in ancient times, although there was initially more interest in the arrangement of the heavens than in their beginnings. In a thesis published in 330BC, Aristotle described the general nature of the cosmos and his belief that the Earth was a stable object around which the Sun, Moon and other planets revolved. He asserted that the 'beginning of anything is the most important part, being indeed half of the whole'. Therefore, a small mistake at the beginning can produce great difficulties in what follows. But shortly after Aristotle, we find that a more immediate interest in the motions of the Sun, Moon and planets saw cosmology steadily subsumed by what became known as astronomy, while the metaphysical concerns that featured in the ancient thinking became the focus of philosophic and religious debate. As little progress could be made on physical observation, the metaphysical component had no reason to develop. In fact, it appears to have stagnated until relatively recent times, when we have observed lively exchanges between scientific findings and their metaphysical interpretation.

In the light of Einstein's theory of relativity in 1915 and Edwin Hubble's detection of galaxies beyond the Milky Way in the 1920s, astronomers such as Georges Lemaitre and later George Gamow proposed the theory that the whole Universe had originated in a great explosion around thirteen billion years ago. It was rather pejoratively termed the 'Big Bang' by British astronomer Fred Hoyle during the 1950s. This was the birth of modern cosmology. In 1988, Stephen Hawking contended that, through 'Big Bang' cosmology, questions relating to the beginning of the Universe had entered the 'realm of science'. John Gribbin claims that these:

> may have been answered by a single highly successful theory that emerged during the 1980s. This theory, which goes by the name of inflation, also explains the incredible flatness of the Universe. At last we come down to the metaphysical nitty-gritty, the questions that used to be regarded as beyond the scope of science. Now, science knows no limitations, and can tackle all of these questions, even if the answers are not complete, or completely understood. We even have a line of attack on the problem of creation itself.[1]

But what was left for philosophers and theologians who also maintained an interest in creation and its beginnings? Fearing physical reductionism, Thomist philosopher William Carroll asks:

> Are we on the verge of a scientific explanation of the very origin of the Universe? The contention of several proponents of the new theories is that the laws of physics are sufficient to account for the origin and existence of the Universe. If this be true, then, in a sense, we live in a Universe which needs no explanation beyond itself, a Universe which has sprung into existence spontaneously from a cosmic nothingness.[2]

According to Stephen Hawking: 'So long as the Universe had a beginning, we suppose it had a creator. But if the Universe is really completely self-contained, having no boundary or edge, it would have neither beginning nor end: it would simply be. What place, then, for a creator?'.[3] Of course, to argue that the Universe has no beginning does not deny the possibility that it has an origin. However, the Australian physicist Paul Davies seems to have a definite answer to Hawking's question, despite a façade of reticence:

> For the first time, a unified description of all creation could be within our grasp. No scientific problem is more fundamental or more daunting than the puzzle of how the Universe came into being. Could this have happened without any supernatural input? Quantum physics seems to provide a loophole to the age-old assumption that 'you can't get something from nothing'. Physics is now talking about the 'self-creating Universe': a cosmos that erupts into existence spontaneously, much as a subnuclear particle sometimes pops out of nowhere in certain high energy processes. The question of whether the details of this theory are right or wrong is not so very important. What matters is that it is now possible to conceive of a scientific explanation for all creation. Has modern physics abolished God altogether?[4]

Curiously, Davies has written elsewhere that:

> Most theologians are prepared to accept that the Universe runs itself without the need for continual supervision by a deity. Instead, the laws of nature are able to regulate all natural activity without supernatural assistance. Nevertheless, God is still invoked to set the system going in the first place. In scientific language, God must be manifesting his powers through the naked singularity that marks the big bang creation. There is certainly no incompatibility between these theological ideas and the scientific version, because the singularity, by definition, transcends the laws of nature. It is the one place in the Universe where there is room, even for the most hard-nosed materialist, to admit God. Yet surely a God that is pushed off the very edge of spacetime is a pale shadow of the deity that most people would wish to accept. In this fascinating subject area, where science mingles with religion and philosophy, the urge to push science to its limits is compulsive.[5]

Heinz Pagels is plainly susceptible to that urge. He is confident that 'from microcosm to macrocosm, from its origin to its end, the Universe is described by physical laws comprehensible to the human mind'.[6] And this, according to Quentin Smith, has enormous philosophical consequences. If Big Bang cosmology is true, Smith argues that 'our Universe exists without cause or without explanation . . . and it exists non-necessarily, improbably and causelessly. It exists *for absolutely no reason at all*' (emphasis in the original).[7] Such a conclusion might seem to abolish the God of Christianity, Judaism and Islam and undermine their claims to revelation of any divine, transcendent and infinite purpose expressed in and through creation. Big Bang cosmology appears to render any belief in God defiant and foolish. But we need to ask the crucial question: is astronomy the new religion, complete with its high priests, sacred texts and holy places?

As religious visions of human life and civilisation have been abandoned by many as being hopelessly lost in superstition and subverted by ideology, science has become the alternative. And the two scientific disciplines most closely associated with a human life and worldviews are biology and astronomy—the first dealing with some of the smallest units of matter and the second with the largest. The biologist is concerned with the smallest building blocks of life and the inevitable demand that practitioners shed some light on the question of whether human biology discloses something of the point of human life and whether human behaviour is determined biologically rather than through the exercise of free will. A demand for such answers can lead to the twin pitfalls of 'determinism' and 'reductionism'—the notion that human identity is determined by genetic factors and that human behaviour can be reduced to biological impulses based on physico-chemical laws. Some 'neo-Darwinist' biologists feel compelled to defend a unified biological theory that accounts for all living things, together with their origins and destiny. Indeed, Richard Dawkins begins his 1986 'neo-Darwinist' work *The Blind Watchmaker* with the grand conviction 'that our own existence once presented the greatest of all mysteries but that it is a mystery no longer because it is solved. Darwin and Wallace solved it, though we shall continue to add footnotes to their solutions for a while yet'. The contrast between Socrates' famous reflection—'All I know is that I know nothing'—and Dawkins' confident assertions could not be more striking.

Clearly, this is a golden era of astronomy, yet it is uncertain whether it will sustain or overturn the Big Bang theory or prove the existence of black holes and dark matter. It is plain that previous eras felt likewise about the advances being made and whether extant explanations would stand the test of time. Edmund Ledger remarked in 1881: 'The progress of astronomical science during the last thirty years has been so rapid as almost to approach the marvellous'.[8] It is impossible to say with any certainty that a bold new theory is at hand or that our received wisdom will be shown to have been folly. As David Leverington has observed:

> A hundred years ago, Newtonian physics ruled supreme. Planck's Quantum Theory, Einstein's Theory of Relativity and Heisenberg's Uncertainty Principle did not exist. The photographic plate and spectroscopy were still relatively modern developments; radio astronomy and space research were both born some decades in the future.[9]

There is not sufficient scope in this book to present a persuasive case for conventional theistic belief, although such a case continues to make religious belief reasonable and rational. The point of touching briefly on cosmology was to highlight the importance of astronomy as a discipline whose consequences go well beyond the physical sciences. Neither cosmology nor astronomy has delivered all the answers to the great existential questions and we are still left asking about the 'why' behind the 'how' that we think we have just begun to understand. It also highlights the need for experts in other academic disciplines, including philosophy and theology, to be attentive to the work being done by astronomers and to the conclusions drawn from the data they collect.

This brings us to the enduring problem of human consciousness—that with which we observe and reason. The philosopher Martin Heidegger argued that 'we must stop looking at beings as beings and start looking at being as being'.

How we observe our external objective world will fail as a method for examining our internal subjective world. The human difficulty is in attempting to understand the Universe of which humanity is a part, while endeavouring to understand itself. While recognising this crucial constraint, there is much that the astronomy and the astrophysical work conducted at the Mount Stromlo and Siding Spring Observatories have yet to reveal to humanity about the Universe of which we are all indeed a part, and from which we gain an understanding of ourselves.

chapter one

A place in the sun, 1905–1923

The earliest European contact with the Australian continent was prompted by astronomical curiosity. Lieutenant James Cook RN was given command of the barque *Endeavour* and ordered on a voyage to the 'South Seas' in order to observe the transit of Venus across the Sun's disc. This was a rare phenomenon of great practical importance at that time: it assisted in determining the Sun's distance from Earth. Astronomers in the eighteenth and nineteenth centuries were aware that the southern skies contained many important objects, especially in the rich southern Milky Way, and would yield many astronomical insights. Astronomy became a feature of British imperial expansion.

The burgeoning interest in objects outside the solar system led in 1820 to the founding by Royal Decree of an observatory at the Cape of Good Hope in South Africa. In the same year, the Governor of New South Wales, Sir Thomas Brisbane, established a private observatory at Parramatta. William Scott founded the Sydney Observatory in 1858. Five years later, a public observatory was established by Robert Ellery at South Yarra in Melbourne on behalf of the Victorian colonial government. It was equipped with a 48-inch speculum metal reflector, making it the largest equatorially mounted telescope in the world. One of its primary tasks was the study of southern nebulae. The Perth Observatory was founded in 1896 with Mr William Ernest Cooke as West Australia's first Government Astronomer. He was later appointed Government Astronomer of New South Wales and Professor of Astronomy at Sydney University, and is credited with building a helio-chronometer, or 'sun-dial', to determine local time and true north.

When B.A. Gould started the National Observatory of the Argentine Republic at Cordoba in 1870, there were observatories in the three southern continents, although astronomical endeavour in the 'Great South Land' remained very fragmented. When the Australian colonies formed a Federal Commonwealth and became an independent nation in 1901, the most significant astronomical institutions remained in the Northern Hemisphere. The Mount Wilson Solar Observatory had been established in Southern California during 1905 under the direction of George Ellery Hale with funding from the Carnegie Institution. The foundation instrument was a horizontally mounted Snow telescope that had previously been used at the Yerkes Observatory. A solar telescope in a 20-metre high tower was then built and used, initially during 1907. A 60-inch reflector was completed with further funding from the Carnegie Institution. When employed for the first time in December 1908, it was the most powerful telescope then in service and made Mount Wilson the world's premier observatory. A 100-inch mirror disc was made for the Hooker telescope, with observations from the new instrument commencing in late 1917.

Despite its early foundation in 1820 as an extension of the Greenwich Observatory, after a major expedition led by Sir John Herschel (son of the astronomer who discovered Uranus, Sir William Herschel), the British observatory at the Cape of Good Hope suffered from a lack of funds, although it possessed a large telescope (an 18-inch reflector). When the Scottish astronomer David Gill became Director in 1879, the observatory enjoyed a substantial rejuvenation. His work on stellar photography in the 1880s was of such quality that many others were convinced of its wider application. In the 1890s, Gill used what was probably the world's best heliometer to measure solar parallax while his *Cape Photographic Durchmusterung* star catalogue was unsurpassed when published in 1900. In 1901, the Observatory was supplied with a 60-centimetre photographic refractor coupled on the same mounting to a 45-centimetre visual refractor. Together with the Dutch astronomer J.C. Kapteyn, and the Director of the Pulkovo Observatory in Russia, Gill persuaded the Transvaal government to establish an observatory on the outskirts of Johannesburg in 1905. When the Union of South Africa was formed in 1910, it became known as the Union Observatory. Although Australia did not possess a comparable institution, many Australians were asking what seemed to them an obvious question given the cost to the newly created nation of establishing a world-class solar observatory: why study the Sun?

There were a host of reasons, quite apart from human curiosity about the most noticeable feature in the sky. It was obvious that the Sun affected the weather. But when telescopes were developed early in the 1600s, many new questions were raised, including what substances did the Sun consist of? Was it solid matter or compressed gas? And after the Jesuit priest Father Christopher Scheiner started observing the Sun during March 1611 in Southern Germany, he asked: why were there so many dark spots on the surface? They had, of course, been noticed visually centuries before. In fact, 'sunspots' were the only solar features visible to the unaided eye. We now know that a medium-sized sunspot has a diameter of 10,000 kilometres (1/140th of the Sun's diameter) and blocks or diverts over 20,000 times more energy than falls on the whole Earth's surface during the same period. Lacking telescopes, it was not surprising that the earliest astronomers thought sunspots were unknown planets close to the Sun. But Scheiner was beginning to believe otherwise and to ponder whether sunspots had any influence on the Earth. It was noteworthy that a period of very few recorded sunspots (from 1640 to 1700), known as the 'Maunder Minimum' after the English astronomer who drew attention to it, coincided with a protracted period of cold weather in Europe described as the 'Little Ice Age'. It was not until 1843 that Heinrich Schwabe was able to prove the existence of a sunspot cycle. He noted changing activity from a minimum level to a maximum, taking approximately ten years. (It was, in fact, really about eleven years in length.) In 1859, Richard Carrington showed that the Sun was rotating faster at the equator than at the middle latitudes. By the turn of the twentieth century, astronomers were very interested in the sunspot cycle, the structure of the Sun's convection zone, the cause of solar flares and the heating mechanisms in the outer solar layers. But what of the Sun itself?

At the end of the eighteenth century, the great William Herschel believed the surface of the Sun was solid underneath its dark atmosphere and that it could be inhabited by living beings. During the 1860s, photographic observations of the Sun's surface showing solar granulation were first made, although there was little consensus on what they revealed. It did appear on first inspection, however, that there were bright clouds over a darker substrate. And then the American astronomer Jonathan Lane argued that the Sun was a spherical concentration of nothing but gas held together by gravitation and having a central energy source. Based on Carrington's observations, Jesuit priest and astronomer Father Angelo Secchi demonstrated that the Sun was indeed gaseous. But how did the Sun generate its heat, and by what means was this heat transmitted? The answers were neither clear nor convincing. Meanwhile, every new discovery led astronomers to realise how little they knew about the Sun. The extent of the research that was needed to overcome this ignorance could only be tackled by a major institution with a large staff and plenty of money. There were many scientists in the emerging international astronomical community hoping that the newly formed Commonwealth of Australia would host such an institution. But where would it be located and who would fund its initial establishment and subsequent operation?

The vision for such an observatory as a public institution originated with one man—Geoffrey Duffield. He would become the driving force behind a campaign that would culminate nearly two decades later in the establishment of the Commonwealth Solar Observatory. The magnitude of the task he faced was truly daunting. As Rosaleen Love has commented:

> Duffield needed more than just the idea of introducing solar physics to Australia. He needed propitious times, both in science and government. He needed good committee technique. He needed good relations with whatever Australian government of the time was in power; and since there were several changes of government between 1905 and 1923, he needed to keep an eye on the opposition even as he courted the government of the day. He needed tenacity. He needed to be a science entrepreneur, in ways that scientists had not had to be in the days when science cost little.[1]

Walter Geoffrey Duffield—always known by his second name of 'Geoffrey'—was born at Gawler in South Australia on 12 August 1879.[2] His family had been pioneers in the Gawler district. His grandfather, Walter Duffield, was a miller who came to Australia in 1839 from Chelmsford, England, and later became a member of the South Australian parliament. The Duffields could rightly refer to themselves as part of the colony's 'establishment', and young Geoffrey came from a highly respected family. As Rosaleen Love has remarked:

> it was a childhood that conferred on [Geoffrey and his brother Kenneth] a sense of personal worth and a social assurance that later stood Geoffrey Duffield in good stead when he called on wealthy and distinguished Australians to contribute to his cause.

The Duffields had also become wealthy by the time Geoffrey was born. After a privileged but undistinguished education at the elite St Peter's Collegiate School in Adelaide, Duffield enrolled in a science degree at the University of Adelaide in 1898. It was a very small intake of undergraduates, making the tuition almost personally tailored to the needs of the students. Duffield's lecturers included Professor William Henry Bragg, who held the Elder Chair in Mathematics and Physics, and who conducted most of the teaching offered to second- and third-year students. Bragg and his son were later awarded the Nobel Prize for their work on X-ray crystallography. Duffield's undergraduate results were as mediocre as they had been at secondary school, and he graduated with third-class honours in physics and mathematics. As the university was poorly staffed and ill-equipped to offer postgraduate programs in science, Duffield decided to continue his studies abroad with the aid of the Angas Engineering Scholarship, which was worth £200 a year for two years. The purpose of the scholarship was to 'encourage the training of scientific men and especially civil engineers, with a view to their settlement in South Australia'. Whereas other scholarships encouraged studies in pure or theoretical science, the Angas Engineering Scholarship sought to foster applied science with practical applications.

Duffield decided to use the scholarship to attempt the undergraduate Mechanical Sciences Tripos at Cambridge University. On completion of his course in mid-1903 and graduation with a Bachelor of Arts degree, he gained a bursary of £50 as the Nobel Research Scholar and the opportunity to work in the engineering laboratories of the British National Physical Laboratory. Having seen the workings of a leading scientific institution at first hand, Duffield sought the McKinnon Scholarship to support his studies as a research student in the Physical Laboratories at Owens College in Manchester, where he gained the degrees of Master of Science (MSc) and later Doctor of Science (DSc) for his work in the field of spectroscopy in 1908. The Physical Laboratory at Manchester also hosted the Central Bureau of the International Union for Cooperation in Solar Research that had been formed after the International Congress of Science met at the St Louis Exposition in 1904.

The idea of a solar observatory being established in Australia first came to Duffield while he was still a research student, during a meeting of the International Union for Cooperation in Solar Research held at Oxford in 1905. This was the real beginning of the movement that led to the creation of a solar observatory in Australia. Astronomers suspected that certain solar brightenings detected with a spectroheliograph could produce a magnetic effect on the Earth. If the Sun were subjected to more close and continuous observation around the world, more solar-terrestrial influences might be detected. However, none of the existing nor proposed observatories was located in the western Pacific area. Although disappointed that solar physics was not being studied in his country of origin, Duffield recognised that Australia was well placed between existing solar observatories in the British colony of India and the United States to host such a facility with numerous suitable observing sites for the one-third of the sky that was inaccessible to northern observatories. There was also the clear advantage of superior viewing conditions in the southern winter, when northern conditions were less than ideal.

Acting on his own initiative, Duffield wrote to his mentor, Professor Bragg, suggesting the possibility of Australia as the site for a new solar observatory. Bragg forwarded a copy of Duffield's letter to the *Adelaide Advertiser*. Duffield claimed that an Australian solar observatory could undertake 'the measurement of solar radiation from different parts of the [Sun's] disc, and from sunspots; atmospheric absorption; sunspot spectra; solar rotation, and the determination of a standard wavelength'. When the text appeared in the 3 April 1907 edition of the newspaper, it provoked a surprising amount of public interest and political attention. The Adelaide Observatory staff was consulted. The State Observatory's Acting Director, George Frederick Dodwell (who would shortly succeed the Government Astronomer, Sir Charles Todd), thought 'the work might with advantage be done in Adelaide, if the necessary means were provided'. Bragg suggested that Duffield also raise the matter with the Premier of South Australia, Mr Thomas Price, because 'the Labor Government has a [budget] surplus, and is a little anxious to be on the side of cultural intelligence'. But, as the federal government was then considering the absorption of all the state observatories into a single Commonwealth department, the South Australian government declared the initiative a federal responsibility and declined to commit any funds.

When the International Union for Solar Research met at Janssen's Solar Observatory at the Chateau de Meudon near Paris during 1907, there was steadily growing overseas support for an Australian facility. A resolution commending the scheme drafted by Duffield and proposed by the widely respected British astronomer Sir Norman Lockyer, discoverer of helium in the Sun and editor of *Nature*, was put to the meeting in the following terms:

> That this International Congress hears with great satisfaction of the proposal to establish a Solar Physics Observatory in Australia and expresses its decided opinion that an observing station in that part of the world would fill a gap which now exists in the system of observatories distributed over the Earth and yield contributions of great value to the study of the solar phenomena.

The resolution was carried unanimously and a copy forwarded to the Colonial Office in London before being relayed to the Governor-General of Australia,

Lord Northcote (Henry Stafford Northcote). Duffield forwarded copies of the resolution to all the major observatories across the world in the hope of broadening support even further.

In the letter to the President of the Solar Union, the Minister for Home Affairs, Senator Joseph Keating, stated that 'it is very desirable that such an observatory should be established . . . but it is considered that it is not desirable at the present juncture to incur the heavy expenditure involved'. The truth of the matter was that the federal government was short of cash after it had invested heavily in the creation of the Commonwealth Bureau of Meteorology. The state observatories had indicated a willingness to undertake solar work but, as the Premier of South Australia again stated: 'My ministers are of the opinion that any system of solar observation and photography which may be established in the Southern Hemisphere should be undertaken by the Commonwealth Government'. It was difficult to argue against this contention, given that astronomy was constitutionally a Commonwealth responsibility.[3] While it was generally accepted that the federal government would eventually provide for an astronomical observatory as a reflection of its constitutional responsibility for this activity (Section 51, Clause viii of the Constitution gave the Commonwealth specific responsibility for 'Astronomical and meteorological observations'), the incumbent government felt that the matter should be delayed until a site for the establishment of a Federal Capital Territory had been selected and developed. This was not likely to happen quickly.

The location of a national capital had been in the minds of those drafting the Australian Constitution. The selection of an appropriate site and its subsequent proclamation were tasks delegated to the federal Parliament. In effect, the 'Federation Fathers' opted for a compromise solution designed to reconcile former colonial rivalries, especially between New South Wales and Victoria, as to which state should host the national capital.[4] Prior to Federation, there was clear and unambiguous consensus: the capital could be neither Sydney nor Melbourne, although the federal Parliament was then sitting in Melbourne. A completely new site would need to be chosen and developed. Australia was, however, not alone in its struggle to create a genuine national capital that was not preoccupied with extant provincial concerns. Many countries have seen once-important historical centres of pre-industrial life eclipsed by rapidly expanding industrial cities with enormous populations, financial wealth and military power. Others have seen great cities with rival claims either refuse to accept the supremacy of the other or ridicule a compromise choice. In what was a thorough compromise, Section 125 of the Australian Constitution empowered the newly created Commonwealth government to consider a site that was in New South Wales, but at least 100 miles from Sydney. The possible locations included Yass, Lake George, Tumut, Tooma, Dalgety and Bombala.

An Advisory Board appointed by the government recommended to the Minister for Home Affairs, Hugh Mahon, that the 'most suitable sites are about Canberra on lands comprised in Duntroon and adjoining holdings'. In October 1909, agreement was reached between Prime Minister Alfred Deakin and New South Wales Premier Charles Wade on the area to be ceded. By the proclamation of the *Seat of Government Acceptance Act* 1909, the area of the newly created Federal Capital Territory was surrendered by the New South Wales government and accepted by the Commonwealth. When there was a change of Commonwealth government in 1910, Andrew Fisher introduced the *Seat of Government (Administration) Bill* into federal Parliament. It became law on 25 November 1910. Having put all the administrative machinery in place, Fisher authorised the issue of a proclamation that the *Seat of Government Acts of Surrender and Acceptance* would come into effect on 1 January 1911. On that date, the Federal Capital Territory came into existence. The Territory thus became a separate political jurisdiction under the exclusive control of the Commonwealth, although the 1909 Act allowed New South Wales state laws to continue 'until other provision is made'. The Territory encompassed 910 square miles of land and a population of fewer than 1,000 people. The southern half of Jervis Bay on the New South Wales south coast was also included in the Federal Capital Territory as a future seaport for Canberra. Where necessary, the Commonwealth government would issue a compulsory acquisition order on all privately held land.

While the Commonwealth government used uncertainty over the site of the national capital to stall for time in the hope of generating more revenue for public expenditures, the recently appointed Commonwealth Meteorologist, Henry Ambrose Hunt, was working energetically behind the scenes to ensure that funding for his fledgling organisation was not diverted to a new solar observatory. Consequently, the matter lapsed for want of federal funds and sufficient political interest. But continuing public interest in the need for such a facility within Australia and overseas meant that the matter would not die. In any event, too much progress had already been made. Offers of donated telescopes for the new observatory had already been made and Duffield had begun receiving and storing them for future use in Australia.

On 12 August 1907, Mr W.E. Wilson, a trustee of the estate of Lord Farnham, had offered a 6-inch Grubb refractor (an instrument that focuses light by means of a lens) with a number of accessories. Farnham's will had stated that this German Equatorial mount telescope should be devoted to 'the best advancement of astronomy'.[5] Lord Farnham (Somerset Henry Maxwell, Bart.) was born on 7 March 1849 in County Wexford and died on 22 November 1900. In addition to having a commission in the army and a substantial land holding, Lord Farnham was an avid amateur astronomer and maintained an interest in natural history and physics.[6] The instrument offered to Duffield was built in 1886 and had been used in His Lordship's own private observatory.

In 1909, Mr James Oddie FRGS of Ballarat offered the Solar Physics Committee a 9-inch Grubb refractor he had been using for observations on Mount Pleasant on the condition that 'it form the nucleus of a solar observatory anywhere in the Commonwealth'. In July of that year, Duffield visited Oddie, a former member of the Victorian Parliament, and inspected the telescope on offer. Although its precise date of manufacture could not be ascertained, it was likely to have been around 1889. The Victorian Government Astronomer, Mr Pietro Baracchi, also examined the instrument. Both agreed that it was acceptable for the envisaged tasks and proposed that Prime Minister Deakin be advised officially of Oddie's offer. In a letter to Oddie, the Prime Minister wrote: 'Permit me now to express, on behalf of the Government and people of Australia, most cordial thanks for your splendid gift'.[7] Baracchi and Duffield both felt that the aged but still very alert Oddie (he was born in 1824) should become a member of the Solar Physics Committee. To keep the momentum going, the Committee published a *Memorandum Upon the Proposed Solar Observatory in Australia*[8] to describe the project to a still curious nation. In the preface, the Committee explained that:

> The study of the Sun is important from two points of view—one is purely scientific, and the Sun is regarded as typical of one stage in the evolution of scientific bodies; the other takes account of *his* unrivalled influence upon the physical and meteorological conditions of the Earth (emphasis added).[9]

One suspects the public were previously unaware that the Sun was best referred to with the masculine pronoun! The principal practical reason for an Australian observatory was the need to keep the Sun under continual observation for 24 hours each day and that this continent was well placed between India and America, with long hours of daylight and clear air. These were ideal conditions for observation.

The memorandum emphasised that there was no such facility in the Southern Hemisphere to cooperate with work then being undertaken by the Smithsonian Institution in Washington on the spectroheliograph and on solar radiation. Furthermore, the initiative was supported by the Mount Wilson Solar Observatory in California, the Royal Observatory of Catania and Etna, and the Society of Italian Spectroscopists. As for the cost:

> the Committee estimates that £10,000 will be required for the equipment, the erection of the necessary buildings, and the housing of the telescopes and apparatus that have been offered. It is further estimated that £1,500 per annum is the minimum sum required for the upkeep of the Observatory. It is hoped that this money will be forthcoming without delay, that the chain of Observatories round the earth may no longer remain incomplete, and that Australia may take her place among the nations by establishing an Observatory for the study of the Sun which will both by encouragement of original research and by its cooperation with other nations, assume an international importance.[10]

On 10 February 1908, the Royal Society in London joined the litany in expressing its strong support for the establishment of a solar observatory in Australia after receiving some skilful lobbying from Duffield. The Society asked Duffield to comment on the resolution that he himself had originally drafted. In a formal statement, the Royal Society was:

> strongly of the opinion that the foundation of a Solar Observatory in Australia is desirable, or else, as an alternative, that provision for systematic solar observations, including an adequate staff, should be made at one of the existing observatories. The latter alternative is suggested on the understanding that the important work on zones of the *International Astrographic Catalogue*, which has been undertaken by three of the four Australian observatories, namely, Melbourne, Sydney and Perth, shall not be disturbed.[11]

The British Association for the Advancement of Science was similarly interested, and at its September 1908 meeting in Dublin formed a committee to cooperate with Australian astronomers in furthering the proposal. The committee was chaired by Sir David Gill, the recently retired Chief Astronomer at the Cape of Good Hope, with Duffield serving as secretary. A grant-in-aid of £50 was made available with Duffield despatched to Australia in late October 1908 with the task of generating public interest and additional funds. The principal purpose of his return to Australia was his forthcoming marriage (which took place in June 1909), but he used the occasion to further his professional interests as well. Generating interest in a solar observatory was still no easy task, despite remarkable astronomical advances being made elsewhere. As Rosaleen Love has remarked:

> the external factors that limited the strategies open in the first decade of the twentieth century included the following: competition for limited financial resources in the newly federated Commonwealth; the conflict of interest between meteorology, with its power base within the Commonwealth Department of Home Affairs, and astronomy, which had bought its autonomy from meteorology in 1907 at the expense of staying a state-funded activity; the ensuing Commonwealth–State rivalry, in this as in other features of Australian political life; and the limitations of the Australian university education system in providing only a basic undergraduate training in science to a few students, so that there was no local pool of excess scientific talent.[12]

In proposing a solar observatory for Australia, Duffield held the view that it was the duty of governments to make provision for science, but he acknowledged that there was substantial scope for non-government bodies to contribute as well. He also used the potentially debilitating federal–state rivalry productively by harnessing the positive effects of state and regional parochial sentiments of pride and prestige. Duffield was one of the first Australians to identify the harmful effects of what has recently been referred to as the 'brain-drain', and the need to create opportunities for young scientists in Australia to prevent them seeking employment abroad. While Duffield himself was not yet a leading physicist or well-known astronomer, he displayed natural ability as a polemicist and advocate for science in the political and public realms.

The Australasian Association for the Advancement of Science met in Brisbane in January 1909. After deliberating on the best way ahead, the Association formed the Australian Solar Physics Committee. The Commonwealth Statistician, Mr George (later Sir George) Knibbs, chaired the committee, with Duffield acting as secretary. Its charter was to lobby government for a solar observatory in Australia as a 'matter of national as well as international importance'. It was agreed that the new solar observatory should be an independent and separate public entity but affiliated with the state observatories and the universities. To establish this vital link, the Professors of Physics at the Universities of Sydney, Melbourne, Adelaide and Tasmania were appointed members of the Committee, together with the two Government Astronomers, Pietro Baracchi and Professor William Cooke, and Senator Joseph Keating and James Oddie. A grant of £50 was also made available.

The Minister for Home Affairs, Hugh Mahon, told a deputation from the Committee on 1 April 1909 that he realised the importance of the plea for an Australian observatory and was sure that the financial aid required was small by comparison with the value of the data to be obtained. He thought the federal Parliament would not be less public spirited than private citizens and would probably give £1 of public money for every £1 raised by public subscription for the erection and equipment fund, and might maintain the observatory after its establishment—'but not at present'. The Minister assured the deputation that its 'requests were very moderate and [that he] was in entire sympathy with the project'. However, the Labor government of Andrew Fisher was replaced by that headed by Alfred Deakin and a fresh approach for government support had to be mounted. But the spirits of the observatory's proponents were not diminished and Prime Minister Deakin was receptive to their pleas.

The proposal was widely discussed at a public meeting held at the Melbourne Town Hall on 19 October 1909, hosted by the Governor-General, Lord Dudley (William Humble Ward). There was additional support from the Governor of Victoria, Sir Thomas Gibson-Carmichael, and a number of representatives from observatories and universities around the country. Lord Dudley remarked that 'it would be little short of a national misfortune if, for the sake of a few thousand pounds, Australia failed to take her place among the nations of the world in scientific research for which her geographical position had marked her out'. The meeting passed a formal resolution that 'the establishment of a solar observatory is desirable and that the federal government be strongly urged to assume responsibility of carrying it into effect'. This was conveyed to Prime Minister Deakin. After considering the matter in some depth, Deakin announced that the Cabinet recognised the Commonwealth's clear obligations under the Constitution. He approved a proposal for formal submission to Parliament that would provide for the maintenance of a solar observatory 'at a suitable post in the interior of the continent'. The initial Commonwealth contribution was to be £1,500 per annum, with scope for an increase as expansion became necessary. However, the cost of establishing and equipping the observatory to adequately perform its basic functions was estimated at £10,000. The government indicated that it was not prepared to provide that much money at that time but suggested that the observatory could be established with the assistance of a number of wealthy private patrons and that, like the Fisher government, it would match the value of private donations (including the value of donated equipment) pound for pound. Thereafter, the government would fund annual maintenance and modest enhancements. Having secured approval in principle, Duffield then stepped up pressure to have the observatory incorporated formally into plans for the new Federal Capital Territory. As a federally funded institution, Duffield argued that it should be affiliated:

> not with one university but with all the universities within the Commonwealth and not with one observatory but with all observatories in the Commonwealth. The matter is essentially one affecting the prestige of Australian science, and not just the science of any particular state (quoted in Love 1985, p. 185).

With the government's announcement, the establishment of the observatory was rapidly becoming a reality. The immediate fundraising target was £5,000. The 'Farnham' 6-inch telescope was valued at £400 and the 'Oddie' 9-inch telescope at £2,000. Franklin Adams offered a coelostat and James Fowler a pyrheliometer at a combined value of £117. As cash donations had already amounted to £1,500, the Committee had reason to be confident that the residue of slightly less than £1,000 could be found in the near future. With this advice, Prime Minister Deakin accepted the offered gifts on behalf of the Commonwealth of Australia and agreed to fund the modification of the 'Farnham' telescope for use in Australian latitudes. He also offered to place a sum sufficient for the inspection of a suitable site on the government's financial estimates. Baracchi was then instructed to take delivery of the 'Oddie' telescope and accessories. However, in July 1910 he arrived at Oddie's Ballarat house to a less than warm welcome: 'Unfortunately, I found the old gentleman [he was then 86 years old] in a capricious and childlike mood, and after two days of ineffectual persuasion had to return to Melbourne empty handed'.[13] Fortunately, Oddie relented and the instrument and accessories were loaded on to a train bound for Melbourne on 15 July 1910. Baracchi then made arrangements for their repair, alteration, addition and storage (the telescope was 'sewn up in cloth') in Melbourne until the site of the new observatory was settled.[14] Oddie's observatory was also dismantled and transported to the capital. Sadly, Oddie did not live to see his observatory or telescope used by the Commonwealth; he died early in 1911.

Deakin then asked Duffield for an inventory of the main requirements of a suitable observing site. The criteria included: low average rainfall and wind velocity; favourable observing conditions during the months of December, January and February; clearness of atmosphere; good elevation; and an absence of vibration, glare, dust and smoke. In January 1910, the government appointed a five-member board to advise on possible observatory sites *within* the recently

The Oddie 9-inch refractor at Melbourne Observatory Photo: MSO

selected Federal Capital Territory. These included Black Mountain, Mount Mugga, Mount Ainslie, the Cotter–Murrumbidgee River junction and Mount Stromlo. The board's members were Robert McDonald (New South Wales Under Secretary for Lands), R.P. Sellors (New South Wales Geodetic Survey), Henry Hunt (Commonwealth Meteorologist), Charles Robert Scrivener (Commonwealth Director of Surveys) and Pietro Baracchi. Several of the sites were not large enough for either present needs or future development. Black Mountain was rejected partly because the 'glare of city lights' would impact adversely on observing conditions. The committee recommended a ridge of land measuring 1.6 kilometres which was situated 200 metres above the surrounding country, approximately 10 kilometres to the west of the Capital centre, that had recently replaced Mount Ainslie as the meridian being used as the origin of the coordinates for the Federal Capital Territory. Early in 1911 a lease for use of the site was signed between the Commonwealth government and the owner, the prominent Canberra grazier, Frederick Campbell. He would receive an annual rent of £26/15/0 for the next two years until ownership of the land passed permanently into the Commonwealth's hands.

In April 1910, Deakin's administration was replaced by another Labor government, again headed by Andrew Fisher; however, government support remained unchanged. In England, the project continued to meet with interest and additional support was sought from the British Empire League. Further pressure was brought to bear when a group of Australian ministers travelled to England for the Coronation celebrations in June 1911. Prime Minister Fisher was met by a deputation consisting of Sir Frank Dyson and Professor Newall from the Royal Astronomical Society, Lord Avebury from the British Empire League and Sir David Gill and Dr Geoffrey Duffield, both representing the British Association for the Advancement of Science. When the Prime Minister was unavailable, the deputation met the Minister for External Affairs, E.L. Batchelor, who promised that he would personally convey to the Prime Minister and his colleagues the importance to international astronomy of establishing a solar observatory in Australia. He was also asked to ensure that adequate funding for the project was maintained in the forward financial estimates. Batchelor pledged to bring the matter to Cabinet on his return to Australia.

While international support continued to grow, Baracchi was instructed to establish a temporary observatory on Mount Stromlo to test the site for a period of at least twelve months 'from an astronomical point of view' and 'to determine definitely whether it answered the requirements of modern scientific research, including astrophysics'. The building constructed to house the 'Oddie' telescope

The Oddie dome at Mount Stromlo in 1926 Photo: NAA

was commenced in March 1911 when the block, on the highest point of the mountain, was marked out with wooden pegs. It was completed in September 1911 at a cost of £600. The Secretary of the Department of Home Affairs, Colonel David Miller, had suggested a wooden hut with a rotating conical roof of canvas. After all, these were simply site tests and Mount Stromlo might have been deemed unsuitable. But Baracchi, who had to live on the mountain, succeeded in gaining approval for a more permanent structure.[15] The 'Oddie' telescope was then transported to Queanbeyan Railway Station where it was discovered that some 'very rough handling' had caused a loose screw to damage the objective lens. There was no alternative but for the instrument to be returned to Melbourne for repairs. On 8 September 1911, the 'Oddie' was finally ready for use on Mount Stromlo. For one week during each month between September 1911 and September 1912, astronomical observations were made from the site by Baracchi and Dr J.M. Baldwin (Honorary Secretary of the Solar Physics Committee and another Melbourne astronomer who served as Baracchi's assistant and later became his successor), who took turns travelling from Melbourne to Yass Junction by train, stopping to open 34 stock gates before reaching Mount Stromlo. The Observatory building consisted of a dome 6 metres in diameter resting on concrete walls.

Four small wings provided a kitchen and accommodation for the caretaker Robert Magill, two bedrooms for Baracchi and Baldwin, and a photographic room. The building was equipped with beds, quilts, tablecloths, serviettes and a range of cooking utensils. There was also an alarm clock and, after January 1913, a wall-mounted telephone connected to an exchange at Queanbeyan.

> During these visits the two made test observations of the Sun, the visibility of the surface detail of the moon and planets, the visibility of the planetary satellites, and double stars down to 0.62 arc seconds of separation. Among deep sky objects, such favourites as Alpha Centauri, Rigel, Kappa Crucis, Omega Centauri, 47 Tucanae, Eta Carina, 30 Doradus, Theta Orionis and the Magellanic Clouds filled Baracchi's observing log over seventy-five long nights at the Oddie.[16]

In a letter to the Reverend A.L. Cortie (who joined the British Association's committee in 1911), Baracchi stated that they found the atmospheric conditions remarkably good and never had less than four fine days and nights out of seven. As the site was prominent and exposed to the weather, strong winds were a concern, although Baracchi thought their effects would be ameliorated by wind breaks formed by a planned timber plantation. This item of correspondence contains the first mention of the name 'Strom', although why the mountain subsequently became known as 'Stromlo' cannot be established. Baracchi noted:

> Mount Strom is the official name of the hill upon which the observatory stands. I intend to carry on observations of the Sun and take star photographs as well as other star observations for testing atmospheric conditions and definition for a complete year, after which I will report to the Government as to the suitability of the locality for a great modern observatory, more particularly solar.

Baracchi's findings were to be submitted for evaluation to the British Association for the Advancement of Science that was due to meet in Australia in 1914. In the meantime, a 3-inch transit telescope was set up in May 1912 in a small hut near the dome to establish the precise longitude of the facility. There was a period in which this instrument defined the Prime Meridian of Australia from which all other lines of longitude were fixed. The Observatory was officially situated at a latitude 35° 19' 5" South and longitude 149° 0' 42" East.[17] The highest point, not far from the 'Oddie' dome, was 780 metres above sea level. West and north of the site lay grazing land, while a deep valley to the west contained the Murrumbidgee River.

Other than site testing, there was little other progress during 1912 or 1913. On behalf of a group of subscribers, Duffield offered the federal government a portion of the essential equipment required for the new observatory in August 1912. The offer was declined on the grounds that the government preferred to wait until the visit of the British Association in late 1914. The government did not want the observatory to be established as a *fait accompli*. Never lacking in enthusiasm, Duffield submitted a memorandum to the federal government in May 1913 concerning Australia's participation in the international scheme for solar research. The government procrastinated again. Ten months later (in March 1914), the federal government finally replied. Its response was carefully timed to precede the departure of the British Association committee members for their Australian meeting. There was a commitment *in principle* to the observatory, but no more. The government stated that:

> in the scheme for the organisation of services in connection with the seat of Government at Canberra, provision has been made for the establishment among general astronomical studies, of a section to be devoted to solar physics in particular.

Some weeks later, the Administrator of the Federal Capital Territory confirmed that the government proposed to establish a permanent Commonwealth observatory, which would be 'adequately equipped, staffed and administered for the highest and most important aims of modern astronomical research'. The new observatory would be located administratively within the Department of the Interior. It was assumed that Mount Stromlo would be the site, but before making a decision the government was still insisting on waiting until the British Association had formally evaluated Baracchi's site report.

At the same time, the Minister for Home Affairs, Joseph Cook, announced that the federal government did not see any need to acquire, nor was it intending to accept any responsibility for, the existing state observatories. These institutions would remain under state jurisdiction and control. This was good news for Duffield and his supporters. For his part, Prime Minister Fisher intimated that he would meet a deputation of overseas astronomers during the visit of the British Association to discuss the nature of the work that needed to be undertaken in Australia in order for solar astronomy to advance. But before the British Association could meet in late 1914, a Serbian radical in Sarajevo assassinated Archduke Franz Ferdinand of Austria on 28 June. This triggered a series of events that plunged Europe into a long and savage war. As a British Dominion, Australia joined Britain in declaring war on Germany. From August 1914 until

November 1918, Australia was committed to supporting Britain in the war against Germany to the 'last man and the last shilling'. The Great War touched every facet of Australian society and most Australian families. More than 59,000 men lost their lives and many hundreds of thousands more were permanently incapacitated as a result of their war service. Hostilities spread throughout Western and Eastern Europe, Asia Minor, the Middle East, North Africa, New Guinea and into China, where Germany maintained a naval squadron, and the Pacific, where Germany possessed several colonies. Given the dire international situation, the British Association could only express its continuing enthusiasm for the project while seeking an assurance that the observatory would be established as soon as possible after the war.

No one was surprised when Prime Minister Fisher refused to make any commitments about the future, although he invited Duffield to discuss with him the offer of astronomical instruments that the government had previously declined to accept. As a consequence of their meeting, the Prime Minister wrote to Duffield on 20 August 1914 and formally accepted the 'Farnham' 6-inch telescope. The government was prepared for immediate delivery of the instrument and made available £100 for its adjustment and transportation to Australia. A spectroheliograph and a pyrheliometer would be purchased from funds already collected, although their acceptance was scheduled for a later date. The federal government was now in the business of astronomy but Fisher's letter ended with a discouraging note:

> No guarantee . . . can be given when a solar observatory will be established, beyond saying that when times are more favourable the instruments will be erected in an appropriate building and observations conducted.[18]

It was obvious that there would be little progress until the war ended although the 'Oddie' telescope was available for limited observing work. Duffield went back to England where he had secured an appointment as Lecturer in Physics and later Dean of the Science Faculty at the University College of Reading. He was a disappointed but not dispirited man.

From 1916 to 1919, while Duffield served with the Royal Flying Corps, little progress was made on detailed planning for the observatory.

Work on the Mount Stromlo site continued very slowly. To rid the hill of rabbits, stabilise the devegetated slopes and improve the 'atmospheric conditions', the Department of Home and Territories (the department was renamed on 14 November 1916) decided, on the advice of horticulturalist

Charles Weston, that a pine plantation was needed.[19] The planting of pines according to a plan devised by Weston began near the Cotter Road in 1915. A number of untested species were tried but it soon became apparent that the 'Monterey pine', also known as *pinus radiata*, was enjoying the greatest success.[20] Weston had already demonstrated the value of Monterey pine as a plantation timber. After initial plantings appeared to meet the need, the site became a permanent pine plantation which reached the ridgeline and the 'Oddie' dome by 1923. By this time, the plantation covered an area of 476 hectares.[21] As John Gray notes:

> Political support for afforestation remained weak in the early 1920s . . . Weston did however continue with Monterey pine planting on Mount Stromlo. In 1924 he made further recommendations about afforestation of the Cotter Catchment area near the recently constructed water supply dam but with no apparent response.[22]

It was not until October 1924, when tenders were first called for construction of the Observatory buildings, that pine trees covering the area to be developed were cut down.

When the war ended on 11 November 1918, attention returned to Mount Stromlo just as the Lick Observatory in the United States made inquiries about establishing its own research facility in Australia. Professor Hussey was despatched to investigate possible locations. He proposed a tentative site on the Canobolas Range near Orange, some 320 kilometres west of Sydney. The notion of Australia assisting well-funded American research began to gain some support. By 1923, astronomers at the Mount Wilson Observatory in California had joined the chorus of those wanting a properly equipped solar observatory in Australia to ensure the Sun was kept under constant observation. While alternative proposals were being considered and lobbying of the Commonwealth government by a range of groups continued, Mount Stromlo was hosting its first permanent residents. The Cotterill family arrived in 1920 and lived in the 'Oddie' building for the next three years. It was still a remote place and young Master Cotterill went to school at Narrabundah on a horse-drawn school bus. Mr J.C. Cotterill was engaged in meteorological observation and bushfire warning and the Mount Stromlo observations were added to those being recorded at the Meteorological Bureau's other facilities at Acton and 'Duntroon'. The newly constructed Cotter Dam was considered an out-station of the Mount Stromlo facility and provided meteorological data to Mount Stromlo for inclusion in the overall weather picture being developed.[23] Provision was also made at Mount Stromlo for the flying of box kites for use in high-level meteorological observation. An octagonal rotatable hut, known as the 'Kite House', was built close to the meteorological site with wide double doors enclosing a powered winch that operated a drum of fine kite wire. However, there is no evidence available to show whether kites were ever flown from the site.

With Australia showing genuine signs of economic recovery from the war by 1920, Duffield (then still working at Reading) wrote to the federal government seeking an indication of its intentions with respect to the observatory in an effort to head off American overtures. The reply was non-committal. The observatory would indeed be built, although when work would commence was uncertain. Duffield returned to Australia in October 1922 and arranged to meet with Prime Minister William Morris 'Billy' Hughes and leading federal ministers. It was not until April 1923, however, that the government gave a firm commitment to establish the Commonwealth Solar Observatory on a site 'high up among the rocks and lizards and miles from anywhere',[24] to 'link up with existing institutions in England, India, America and certain European countries'.[25] The American aspirations had been thwarted and Australia would become an international player in astronomy. The Commonwealth Meteorologist agreed with the choice of site, advising the Secretary of the Department of Home Affairs that 'the crest of the Mount Stromlo ridge is an ideal site for a scientific community'. But what about the social needs of the community that would be created by the observatory? The Commonwealth Meteorologist claimed some knowledge of human fallibility when he remarked:

> The exclusion of the dwellings, which would be situated on the slopes without the compound, is also important as difficulties are always liable to arise from purely personal reasons, particularly where the families of married officials are concerned.

He was also concerned about the deleterious effects that staff housing would have on the scientific work: 'the roofs of dwelling houses should not rise above the level of the crest of the hill, and the houses themselves should be situated preferably on the south-eastern slopes'. It was resolved that the accommodation would consist of:

> married quarters for one director; married quarters for two assistants; possibly married quarters for one senior mechanic; single quarters for research fellows, one professional officer, two clerks, one apprentice, one janitor, one chauffeur and one cook . . . [while] the layout and planning of the buildings should be such as to allow for expansion.[26]

All of this development would require coordination and leadership. To select a Director for the new facility, the Australian government asked a committee of British astronomers, including Sir Frank Dyson and H.H. Turner, to recommend a suitable candidate. The most obvious choice was Dr Geoffrey Duffield. H.H. Turner later wrote:

> It was as natural that Duffield should be the first Director as that [John] Flamsteed should be appointed the first Astronomer Royal. 'Who is to take charge of the new Greenwich Observatory?' said King Charles II. 'Why, the man who told you there was need of it.'[27]

Duffield was back in Australia lobbying for work to begin at Mount Stromlo when his name emerged as the best candidate for the directorship. But his acceptance was apparently by no means certain. As his daughter remarked:

> My father was not at all anxious to return to Australia in a permanent capacity as he was too much involved with his own work at Reading University and dearly loved England. He was also a very bad asthmatic, having suffered from this condition since the age of four.[28]

However, there was never any serious suggestion that he would decline the offer if and when it came formally from the Commonwealth government. As part of the negotiations, he was offered a salary of £1,350 with a further £700 to be made available for the purchase of equipment.

Duffield spent a week in Canberra during July 1923 and met with Colonel P. Owen, the Director of Public Works, to discuss the general layout of the observatory and additional specific requirements. The residential area was to be on the eastern side of Stromlo ridge to give the dwellings some shelter from the coldest and strongest winds. Sites were also chosen for a laboratory, a workshop and administrative offices near to the proposed location of the telescopes on the summit of the ridge. Provision was also made for a water supply, and for electric light and power to be connected to the residential area. Colonel J.T.H. Goodwin, the Commonwealth Surveyor General, arranged for an immediate survey of the construction area, while Duffield examined and approved plans for the houses for the Director, First and Second Assistants and the Bachelors' Quarters in addition to the laboratory, workshop and office. On 23 August 1923, Duffield returned to the University College of Reading until a successor could be found and the Commonwealth government confirmed his appointment to the proposed 'Commonwealth Solar Observatory', as it would be known.

If it was difficult to underestimate the magnitude of the challenge Duffield faced in 1905, it is hard to overstate the magnitude of his achievement in seeing the Commonwealth Solar Observatory established in 1923. As Rosaleen Love concluded:

> Both consciously and unconsciously, Geoffrey Duffield manipulated the public image of science in a remarkably clever way . . . how one man was able to masquerade as a pressure group; how the principal actor in this scientific drama chose to recast dramatic events at various points and present his calculated actions as the spontaneous actions of others; how the passive voice of proper scientific discourse can be used to conceal motives and actions of the individual, and how the advancement of science can be presented to government as the playing out of some kind of inevitable natural drama where it is not the fate of the gods that hangs in the balance but the fate of the Empire or of nations, should they choose to obstruct the unfolding of inevitable public support for scientific institutions.[29]

Duffield's vision was about to become a reality.

chapter two

The road to Mount Stromlo, 1924–1930

HAVING NEGOTIATED A RELEASE FROM Reading University, Duffield formally became the Director of the Observatory from 1 January 1924. His appointment date coincided with the official establishment of Mount Stromlo as the Commonwealth Solar Observatory. Within three months of assuming office, and while he was still in England preparing for his departure and gathering instruments for the new Observatory, John H. Reynolds, a distinguished British amateur astronomer from Birmingham and later President of the Royal Astronomical Society in London, offered Duffield a 30-inch reflecting telescope with equatorial mounting and clock drive for use on Mount Stromlo. Reynolds had acquired several reflecting telescopes and had earlier (1905) given a similar instrument to the Helwan Observatory in Egypt. Reynolds was born on 8 February 1842 and died on 17 July 1927.[1] He held an MSc and was Director of Higher Education and Principal of the Municipal School of Technology in Manchester. His other appointments included President of the Association of Technical Institutes and Member of the British Science Guild. The telescope he offered Duffield had originally been fitted with a 28-inch mirror (replaced by a 30-inch mirror) and had been made for Reynolds' private observatory at Low Wood in Harborne near Birmingham. Reynolds was also prepared to have the telescope adapted for the latitude of Canberra and packed and delivered to a British overseas port on condition that the Australian government cover the costs of its packing and transportation to Mount Stromlo. This was agreed and the 'Reynolds' telescope was accepted.[2] The cost of transportation—£20—was exceeded by expenditure on insurance—£30. The disassembled 'Reynolds' telescope would finally arrive on Mount Stromlo during 1929. To coincide with its arrival, a telescope building and 26-foot dome were completed. Although the telescope would be available for use in 1930, there were a number of technical difficulties (detailed in Chapter 3) which delayed its efficient use for a number of years.

En route to Australia in 1924, Duffield visited observatories in Canada and the United States, making contacts and seeking advice on the most ideal work to be carried out in Australia. The new Director and his family finally arrived in Australia in November 1924, expecting that the Observatory buildings would be ready for occupation. Although the plans had been approved twelve months earlier, there had been a series of delays in construction. Light and power cables had been extended as far as the residential dwellings, but work on the Observatory buildings had not even begun. Duffield was, rightly, angry at this inauspicious start. In Sydney, Duffield acquired 'a huge Crossley motor car which had a canvas roof and no side flaps to keep out draughts and Canberra's icy-cold winds. This car had belonged to a fleet of cars used by Edward, Prince of Wales, when he visited Australia'.[3] Together with his wife, daughter Joan, sons Peter and Michael, and a secretary, Miss Dorothy Hawkins, the family made their first trip to Canberra. The Crossley struck a straying cow en route near Bungendore during the night, which delayed their arrival until 2.00 a.m.

W.G. Duffield, Pioneer Director of the observatory at Mount Stromlo
Photo: MSO

In time for Christmas 1924, Duffield and his family took up residence in the new Hotel Canberra. As a sign of things to come, the hotel itself was still unfinished, although it was to be the Duffield family home for the next two-and-a-half years. The Duffields were the first to sign the guest register. Three months later, three rooms in one wing of the hotel were made available as a temporary observatory and as a place to receive and assemble new equipment.

In his first annual report dated 2 January 1925, Duffield stated:

> As large a portion as possible of the apparatus required for the new instruments should be made in Australia. It is my earnest desire that we should take our place among the great observatories of the world. It is even possible that we may lay the foundation of a lens-making industry in Australia. This might have far reaching importance in the event of our ever becoming isolated, and the men trained in the works might ultimately develop an industry which would make periscopes for submarines, range-finders, theodolites and field glasses.

These were prophetic words. They were also eminently practical because it would be entirely impractical for the Observatory to be totally reliant on the supply of British equipment. A few days after the first annual report was completed, the *Sydney Morning Herald* covered the arrival in Australia of the 'Reynolds' telescope, quoting a statement from Duffield that the instrument would be used 'for examining spectra of the stars and obtaining information as to the distribution of energy from which it is hoped the temperature of the stars can

The Hotel Canberra pavilion in which the Commonwealth Solar Observatory was first housed Photo: MSO

be ascertained'.[4] As soon as the Observatory was established on Mount Stromlo, real scientific progress could be made and the Australian parliament and people shown the fruits of their investment in astronomical research. But construction work all over Canberra was proceeding very slowly.

The need to liaise with staff of the Department of Home and Territories based in Melbourne, and particularly the absence of any technical support in Canberra, hindered the development of the Observatory site. Duffield spent many days and nights during 1925 travelling by train to and from the Victorian capital. However, this did allow him to see his children, who were studying at boarding schools in Melbourne. Throughout his first year in Australia, Duffield recorded the difficulties he faced in having the facilities constructed to his exact specifications. His diary entry for 23 July 1925 gives an insight:

> Inspected Administrative Block. Contrary to my previous representations to the Chief Architect there was no water in the rooms to be used temporarily as workshop and laboratory! There was no alteration to the number of windows in the Library and Director's rooms as had been requested. They will be impossibly draughty. No progress has been made in taking the electric light and power up to the Oddie building although this had been promised by the Electrical Department of the Federal Capital Commission some months ago.[5]

Duffield later met with the Chief Commissioner of the Federal Capital Commission and conveyed a general lament about 'the lack of interest taken by the Commission in the Observatory development and their repeated failure to take action'.[6] Duffield believed the Commissioner had more delegated authority from the Minister than he was prepared to exercise, and this frustrated construction of the Observatory. There were occasions on which the Commissioner seemed reluctant to authorise even minor expenditures, such as on office furniture, without seeking specific ministerial approval. To make matters worse, the Observatory staff were frequently asked to vacate some of their rooms when the hotel found guests willing to pay a higher tariff than the Department of Home and Territories. One such occasion involved a party of polo players, leading Duffield to comment in his diary: 'An instance of the sacrifice of scientific interests for pecuniary gain'.[7] These practical difficulties were exacerbated by problems with the Observatory's funding arrangements. Some of the monies allocated to site development had been absorbed back into Commonwealth consolidated revenue when, due to building delays, they had not been expended. Furthermore, Duffield believed there was insufficient legislation to support the Observatory's operations and future expansion. He felt very much at the whim of government and a fickle political climate. Although the Observatory had been an expensive undertaking for the Commonwealth government, the money originally donated by the public to assist in the establishment of the facility had not been spent and now amounted to more than £2,000. Rather than expending the money rashly, Duffield decided to create a 'Foundation and Endowment Fund'. Notwithstanding numerous administrative hurdles, the Fund was finally constituted by an Act of the federal Parliament in 1930. The legislation provided for the appointment of three trustees who would make the interest generated by the invested capital available to the Director every year to 'expend for the purpose of the Observatory'. In 1925, however, Commonwealth funding was assured for the immediate future.

It was now time to begin attracting the Observatory's foundation staff. The positions of Director, First Assistant, Second Assistant, Research Fellow and Mechanic were soon to be formally created, although Duffield had authority to interview applicants in anticipation. It is noteworthy that he was embarking on a recruiting drive at the very time there was a dearth of solar astronomers. As he did not have the opportunity or the capacity to nurture young local astronomers, Duffield had to attract staff from overseas institutions. His first recruit was William Bolton Rimmer from the Norman Lockyer Observatory (also known as the 'Sidmouth Observatory'), a small private astrophysical institution in Devon, England. 'Bill' Rimmer was born in Derby, Lancashire in 1882 and studied at the University of Liverpool. He began his working life as a surveyor in British Columbia and joined the Royal Engineers when war broke out in 1914. He was commissioned and held the rank of captain. Once demobilised and at the age of 36, Rimmer decided to become an astronomer. He attended courses at Imperial College, London from 1919 until 1921 when he gained employment as a Research Assistant at the Lockyer Observatory. When he was recruited in 1924, he was the only trained astronomer that Duffield planned to engage. The other members would be geophysicists or physicists with some interest in astronomy. Rimmer was appointed Assistant Solar Observer in 1924 with a salary of £816. He married Kathleen Hayes of Queanbeyan in 1927 and was to serve at Mount Stromlo until his death on 2 July 1945. There is evidence to suggest that the relationship between Duffield and his 'deputy' had deteriorated by 1927 and that this in part contributed to Duffield's later decision to recruit Arthur Hogg at a comparable level to Rimmer. But whereas Rimmer was largely indifferent to status and position, Hogg was more ambitious and conscious of his standing at Mount Stromlo.

B.P. Clark was employed as a mechanic. Although he was a married man, both he and Rimmer lived in the 'Bachelors' Quarters' in Acton until residences were completed on Mount Stromlo. Later, Clark would fall and die in a Queanbeyan street (on 23 February 1933). H.J. 'Jim' Banham was recruited in July 1925 as junior mechanic and later workshop foreman. There was also a clerk, W. Blackman, and a groundsman, Zuriel H. Nowland. By the end of 1926, Duffield had approval to appoint two groundsmen. Their duties included 'cleaning, attention to grounds and gardens, drive van for mail, supplies, goods etc., and to attend to water supply'.[8]

Duffield then recruited additional professional staff: A.L. Kennedy (in September 1925), Arthur John Higgs (in September 1926) and William Eric Duncanson (a Sydney University physics graduate who would leave Australia in 1930 and spend the rest of his life teaching physics abroad). Higgs was born in Sydney in 1904 and was another physics graduate with a special interest in the properties of the ionospheric layers and radio propagation. He would later build equipment for use in photoelectric photometry and remain at Mount Stromlo until 1940 when he joined the CSIR's Radiophysics group where he remained until the end of his working life. Duffield also arranged for W.H. Love to be appointed as a Research Fellow. This particular appointment proved to be an unhappy arrangement on both sides.

On 25 August 1925, Love was 'reprimanded for having a filthy collar'. Duffield remarked: 'Unless this boy becomes cleaner and tidier in his person it will not be possible to retain him on the staff'. The following day he 'again appeared in a collar which would disgrace a labourer'. Love was again reprimanded by Duffield and 'ordered back to his lodgings [Hotel Ainslie] and put on a clean one'. In response, Love submitted his resignation 'as the result of his being insulted'. Two days later, Duffield accepted Love's resignation with some regret. In his diary entry for 28 August 1925, Duffield lamented:

> He seemed surprised that any exception should have been taken to his personal habits. I told him his collar had become notorious and that it was a matter of frequent comment . . . I fancy he is quite unaware of his dirtiness and the bad impression he creates.

Clabon Walter Allen replaced W.H. Love as a Research Fellow in January 1926. Allen, known universally as 'Cla', was born in Perth on 28 December 1904. His father was a lecturer in physics at Perth Technical School, but died when Cla was aged just seven. Cla enrolled in a science degree at the University of Western Australia in 1923 and graduated three years later. Because he gained a distinction in physics in his degree examinations and his name came to Duffield's attention, Allen was awarded a research fellowship at the Observatory in January 1926 and became an assistant in 1928.

Shortly after he arrived at Stromlo, Allen began spectroscopy work using some uncalibrated plates of the copper-arc spectrum taken earlier by Duffield in Manchester.[9] He was soon obtaining his own copper-arc spectra, a dangerous exercise during which he nearly killed himself when he touched the live terminals and was saved only by throwing himself forward until he broke the circuit. On completion of the solar telescope tower on Mount Stromlo in 1931, Allen began studying the intensity of Fraunhofer lines—dark lines that cross the solar spectrum and represent the absorption of certain wavelengths of radiation by specific elements in the Sun's atmosphere. His photometric results provided large-scale tests of the then current theory of the emergence of light from the Sun's atmosphere and contributed directly to analysis of stellar spectra, thereby having an important influence on the development of stellar spectroscopy generally. He completed his honours examinations and thesis while working at Stromlo, showing a wonderful aptitude for research, and then went on to complete his MSc and his doctorate, based on research and papers written for publication in the Observatory's *Memoirs*.

As a result of his work, Allen was awarded a Hackett Research Scholarship for two years' work at the Solar Physics Observatory in Cambridge and later at Mount Wilson. As the Commonwealth Public Service allowed only a year's leave of absence, an Act of the federal Parliament was required to permit Allen to accept the scholarship while retaining his position at Stromlo. While overseas, Allen familiarised himself with astronomical work then underway in Britain and managed to re-acquaint himself with Lord Rutherford, whom he had met briefly in 1925 after a public lecture in Perth. On returning from overseas, Allen produced a comprehensive photometric atlas of the solar spectrum, which immediately gave the Observatory international status. It is widely considered to be the Observatory's most important achievement during its first fifteen years.

The astronomers worked as a team and frequently assisted with each other's research projects. Rimmer and Kennedy, from the Adelaide Observatory, were both concerned with stellar problems rather than solar research. Rimmer had experience in stellar spectroscopic parallax observations and would adapt the 'Oddie' telescope for similar work. Kennedy's experiences with Douglas Mawson in the Antarctic and in surveys across Australia were crucial in establishing the new observatory. However, Kennedy never felt settled at the Observatory and would leave Mount Stromlo in November 1927. Allen was embarking on a research fellowship with a special interest in spectrum-line photometry at Duffield's suggestion. Arthur Higgs had a similar background to Allen and would prepare equipment for ionospheric studies. With his postgraduate qualifications

C.W. Allen at the Sun Telescope
Photo: MSO

in chemistry and valuable industrial research experience, it made sense to appoint another new staff member, Arthur Hogg, to concentrate on atmospheric ion problems.

Dr Arthur Robert Hogg was born at Creswick in Victoria on 25 November 1903.[10] After studying at the Royal Melbourne Institute of Technology, he graduated from the University of Melbourne in 1923 with first-class honours in chemistry and the Dixson Scholarship. He gained an MSc two years later with the Kernot Scholarship. His first employment as an industrial research chemist was with Broken Hill Associated Smelters at Port Pirie in South Australia, where he served as Assistant Superintendent of Research until 1929 when he was recruited by Duffield to work at Mount Stromlo. Hogg would work for the rest of his life at Mount Stromlo, where he lived with his wife, Irene Doris Randell, whom he married in 1933, and two sons and a daughter. On arriving at the Observatory, Hogg began research into a number of atmospheric electrical phenomena, particularly the conductivity, ionic mobilities and ion balance in the lower atmosphere. He was a very practical scientist and certainly had the greatest expertise in electronics at the time of his appointment.

Although this diverse group of scientists showed a readiness and a willingness to work together, the consensus of the staff was that Duffield was not a naturally gifted organiser nor an especially talented leader of men. Allen kept a diary and commented frequently on the Director's traits and the impact of his family on Observatory life:

> Today the Dr [Duffield] set me going on some hair-brained arrangement of lenses, which simply can't be done with the apparatus in stock . . . The Observatory is getting into a frightful state of disorder and the Dr's opinion never coincides with anyone else's . . . Mrs Duffield wanted to count the linen in our rooms [in the Bachelors' Quarters] but we locked our doors and she had to do without.[11]

This criticism is probably somewhat overstated, however. Duffield was obliged to attend to diverse matters and to contend with a wide range of competing interests. By way of example, the arterial road from Canberra to Mount Stromlo and the roads formed on the summit were never adequate or safe. They were unsealed and unstable and access was unreliable. In heavy rains, the road around the incline was either washed away or left with corrugations and deep gutters. The 'Oddie' building, which had already been standing for nearly fifteen years, also required considerable attention before it could be used. The dome had to be overhauled, the windows fitted with shutters, the water tank repaired and the interior cleaned and painted. It was then discovered that the instrument was not sufficiently rigid for astrophysical work and would have to be remounted on a firm base and a portion of the standard recast to suit the latitude of Mount Stromlo. This was only realised after the telescope toppled over on 16 September 1925. Duffield noted in his diary:

> At 3.30pm I was called to the telephone by Mr Rimmer . . . requesting my immediate presence as there had been an accident. No one hurt but the instrument had overturned. He was too agitated to tell me the extent of the damage . . . On arrival at the dome in pouring rain, Banham and Clark were standing like morons by the overturned telescope. The section holding the lens and prism was dangling mid air at the end of the pulley block. Thank God. The telescope had fallen towards the east wall and part of the main castings were broken and the declination axis also smashed beyond the bearing . . . Fortunately no damage to any of the three workers and no damage which cannot in time, I think, be repaired.

Most of the repairs were effected by the end of 1925. Additionally, a camera replaced the eyepiece in the 'Oddie' and a prism was fitted in front of the objective. The telescope was finally ready for use in February 1926 and a program of spectroscopic parallaxes of southern stars was begun. In 1929, a Dallmeyer 6-inch astrographic camera was mounted on the telescope. The intention was that it should be available for incidental purposes, such as the photography of comets. In the period January–March 1931, 68 plates were taken of the minor planet Eros.

Given the many demands on Duffield's time and attention, it was to be expected that some confusion and a lack of clarity among the staff about their duties would result. With the Observatory buildings still under construction and the staff arriving in Canberra, assembling equipment and carrying out calibration trials at the Hotel Canberra kept everyone fully occupied. It was then discovered that some of the equipment was either damaged or defective as a consequence of its transportation from England or prolonged storage. Several key items had rusted and progress was slow. But Rimmer and Clark still managed to arrange a workshop and laboratory. The first time service was based on two slave Shortt clocks, one master free-pendulum and two chronometers. These were all set up and in use by 1927, with the sole exception of the master free-pendulum which was installed in 1929. All clocks were set to Eastern Standard Time (EST), with one of the slave clocks set at Sidereal Time. When the master–slave combination was operating in 1929, a set of radio time comparisons was initiated. The main comparison stations were Sydney, Melbourne, Adelaide, Honolulu, Germany and Malabar. A new long-wave receiver for time signals entered service in December 1931. A Shortt clock No. 31 was also added to the Observatory's time service equipment in 1931 and erected in a temperature-controlled room of its own in the basement. A new short-wave receiving set was completed for the regular reception of overseas signals for comparison with it.

Duffield was pleased with what had been achieved in scientific research, given the many practical difficulties associated with creating a modern observatory from scratch. He was also encouraged by the number and calibre of the staff he was able to employ. Even as his institution began to make progress, there was a bigger agenda. Duffield remained concerned about the cause of astronomy elsewhere in Australia. In October 1925, he wrote a letter of complaint concerning the proposal of Jack Lang's New South Wales Labor government to close the Sydney Observatory. Although astronomy was a Commonwealth responsibility, Duffield believed that useful work was being done at the Sydney Observatory and that the state government (Minister for Finance) and the University of Sydney rather than the Commonwealth government should be prevailed upon to support the work. He might also have feared the possibility of some of his funds being diverted to a state observatory. However, inasmuch as Duffield was devoid of parochial sentiment, it is most likely that his support for the Sydney Observatory was an expression of genuine solidarity.

With the basic requirements of an observatory in place, the staff could turn to their primary function: the study of solar radiation using the 45-foot solar tower telescope. It was comparable to any similar instrument in service anywhere else in the world. The Sun's disc was examined and records of sunspots were maintained. The pyranometer despatched from the laboratory of Professor Ångström of Uppsala was erected on one of the chimneys of the Hotel Canberra with its photographic recording facility arranged in a screened-off portion of a passage. The pyrheliometer was later delivered, enabling the records to be standardised.

A Benndorf electrometer was installed in one room and continuous records of atmospheric electric potential gradient (electric field) commenced. In another room, the intensity and breadth of spectrum lines on plates Duffield had recorded in England were measured with a Moll microphotometer. When a fourth room was secured, it was used to accommodate a Hilger quartz spectrograph. Duffield provided his own grating spectra of pressurised metallic arcs he had brought from England. In 1926, atomic spectroscopy had reached the stage where spectra were being analysed into multiplets, and certain 'sum-rules' were being suggested from which to calculate the relative intensities of lines within multiplets. Allen started to measure the intensities of any lines that could be produced easily and which were in multiplets that appeared reliably classifiable.

The readily available copper-arc spectrum included recently identified multiplets suitable for measurement. However, these multiplets were found to be quite abnormal. Within all other multiplets, the character of the lines was the same throughout. In the copper-arc multiplets, about half the lines were broad and the other half sharp. Moreover, the broad-line group changed intensities with respect to the sharp-line group when there was a change of arc conditions. The multiplets plainly deserved further study. First, the relative line intensities were measured with the help of the quartz spectrograph and the microphotometer. In 1929, the 3-prism tunnel spectrograph became available and line breadths were measured. It was found that the broad lines retained their breadth when the arc pressure was lowered although they tended to disappear. Two papers were written giving the full experimental results. It was then revealed that the anomalies could be explained by A. Shenstone's new theoretical concept of auto-ionisation. The copper spectrum measurements were the only ones that could illustrate the effects quantitatively and, as such, were later quoted in spectrographic textbooks.

The type of work commenced in 1925–26 established a pattern that was followed as the Mount Stromlo site gradually became available for scientific research. By January 1926, two years after its creation, the Observatory occupied a suite of six rooms at the Hotel Canberra. To his great relief, Duffield was advised that permanent astronomical work could commence at Mount Stromlo, where the buildings would be available for occupation by the end of 1926. However, problems on the mountain remained. Building was still proceeding slowly, the sewerage system was deemed inadequate and additional work was authorised. Happily, Duffield was satisfied with the substantial reservoir with a capacity of over 4,500 litres that had been built near the summit. He told the Minister: 'It will give the settlement an ample supply of water for many years to come'.[12] Some progress had also been made on the staff tennis court. This was practically the only provision made for staff recreation at the Observatory. It was built with Commonwealth funds and would be completed and available for use in January 1930.

Preparations were then made for the relocation to Mount Stromlo from the Hotel Canberra. All of the temporary equipment was loaded on to a horse-drawn cart for the short but difficult journey up the hill. Mount Stromlo was 11 kilometres from the Hotel Canberra and more than 200 metres higher in altitude. The main Observatory building and staff houses, located a short distance away, were first occupied in December 1926. In addition to the equipment, the central Observatory block contained offices and workshops which were available shortly afterwards. The 'Oddie' dome continued to house the 'Oddie' telescope and the recording pyrometer. The 'Kite Room' contained the meteorological equipment and was the base for potential gradient observations. It was secured in a fixed position in 1927 and the winch removed. Building continued into 1927, with the addition of the eastern and western wings of the main Observatory building.

The Commonwealth Solar Observatory, c. 1928 Photo: NAA

The solar laboratory was located in the east wing of the main Observatory building, while the laboratory itself was situated in the basement. It was available for all spectroscopic uses by the end of 1928. This part of the building also contained a covered tunnel for a large spectrograph. The ground level included a library and lecture room. The library contained a vertical concrete column that would accommodate the Sun-telescope beam when that instrument was finally available for use. The west wing contained the workshops, an electrical substation with transformers, a DC generator and a storage battery set. This wing was fully in use by the close of 1928. The dome for the 'Farnham' 6-inch telescope surmounted the electrical sub-station and included a photographic dark room. Although the dome was not completed and assessed as adequate until October 1928, the instrument was installed during July 1928 and a series of tests conducted.[13]

A photoelectric program was planned for the 'Farnham' telescope. It was to include a study of variable stars and photometry of planets and satellites with the overall objective of gathering independent evidence on solar variability. A Gunther & Tegetmeyer stellar photometer unit with cells and filters was acquired for these purposes, but it proved to be less than well suited for use with the small, 6-inch telescope, although considerable time and ingenuity were invested in the investigation and elimination of factors tending to produce errors. Moreover, its light-gathering power in relation to the cell sensitivities was not sufficient for any extensive development of the programs. The Observatory staff decided in 1930 that the photometer was too limited in its usefulness and it was removed from the telescope.

As the staff began to deal with the many technical and logistic problems associated with occupying the new building and operating the equipment, the Duffield family relocated to Mount Stromlo. They initially camped in the First Assistant's house until the Director's spacious two-storey residence known as 'Observatory House' was completed. Joan Duffield recalls: 'My mother wasted no time in getting the garden growing. Likewise, we all put our hands to the wheel and made an orchard on a small plot of vacant land. Here we planted apples, pears, oranges, lemons, peaches, apricots, plums and nectarines'. Her parents were also thoroughly involved in Canberra's emerging community life:

> My father was a foundation member of Canberra's first Rotary Club and was on the Board of the Canberra University College. My mother was a foundation member of Canberra's Mothercraft Society and a committee member of the YWCA. My parents were staunch supporters of the very beautiful little Anglican Church of St John the Baptist in Reid and

> my father was a member of the Parochial Council. Each Sunday my mother drove down from Mount Stromlo to play the organ at both morning and evening services. They were both devoted to music; my mother played the violin and my father the double bass. They were instrumental in forming what they called the 'Stromberra Quintet'. Each week they would trundle down the mountain in the old car with the double bass slung between the front and back seats, and spend an evening with other members of the Quintet rehearsing for an upcoming concert. These concerts were usually held in the old Causeway Hall in Kingston or in the Albert Hall when it was completed in 1928. The proceeds went to various charities, buildings or church organ funds.[14]

When 'Observatory House' was finally occupied, the Duffields hosted musical evenings for staff and visitors with recitals from a selection of instruments including the family's Steinway grand piano. In addition to those invited to Mount Stromlo for social occasions, the Observatory attracted a procession of official guests once scientific work began. Prime Minister Joseph Scullin visited with his wife; the Prince of Siam, Purachartra, and his entourage also travelled to Mount Stromlo. In fact, there was such interest in the Observatory that Cla Allen remarked in his diary that 'these visitors are taking up a lot of our time'.[15]

Dr and Mrs Duffield played double bass and violin in an early Canberra quintet
Photo: Australian Academy of Science

Despite the number of new residents, Mount Stromlo was still a remote place. Domestic supplies were obtained weekly from Queanbeyan, while the groundsman also delivered bread and milk that was distributed among the residents. At the Bachelors' Quarters, a residents' committee was formed and a cook-cleaner was employed. Black Orpington fowls were one of the varieties of domestic animals introduced to the mountain. The staff used their recreational time to play ping-pong, billiards, mah-jong, bridge and chess, as well as playing golf and bushwalking. But for the single men life was frequently uneventful and boring. Allen recorded in his diary: 'The monotony of this place led Jim, Dave and I to walk down to the gate and back in our pyjamas just to have something to do'.

By 1929, Duffield's vision had taken concrete form and the foundation of the new institution had been established. The equipment had been installed and the staff had been recruited and accommodated. Observing at Mount Stromlo commenced immediately according to the pattern begun at the Hotel Canberra, with meteorological observing and recording the only additions to the staff's duties. Solar radiation observations were made daily and records of sunspots continued to be kept. In early 1928, two Abbott silver disc pyrheliometers were obtained and adopted as standards for calibrating the Ångström instrument to the Smithsonian scale. Atmospheric potential gradient measurements with the Benndorf electrometer proceeded. The work on the luminosity of the night sky went on.

The weather data first recorded by J.C. Cotterill in 1920 were also continued with the maintenance of a full set of automatic meteorological instruments. Both the Cotter Dam authorities and the Mount Stromlo foresters were interested in rainfall and evaporation. From 1927, the instruments maintained at the Observatory included continuous thermograph, barograph, bright sunshine and rain recordings. Each day at 9.00 a.m, 3.00 p.m and 9.00 p.m, regular readings were taken of temperature (wet bulb, dry bulb and ground), wind speed and direction (using a cup anemometer), rainfall and descriptive conditions including cloud, wind, visibility and special features. In 1929, a Halliwell-Dines non-integrating Anemo-Biagraph recorder was installed. Hourly readings were scaled and tabulated from these continuously recording instruments with the data sent on a monthly record sheet to the Meteorological Bureau. The meteorological observations were of value in processing other data. Then tragedy struck.

The winter of 1929 was very cold and snow lay thick on the ground at Mount Stromlo. A severe influenza epidemic made its way to the Observatory to which Duffield fell victim. He could not be moved owing to the snow and the frozen roads. After a nine-day illness during which his daughter was summoned home from her boarding school in Frankston, Walter Geoffrey Duffield died on 1 August 1929. The staff were devastated. It had been his vision that had led to the establishment of the Observatory and his persistence that was so crucial in ensuring the Commonwealth government completed the project. Duffield played his Director's part very well. He decided in general terms what types of observation, experiment or analysis should be made and then acquired the necessary equipment. He also had to plan the essential buildings for such activities and to get them started before the Observatory would begin. It was Duffield who decided that the Observatory work should be expanded to include studies of the Sun, geo-physics, spectroscopy and southern sky astronomy.

Most appropriately, Duffield was buried on Mount Stromlo in a grave overlooking the evolving Observatory to which he had contributed so much. The impromptu establishment of a cemetery on Mount Stromlo caused a controversy in the Department of Home Affairs (as the Department had again been renamed on 15 December 1928). Although the Duffield family wanted to secure custody of the burial ground for future use and permanent preservation, Departmental officers were concerned about its status and protection as Commonwealth territory. The transfer of title to the land was not a straightforward matter, but eventually the Commonwealth agreed to the creation of a Board of Trustees to

Pall bearers at the funeral of W.G. Duffield, 5 August 1929 Photo: NAA

oversee the land. There was also an ecclesiastical issue. Was the site a duly consecrated burial ground? The question was settled by the Anglican Bishop of Goulburn, the Right Reverend Lewis Bostock Radford, who wrote to the Minister for the Interior on 24 August 1933: 'For the permanent information of your Department, I hereby certify that on 5 August 1929 I consecrated on Mount Stromlo a space of ground for the burial of the late Dr Duffield, large enough for and intended for the burial of his widow when the time shall come'.[16] Doris Tennant Duffield was buried on the site after her death in 1956.

Geoffrey Duffield's headstone carried an inscription from *The Torchbearers* by English poet Alfred Noyes: 'Take thou the torch. Carry it out of sight into the great new age I must not know, into the great new realm I must not tread'. A cross was inscribed with the motto of the Royal Flying Corps—*per ardua ad astra*. A memorial plaque in the Church of St John the Baptist was eventually dedicated in 1954. The memorial bears a representation of the Sun telescope mounted on the Observatory library with symbols of heavenly bodies in each corner. Duffield's published work had covered a range of scientific concerns from applied physics to theoretical astronomy. His earliest interests had been the effect of pressure upon arc spectra, the consumption of carbon in the electric arc, the determination of gravity at sea and the luminosity of the night sky.

Duffield's own researches as Director were spectroscopic, and he correctly judged that the spectroscopy needed for the Observatory would involve the photometry of spectrum lines. It was on this basis that he acquired the recently invented Moll microphotometers, the quartz spectrograph for laboratory studies and the 3-prism spectrograph for the solar laboratory. These made possible the spectroscopic researches already described. With a mind to future study of the southern skies, he had accepted the three telescopes when they were offered. For solar studies, he ordered the solar tower telescope, the spectrohelioscope and many pyrheliometers. Although his colleagues did not regard him as a genius or a physicist who would break new ground, he was a highly intelligent, industrious, creative and gifted man.

The staff continued without Duffield. However, they were to suffer their next setback within four months when the Great Depression followed the Wall Street stockmarket crash of October 1929. The effects were felt almost immediately in Australia. Unemployment, already high at 10 per cent, jumped by another 11 per cent within twelve months before hitting 32 per cent in mid-1932. Factory output fell by 10 per cent in 1929–30 and a further 30 per cent in 1930–31. With unemployment increasing and production decreasing, the government's income and revenue base was rapidly diminishing. In Canberra,

Commonwealth construction programs came to an immediate halt and the Federal Capital Commission, the body charged with oversight of Canberra's development, was abolished. Workers were laid off and new projects were abandoned as the state governments struggled even to understand the causes, let alone the remedies, for this cataclysmic upheaval in the capitalist economy. The Hotel Canberra and the Hotel Kurrajong were the only two of Canberra's five hotels to remain open. In this context of social disruption, poverty and despair, all official expenditure was reviewed and reductions achieved where possible, as governments tried to alleviate suffering while limiting the size of their deficit budgets. By 1930, many people were desperate just to survive, although there would be a respite for some in 1931 when the Federal Highway, linking Canberra to Goulburn via Collector, was completed by unemployed men working for 'relief money'. With a population of just 7,000 people, many of whom had been dependent on government for their income, life in Canberra became subdued and sombre.

The Commonwealth government's quest to reduce its financial outlay affected the Observatory in several ways, including the appointment of Bill Rimmer, the First Assistant, as 'Officer-in-Charge' rather than as Director following Duffield's untimely death. Had the Depression continued much beyond 1932, it was quite likely that Mount Stromlo's operations might have been suspended. Thankfully, the first fruit of the Observatory's staff and facilities was an official report, known as *Memoire Number 1*, produced in mid-1929. It was entitled 'The Luminosity of the Night Sky'. Owing to the financial stringencies of the Great Depression, the *Memoire* was not published until 1935. It also contained five previously unpublished reports that were produced between 1927 and 1934. The first report was dedicated to:

> those generous benefactors who have made valuable gifts of telescopes to the Observatory; those generous contributors to the Foundation Fund of the Observatory; and, those leaders of scientific thought in many countries who for so many years gave their influential support and encouragement to the movement to found this Observatory.[17]

Prompted by a request from Lord Rayleigh and based on a series of observations at the Hotel Canberra between October 1925 and October 1926, and at Mount Stromlo between October 1926 and December 1927, the luminosity study commenced at a time in Canberra when there was very little interference from domestic illumination. The astronomers made note of the difficulties of working even temporarily at the Hotel Canberra:

> There were then no street lights, but the head lights of motor cars, as they swung round a certain bend in the road, became an increasing source of trouble and delay as the city developed, but by waiting long enough it was always possible to avoid them.[18]

This was in contrast to the Mount Stromlo site, 'where observations could be made under excellent conditions; only to east, looking towards Canberra, is there any habitation whose light can be seen'.[19]

The instrument used in this study was a 'Rayleigh Photometer'—one of the instruments personally designed and supplied by Lord Rayleigh to observatories throughout the world for the purpose of determining the distribution of the intensity of the auroral radiation over the globe—in which a block of self-luminous uranium salt acted as a light source compared with the sky by means of a mirror. The two sources were equalised by interposing neutral filters. Colour filters labelled blue, auroral and red were introduced in the hope of isolating the light from the auroral line. Kennedy made regular observations on moonless, cloudless nights until October 1927 and these were later continued by Higgs. On each occasion, the photometer was pointed to the part of the sky that seemed least influenced by the Milky Way, bright stars, zodiacal light or light from human settlement. The general trend of the night-sky brightness was a maximum in April–May for both Canberra and Mount Stromlo. No clear reason emerged at the time for this maximum. An attempt was made to correct for unwanted sunlight and starlight by studying the effects of an admixture of moon-light. In this way, certain fluctuations in the data were decreased but the main features were unchanged.

The Australian results were generally consistent with those from London although, as Duffield noted, some of the Australian readings were based on more data than those in England. The astronomers at Mount Stromlo concluded that the luminosity of the night sky originated from sunlight rather than moonlight. However, they acknowledged the possibility that starlight may also have played a part. The most significant outcome was the provision of additional grounds for accepting Rayleigh's division of the aurora into two types: polar and non-polar. The observations also pointed to the need for more detailed work on the fluctuations of night sky luminosities in southern latitudes.

But perhaps the real significance of these observations in Australia was not in the results—although they were important—but in the fact of a published report that showed the determination of Duffield and his team to produce observational data of interest to a wider scientific community. The Observatory's first *Memoire* proved its worth and demonstrated its potential to make significant contributions to astronomical knowledge, particularly in relation to the Sun. From 1927 to 1938, similar observations continued, mainly undertaken by Higgs. The results were initially conveyed to Lord Rayleigh, but from 1934 they were forwarded to the Astronomer Royal, Sir Harold Spencer Jones, for analysis with data collected elsewhere.

The second *Memoire*, 'The Luminosities and Parallaxes of 350 Stars of Spectral Type B', was prepared shortly after the first in February 1930. It was based on Rimmer's determination of spectroscopic parallaxes of the brighter stars in the southern sky. Astronomers used the word 'parallax' as a synonym for 'distance'. This is because the earliest stellar distances were measured by a trigonometrical technique using the diameter of the Earth's orbit around the Sun as the baseline. The six-month change in a star's apparent position in the sky is called its parallax, and measuring this angular movement yields its distance. In the 'spectroscopic parallax' technique, a star's *intrinsic* luminosity is inferred from its spectrum, and observing its *apparent* luminosity then yields its distance.

This *Memoire* was the culmination of a four-year study. From the opening of the Observatory, Duffield stated that Rimmer would have the use of the 'Oddie' 9-inch refractor to continue his studies on spectroscopic parallaxes. The telescope was prepared for the work by the addition of a 32° 40' dense-flint-glass objective prism that fitted over its 9-inch objective to form spectra of stars in the focal plane where they could be photographed. Rimmer photographed suitable spectra of around 530 stars during 1926–29. The procedures for determining the parallaxes were dependent entirely on the comparison of one spectrum with another by visual inspection, mainly the appearance of the hydrogen lines in their spectra. It was known that in high luminosity stars these lines are sharp and well defined while in fainter stars they become progressively wider and more blurred. Attention was first given to the *B* stars. Rimmer's plates were then stored before being studied much later by Mary Lee Woods. By that time, the 'Yerkes' system of stellar classification by both temperature and luminosity had become available. The latter studies published by Woods in 1955 evaluated 520 stars of types *O* to *M*. They were not actually converted into parallaxes, but this could be done in a straightforward manner from known information.

Just as the future of the Observatory looked bleak, an 'atmospherics' research station was established at Mount Stromlo in conjunction with the Radio Research Board (RRB) of the Council for Scientific and Industrial Research (CSIR) in September 1930. The purpose of the station was to investigate the nature and distribution of sources of atmospherics, the characteristics of their propagation and their relation to meteorological and other phenomena, with special reference to their value in weather forecasting. Three huts were erected on 'Radio Hill' north of the 'Oddie' dome to house the equipment that consisted principally of a cathode ray direction finder and a narrow sector continuous recorder, both designed and supplied by the DSIR in London. An RRB officer, L.G.H. Huxley, operated the station until 1 May 1931 when the Observatory assumed responsibility for its operation. In 1935, the equipment was moved nearer to the main buildings at the Observatory to promote more efficient maintenance and simply to save walking time. A new cathode ray oscillograph was installed and the site errors for the new location determined. Continuous records were maintained until 6 December 1937, when they were suspended temporarily. Improved ionospheric sounding equipment was constructed during 1938 by Arthur Higgs and brought into operation in January 1939. With this equipment, a 'sounding' of the ionosphere took 2.5 minutes and was repeated automatically eight times every hour. The data obtained for each half-hour of the day and night were regularly supplied to the RRB.

While some thought the first five years of the Observatory's work lacked coherence and an organising focus—a situation that could be explained by the Observatory finding its feet and locating its scientific niche—the next decade would reveal a more urgent quest for consistency and purpose.

chapter three

Problems and progress, 1931–1939

Observatories are intentionally located away from major human settlements to avoid the effects of 'light pollution'. This meant that those who staffed the Observatory at Mount Stromlo not only had to work closely together and remain in the immediate proximity of the telescopes to take advantage of the best viewing conditions, they also had to live together. The Depression took its toll indirectly on the small Mount Stromlo community. Although there was no direct impact on the Observatory's core activities, the many destructive effects of the Depression on Canberra's social life affected individual staff members and influenced both the quantity and quality of the work that was conducted. The Commonwealth Attorney General, Mr Robert Menzies, referred to Canberra in 1934 as a 'place of exile'. It was remote from the other capital cities; a regular air service to the fledgling capital did not begin until 1936 when a grass airstrip was developed. It was not until the mid-1930s that money was again available for government construction programs in Canberra when both the Australian War Memorial and the National Library in Kings Avenue (since demolished) were commenced.

Despite Canberra's isolation and lack of infrastructure, the Observatory was technologically and administratively well established and a program of research was proceeding smoothly when Duffield died in 1929. Bill Rimmer's appointment as Officer-in-Charge rather than Director meant he was obliged to deal with a heavy load of administration that had deleterious consequences for his own research. The two incomplete projects that demanded his immediate attention were the spectrohelioscope and the Sun telescope, while his research was limited to two long-running programs concerning the Sun: the first involved use of a pyrheliometer to determine whether the Sun's radiation was constant or varied, and the second employed a pyranometer to make a continuous record of the heating of the Earth from energy supplied by the Sun, sky and clouds. The Observatory undertook this work while the RRB atmospherics research station concentrated on the upper atmosphere and the influence of the Sun's ultraviolet radiation.

The existence of sunspots had been known long before the establishment of the Observatory but any terrestrial effects that might be associated with them were vague and uncertain. It was therefore clear that the Observatory would require daily observations of sunspots for correlation with other phenomena. Daily sunspot sketches were made from early 1926 by projecting a solar image from a 3-inch telescope on to paper. From the sketches, the positions and sizes of the sunspots and spot groups were tabulated. These tabulations supplemented the standardised and very useful sunspot number data from Zurich.

The development of the spectrohelioscope by George Ellery Hale and others demonstrated that there were many rapidly variable features on the Sun that could not be seen with a normal white light telescope. As mentioned earlier, there had been hints from as early as 1906 that some of these features could have effects on the Earth. Duffield had intended to develop similar work at Mount Stromlo. He was undecided whether to erect a spectroheliograph, which was slow working and needed photography, or a spectrohelioscope in which variations could be visually observed. He decided on the latter. A spectroheliograph, however, became available and one of these instruments was delivered to Mount Stromlo in 1929. The installation of a spectroheliograph commenced in 1933 with the construction of a purpose-made building. The instrument was not originally provided with a suitable diffraction grating and it was several years before one could be acquired. A small building for the instrument was completed in June 1936, allowing the spectroheliograph to be brought into operation.

To prepare and operate the spectroheliograph, Sydney Williams (April 1936–May 1937) and then Ronald Giovanelli (May 1937–August 1939) were appointed. They were both physics graduates. Sydney Ernest Williams was educated at the University of Sydney and graduated with a BSc in physics.[1]

He completed a Master of Science in 1932 and his PhD at Bristol in 1936. His time at Mount Stromlo was short as he returned to Sydney University briefly in 1937 before being appointed Lecturer in Physics at the University of Western Australia in 1940. He was employed at the university for the remainder of his working life, becoming Reader in Physics in 1961. He died in 1979.

Ronald Gordon Giovanelli was born in Grafton on 30 April 1915. His father was a mathematics teacher and later headmaster. Giovanelli attended Fort Street Boys' High School and won a Public Exhibition to attend the University of Sydney from which he graduated with First Class Honours in Physics in 1937. He was offered a Research Fellowship at Mount Stromlo in 1937 and remained on the staff there until 1939. His early work was concerned with solar flares and related phenomena occurring within the vicinity of sunspots. As Piddington notes:

> The equipment available at Mount Stromlo was not of outstanding quality and so Giovanelli concentrated on obtaining sufficient data to allow good statistical results to be determined. These included correlations of flares with geophysical effects such as auroras, and the main structural characteristics of the optical flares and their relationships with sunspots. From all these results, Giovanelli concluded that flares are basically an electromagnetic phenomena [*sic*] in which electrons are accelerated by electric fields induced by changing magnetic fields.[2]

Giovanelli left Mount Stromlo after his Research Fellowship to take up a CSIR Science and Industry Scholarship for training at the National Physical Laboratory in Britain.

Giovanelli introduced various estimates of eruption characteristics and related these to ionospheric characteristics. He demonstrated that the ultra-violet radiation affecting the ionosphere has its origin at the seat of the eruption.[3] Statistics for the eruptions and their relationships with the associated sunspots were also studied and reported by Giovanelli,[4] while the association of eruptions with radio-fade-outs was the subject of a paper published in 1939 by Giovanelli and Higgs.[5] It was shown that an eruption does not cause fade-out unless the Sun is more than 15° above the horizon. Giovanelli developed eruption theories dealing with the motion of eruptive prominences in relation to radiation pressure. He suggested that the Lyman-α line of hydrogen provided the radiation pressure.[6]

With the spectrohelioscope in use, a watch on solar flares, known in this period as chromospheric eruptions, was commenced. Routine visual surveys of the Sun's surface were made on all days when the weather permitted and monthly records of chromospheric eruptions were sent to Paris for inclusion in the *Quarterly Bulletin of Solar Character Figures*. From the beginning of 1937, reports of solar disturbances were supplied to the RRB in Sydney. A cause-and-effect relation between solar flares and radio fade-outs had been suggested during 1937 by J.H. Dellinger and was soon confirmed at Mount Stromlo and in Sydney by Martyn, Munro, Higgs and Williams.[7] This relation was the subject of later research. During 1939 the spectra of very bright eruptions were obtained with the Sun telescope and a 3-prism spectrograph. Only 116 emission lines could be detected in the whole complex spectra and these were mainly Fe^{+} and Fe lines. Allen could not find a satisfactory explanation for this surprising result.[8] Two years later, the sunspot cycle was close to maximum and flare lists were becoming extensive. The Mount Stromlo lists soon dominated the reports from this longitude. It was a notable and politically strategic achievement, as the Commonwealth Solar Observatory had begun to justify its existence by fulfilling its original charter to provide solar data in the Western Pacific longitudinal gap.

Sun telescope

Solar research was the primary reason for the creation of a solar observatory in Australia, although it would be nearly seven years from the date of the Observatory's establishment before the Sun telescope was brought into service. The delay was, however, attributable to a series of factors and mistakes beyond the control of Observatory staff. Tenders for construction of the Solar Tower to be built on the east end of the main Observatory building were sought in January 1926,[9] with the successful contractor asking in April 1927 for the completion date to be delayed as excavation of the ground for the basement proved to be more difficult than envisaged.[10] The breaking of windows and furniture in the Administration Block by flying rocks and rubble after excavation blasting was a practical demonstration of the hard ground at Mount Stromlo. By November 1927, the Queanbeyan firm of W.H. Mason conceded that excavation of the ground for the Solar Tower's basement was beyond its capabilities and the work was surrendered to the Federal Capital Commission. Nearly a year later, the detached vertical ferro-concrete dome (based on the Einstein Tower at the Mount Wilson Observatory) was still not ready.[11] The proposed shutter opening system was considered too expensive, with the shutter wheels in the original design replaced with ball-bearing wheels.[12] The Observatory staff proposed the acquisition of a 1 horsepower rotation motor for the dome at an estimated cost of £80. At the end of 1929 the major parts of the instrument had been received. The remainder of the equipment required was constructed during 1930 at the Maribyrnong ordnance factory before being transported to Mount Stromlo for assembly. By October 1931, the mechanical and electrical parts of the instrument had been tested, the optical parts fitted and the final adjustments made. When it finally entered service in November 1931, it was considered comparable in capacity to the best in use anywhere in the world.

The Sun's image was reflected by a mirror system in the dome consisting of a coelostat and a mirror (both 46 centimetres in diameter) vertically downwards to the 30-centimetre objective lens located at the top of the tower immediately below the secondary mirror. As the Sun moved through the sky, the orientation of the mirrors was controlled electrically. A 25-centimetre mirror in the basement directed the Sun's image horizontally on to the slit of a solar spectrograph. Manufactured by Hilgers of London, this instrument was mounted in a tunnel extending away from the basement. It contained three glass prisms in a Littrow mounting. Intensity-calibrated solar spectra were obtained by either the slit-width method or the perforated screen method, with their images recorded on photographic plates. A photometric study was made of all Fraunhofer lines that could be measured in the available spectral region (3900–6600 Ångströms (Å)). For each line the main measurement was the observed central intensity relative to the continuum. Instrumental effects were then eliminated so that the measurements could be converted into line central intensities and into equivalent widths. More detailed contours were measured for specially selected lines. A complete atlas of the solar spectrum over the region 3650–8000 Å was photographed by Cla Allen using the 3-prism spectrograph and published in two of the Observatory's *Memoires* (1934 and 1938); it was a forerunner to the classical *Photometric Atlas of the Solar Spectrum* published by Minnaert, Mulders and Houtgast in 1940.[13] In 1934, the theory of emergence of light from the Sun's atmosphere was well developed and could be used to analyse both the contours and relative intensities of Fraunhofer lines. The Allen data provided opportunities for large-scale tests of such theories. The results indicated what values should be used for various poorly known parameters and showed when and how certain empirical rules could be applied. The 'curve-of-growth' method of interpreting stellar line photometry was an outcome of this Fraunhofer line intensity analysis, and became the standard method throughout the world for analysing stellar spectra for more than 40 years. This was easily the most important work carried out at the Observatory during the 1930s.

With the Sun telescope operational and producing excellent results, attention could finally be turned to the 'Reynolds' telescope. The Observatory staff intended to use the 'Reynolds' telescope for the photography of southern nebulae. However, as they had many other tasks to complete, its unavailability was not of serious immediate concern. Plans for the dome building were developed in September 1925 after Duffield had spent six months determining the construction specifications. He was insistent that the 'materials and workmanship are to be the best of their respective kinds', with plate steel and even the individual rivets used to be quality surveyed before acceptance.[14] Not surprisingly, the cost was initially estimated at £4,500 in March 1925, but was reduced (with the inclusion of certain economies) to £2,250 by March 1927.[15] W.H. Mason was awarded the contract for construction of the building with a completion date of 20 August 1927 (subsequently delayed to 31 January 1928), while the James Steel Engineering Company won the contract for construction of the dome.[16] The main body of the telescope had arrived in 1928, although the final piece was not received from Britain until late 1929. A number of modifications were made including a chain drive to operate the clock winding gear; additional teeth to be cut on the Wray declination spur

The 'Reynolds' dome under construction, 1927 Photo: NAA

wheel; and re-silvering of the mirror. With the Sun telescope in service by the end of 1931, most of the main installation work for the 'Reynolds', including erection of a steel dome 26 feet in diameter, was completed by 1932. The mechanical parts were then assembled and the preliminary adjustment of the polar axis carried out. But several structural and adjustment difficulties were still to be overcome. The delays were prolonged by the deployment of the staff on other projects.

While Rimmer's own specialty was the maintenance of telescopes, he was unable to give the time necessary for the night trials that an unfinished telescope required. The first photographs (65 in all) were taken with the telescope in 1933. Six years later, the telescope was converted to a modified Cassegrain configuration when a program of photoelectric photometry and stellar spectroscopy was being developed, although telescope flexure made these projects problematic. In 1939, the telescope was repainted and overhauled with the intention that it would be used to examine bright southern stars to detect previously undiscovered brightness variations. With the intervention of the war in 1939, however, productive use of the 'Reynolds' telescope was suspended. The only changes made to the instrument during the war years were the refiguring of the mirror and bracing of the tube assembly with wire for greater rigidity. While plenty of attention was devoted to the telescope in the second half of the 1930s, it did not produce results commensurate with the investment of time, energy or money. The best days for the 'Reynolds' telescope were still to come.

As for the 'Farnham' telescope, the staff had decided in late 1930 despite its limited use for photoelectric photometry, to keep the instrument maintained and operational in the event that it is 'required for observations of special objects appearing in the sky. No occasions have arisen for its use for this purpose'.[17] By the end of the following year, its potential use had been further limited to 'ordinary visual observations, including showing visitors the night sky'.[18] In 1936, a solar disc camera was made and fitted to the telescope as the first in a series of enhancements and improvisations that would give the instrument a new lease of life after 1945.

Solar radiation and energy

The Sun's energy output is almost entirely in the form of electro-magnetic radiation of which about 20–30 per cent is lost on transmission through the Earth's atmosphere even on a clear day. Measurements of the Sun's energy actually received on Earth were considered an essential part of the Observatory's observing program. From such observations, a large and difficult correction is required in order to derive the Sun's total energy emission, sometimes called the 'solar constant'. The important question to be decided was whether the Sun's total emission *was* actually constant. During the 1930s it was suspected that the total emission might vary with the sunspot cycle. The instruments available for study of solar radiation were pyrheliometers, which pointed to the Sun and measured the radiation received from it, and pyranometers, which remained horizontal and measured the vertical component of the Sun and sky and bright cloud radiation.

Pyrheliometer measurements were made at noon on all sunny days at Mount Stromlo, and occasionally a series of measurements would be made throughout a clear day as a check on the atmospheric absorption. The regular measuring instruments were Ångström pyrheliometers that were standardised by measurements with Smithsonian pyrheliometers. The unit of measurement was calories per square centimetre per second. It was recognised that the values were influenced by the water vapour in the atmosphere, the dust content of the atmosphere and, of course, the altitude of the Sun. Three methods were developed for measuring atmospheric water vapour. The first involved measuring the atmospheric humidity with a pyschrometer and converting the humidity into total water vapour in the path by using Hann's empirical equation. The second measured the absorption by the infrared water vapour bands of the solar spectrum using a Kipp and Zonen infrared spectrometer. The third used photometry of the intensities of the yellow water vapour absorption lines, with the adjacent Fraunhofer lines being used as intensity standards.

The dust content of the atmosphere was estimated by Allen from the intensity of the Sun's dust halo.[19] These detailed studies enabled certain adjustments to be made to the primary pyrheliometer readings whereby estimates of the total solar radiation outside the atmosphere could be made. Such evaluations showed very small variations throughout the years, and much significance was placed on establishing the reality of those small effects. Any real variability could only be decided by comparing Mount Stromlo observations with analogous data from the Smithsonian Institute. The analysis of the Mount Stromlo data was the responsibility of Rimmer until 1945. The results were later prepared and published.[20] The main conclusion was that no evidence could be found of any real variability (greater than 0.1 per cent) of solar radiation.[21] It was 30 years later, when radiometers were carried above the Earth's atmosphere on orbiting satellites, that it was finally established that the

Sun's irradiance does indeed exhibit changes of 0.1–0.2 per cent.[22] Some last typically for several days and are closely associated with the passage of sunspot groups across the solar disc. Variations due to several other causes have also been detected.[23]

An early suggestion was made that the solar variability could be estimated by studying the brightness of Sun-illuminated planets and satellites. The observations could be photometrically standardised against stable stars. Such measurements were planned for the photoelectric photometer but were not actually undertaken. An estimate of 1.97 cal cm^{-2} s^{-1} for the total solar radiation outside the atmosphere (at mean solar distance) was made,[24] and this might be compared with a much later evaluation of 1.950 cal cm^{-2} s^{-1}.[25] There are now known to be very large variations in the solar X-ray, far ultra-violet and radio-wave energies emitted by the Sun, but these spectral regions contain a quite negligible portion of the Sun's total emitted energy. The regular pyrheliometer measurements of 1932–33 were used to estimate the radiation loss observable in Canberra as a result of the Andes volcanic eruption of 18 April 1932. Evidence of this radiation loss was detected in Canberra from early May to November 1932. The loss was approximately 4 per cent for most of this time, but became undetectable by early 1933.[26]

Pyranometry was just as unspectacular and the results as unexciting. There were daily continuous pyranometer recordings of the combined energies of the Sun, the sky and the clouds received on a horizontal surface. The results were summarised by Rimmer and Allen.[27] Perhaps the most useful form for expressing the results was a plot for each month of the daily Sun, sky and cloud energies against the hours of bright sunshine. Such data can be used to express the solar heating of the Earth for any day as a function of only the time of year and the hours of bright sunshine. An incidental outcome of the solar radiation work was the preparation of a table giving the transmission of the Earth's atmosphere to total solar radiation.[28] The table can be applied fairly directly to the practical use of solar energy. Measurements of the ultra-violet component (wavelength <3200 Å) of solar radiation were made by Hogg with an Elster and Geitel cadmium photoelectric cell.[29] These measurements showed a large variation—a factor of five—between summer and winter in agreement with approximate calculations. The results were relevant to the erythemal and anti-rachitic effects of solar radiation. With hindsight, these were not particularly significant programs although they did make use of the available instruments and kept the staff occupied. There was no surprise and little sorrow when this work was discontinued in 1940.

The upper atmosphere

The reasons for studying the upper atmosphere at a solar observatory were self-evident. These layers are illuminated by the Sun's ultraviolet radiation and were the subject of special interest to the Radio Research Board (RRB).

Ozone is easily the largest absorber of solar radiation in the wavelength range 2000–3000 Å. The world distribution of ozone and its annual variation was of interest long before scientists established a connection between ozone depletion and global warming. From 1929 a Dobson photographic spectrograph, with fittings for the photometry of the solar spectrum in the 3000–3300 Å band, was put to daily use at the Observatory for measuring atmospheric ozone. The photometry gave a measurement of the amount of ozone above Mount Stromlo, the results being expressed in millimetres of ozone at standard temperature and pressure. The first results were published by Higgs in 1934 and covered the period 1929–32.[30] The values varied fairly regularly from a minimum of 2.45 millimetres in April–May to a maximum of 3 millimetres in September–October. There was no evidence of a diurnal variation. Indeed, by assuming there was no such variation, it became possible to use exposures made throughout a clear day to estimate the height of the ozone layer. Higgs' measurements led to a mean height of 56 kilometres, which is generally consistent with the 45–50 kilometres found elsewhere.

Observation continued until 1939, by which time eleven years' worth of data were available. Preparations for publication of the results in 1940 were underway when war intervened and Higgs was seconded to the Radiophysics Laboratory in Sydney. The data have never been retrieved and their potential benefit remains unexploited.

Radio atmospherics

The ionosphere is produced by solar ionising radiation and, as a consequence, there are certain characteristics of the ionosphere that are related to solar activity. Ionospheric problems are usually regarded as part of radio-science because the ionosphere itself can best be studied by radio-reflection methods. Arthur Higgs worked with L.G.H. Huxley in setting up a cathode-ray direction finder (CRDF) and a narrow sector directional recorder (NSDR) for observing and recording atmospherics. He assumed responsibility for this research when Huxley returned to England in 1931. Similar observing stations were established in Melbourne and Toowoomba, and by the combined study of individual atmospherics it became possible to locate the sources. This study confirmed that all atmospherics originated in lightning discharges. Intensity-wavelength characteristics were determined. As expected, the sources were located in thunderstorm areas and barometric lows. By operating the equipment at low frequency (10 kc/s), it became possible to make regular recordings of the major thunderstorms in central Africa and the Malay Archipelago.

The expected dominance of major thunderstorms in central Africa was to some extent confirmed. If late-afternoon African storms can be considered to determine global activity, a world maximum should be expected at about eighteen hours (universal time). This would be in phase with the world maximum of potential gradient and the local maximum of air–Earth current. The Mount Stromlo publications did not direct attention to the possibility that there may be a physical connection between the results of the atmospheric electrical observations described here.

In 1934, Higgs became interested in the fading signals observed in radio circuits and set up intensity recorders (receivers) on fixed transmitting stations to provide evidence of 'fade-outs' reported on short-wave stations. In 1936, a pulsed transmitter–receiver combination, capable of sweeping automatically over a range of frequencies to record the variations of height from which echoes were received, was provided by the RRB. The results proved so revealing that Higgs designed and built improved equipment covering a wider frequency range in a much shorter time. This made a virtually continuous watch on the ionosphere available for the first time.

The regular data extracted from these records included the height, electron density and thickness of the E, F1 and F2 ionospheric layers, and their variations with time of day, season and sunspot cycle. Thus this information provided a quantitative picture of the daily and seasonal variations of the ionosphere. As data accumulated, it became possible to carry out a quantitative evaluation of the ionising radiation emitted by the Sun and its variation with the sunspot cycle; the basic information needed for predicting the maximum and optimum radio frequencies which should be used for communication by radio over specified circuits also became available. This would be a critical application during wartime.

One of the first results to emerge from use of the ionospheric sounding (P'-f) equipment was confirmation that the sudden ionospheric disturbances (SID) reported by Dellinger in 1937[31] and now referred to as radio 'fade-outs' or 'black-outs', are closely associated with major solar flares. Martyn, Munro, Higgs and Williams were able to show in 1937 that an effect on the ionosphere could be detected with every well-observed flare.[32]

Atmospheric potential gradient

There were no solar influences found in the atmospheric electrical variations and none were really expected. Yet there was some evidence of a worldwide influence.

Interest in the atmospheric field derived from the fact that the field persisted and, in fine weather, was constantly in one direction over the whole Earth. The observations became part of the routine on Mount Stromlo, although they were of marginal interest to a *solar* observatory. Nonetheless, a Benndorf electrometer connected to an ionising collector situated about 2 metres above a level piece of ground gave a continuous recording of the potential gradient of the atmospheric electrical field throughout the period 1926–39. The initial installation and control were in the hands of A.L. Kennedy, while later routines were undertaken by Allen and then published.[33] They provide monthly tables of a number of factors, including diurnal variations and daily means—although only in fair weather conditions. In wet and unsettled weather, and during thunderstorms, the field varies erratically in both directions.

A systematic effect could be detected on the rare occasions when a significant wind persisted for several days. At such times the diurnal variation was smoother and had a maximum at five hours (local time) or nineteen hours (universal time). This variation is in phase with the fairly regular fields observed widely over the oceans.[34] The implication is that during normal fine conditions on Mount Stromlo the worldwide variation is swamped by the stronger local effects.

Conductivity and atmospheric ions

From 1931, observations of atmospheric electricity were undertaken by Arthur Hogg. Measurements were made of atmospheric electric conductivity, condensation nucleii, large and intermediate ions and the rate of atmospheric ionisation so as to give as complete a picture as possible of the ionisation equilibria in the lower atmosphere.

These inquiries began when Hogg used a Gerdien apparatus in 1929 to initiate extensive investigations into the electrical conductivity of the lower atmosphere.[35] He was interested in all the physical causes. In 1930 he made nuclei counts with an Owens dust counter which led, in turn, to ion counts. By applying electrical fields, he derived mobilities. However, he found it was necessary to differentiate between three types of atmospheric ion: small, intermediate and large. Each differed in their number-densities, mobilities and formation rates (average life or recombination coefficient) and in their contributions to the total conductivity.

From the conductivity and the potential gradient, the air–Earth current could be derived. This current proved to have a diurnal variation similar to, and in phase with, the potential gradient observed on the oceans.

Ion number-densities are associated with a physical balance between the rates of ionisation, the mean life of ions and their recombination coefficients. All were studied by Hogg and their numerical values determined.[36] The production of small ions was found to be due to radioactive matter in the air. The diurnal variations of ionisation rates were closely related to worldwide potential gradient variations. The ion mobilities were obtained by careful observations of conductivity and of ion numbers. These experiments were made in a cylindrical condenser which had an axial electrode through which was sent a smooth stream of air.[37] The values obtained were in close agreement with many other observations from across the Earth. Since intermediate ions are due to air pollution, Hogg studied them at Kew Observatory in London during 1937–38.[38] While at Kew, he also examined the effects of space charges on the potential gradient very close to the Earth. He found the expected effects to be very small and discovered that the gradient at the surface was the same as at a height of 1 metre.[39] These results demonstrated Hogg's practical aptitude as much as his theoretical insights. As Gascoigne notes:

> Most of this equipment he designed and built himself. Hogg was a skilful and ingenious experimenter, a facility which stood both him and the Observatory in good stead throughout his life. He had a meticulous eye for detail, and took the greatest pains to eliminate systematic effects from his measurements. His early work, on the diurnal variation of conductivity, and on the average lives, rates of production and mobility of small ions, attracted the attention of Whittle, who invited him to work at Kew Observatory. Hogg went there in 1937–38, taking with him a beautiful piece of equipment he had built in Canberra, with which he made some of the first measurements of the mobilities of the intermediate atmospheric ions, relatively frequent in the industrial atmosphere of London. In this important piece of work he was able to show conclusively that these ions existed in discrete groups, composed of one or more droplets of sulphuric acid, each containing about 2,200 molecules.[40]

Hogg's other research interest was in cosmic rays. During the 1930 meeting of the Australian & New Zealand Association for the Advancement of Science (ANZAAS), held in Brisbane, a group of scientists particularly interested in physics agreed that the much-discussed cosmic ray observations of the day were not being interpreted properly. The meeting recommended that 'the

Commonwealth Solar Observatory be asked to carry out continuous observations on the intensity of the so-called Penetrating Radiation (cosmic rays)'. Hogg was already familiar with cosmic rays and, from 1932, had observed them with aligned Geiger-Müller counting tubes. As he had already studied the effects of cosmic rays on ionic equilibrium in the atmosphere, he took up the challenge.

Hogg decided on the ionisation method of observing. He then gathered equipment with the intention that it would operate for several years without interruption. In 1934, a special double-walled building was erected that could maintain the whole apparatus in fine thermal equilibrium. The equipment was filled with CO_2 at 10 atmospheres pressure and was surrounded with 10 centimetres of lead shielding. It was maintained at 25.0 ± 0.3°C. The electrode was connected to a Compton recording electrometer. The equipment began recording both hard and soft cosmic ray components from September 1935. Observations ended in September 1940. While an analysis of the data was precluded by the war years, the results finally appeared during July 1949 as *Memoire Number 10*.[41] The absolute intensity was found to be 1.91*I* in standard cosmic ray units. The variations could be traced to solar and geomagnetic influences, and large bursts showed a barometric effect.

Review

The main scientific results of work carried out by 1939[42] were: parallaxes of *B*-Stars (Rimmer); the proof that total solar radiation variability was no greater than about 0.1 per cent (Rimmer, Allen); the establishment of relationships between solar flares and radio fade-outs (Williams, Giovanelli and Higgs); physical studies of solar flares (Giovanelli); quantitative measurements of auto-ionisation effects in the spectrum (Allen); full quantitative measurement and analysis of the intensities of Fraunhofer lines (Allen); detailed measurement and analysis of lower atmosphere electrical factors from experimental measurements—conductivity, nuclei counts; small, intermediate and large ions; ion mobilities; recombination coefficients; air–Earth currents (Hogg); extended study of radio atmospherics and establishment of their source in lightning discharges (Higgs); construction of *P'-f* equipment and regular measurements of ionospheric electron densities and heights of ionospheric layers (Higgs); and a full five-year study of hard and soft cosmic ray components with analysis (Hogg). Given the difficulty associated with scientific research in the decade that began with the Great Depression, the limited equipment available for use at Mount Stromlo—the principal donations to the Observatory consisted of telescopes that had no special relevance to solar work—and the small number of staff supplemented by scientists such as D.P. Mellor (1928–29) and F.J.P. Wood (1935–37) who were seconded temporarily to Mount Stromlo, the Observatory's achievements during the 1930s were not inconsiderable. But the institution lacked overall direction and a powerful advocate to promote the importance of astronomy and the need for Mount Stromlo to acquire new equipment.

In the context of Mount Stromlo's future, the question of the Observatory's Directorship was raised in May 1937 when the Secretary of the Department of the Interior (the Department was again renamed on 12 April 1932) requested that Dr A.C.D. (later Sir David) Rivett, Chief Executive Officer of the CSIR, and Professor Oscar U. Vonwiller, Head of the Department of Physics at Sydney University, report on the scientific value of the work being conducted at the Observatory, the adequacy of the equipment and whether a fully qualified Director needed to be appointed.

After visiting the Observatory, Rivett and Vonwiller confirmed the usefulness of the work in progress and the general sufficiency of the equipment for the kind of work being undertaken. They were particularly impressed with the work of Cla Allen, which probably tipped the balance in favour of the Observatory continuing. They pointed out, however, that the successful completion of some of the programs would require the acquisition of supplementary instruments, while several items of existing equipment—most notably the Reynolds reflector—had been standing idle for some time owing to a shortage of staff. The committee recommended retaining Rimmer as Officer-in-Charge while recognising his service with promotion to First Assistant rather than appointment as Director. In terms of aptitude and scientific ability, Rimmer was not Director material. Therefore, Rivett and Vonwiller also suggested the initiation of a search for a suitable Director throughout the British Empire and the creation of a Board of Visitors similar to that operating with the Royal Observatory at Greenwich to 'advise upon but not to control the work of the Observatory'.

All three recommendations were adopted. Rimmer was retained and promoted. He did not feel slighted by the decision to appoint a new Director, as he realised the limitations of his own training and that his own appointment had been both an interim measure and a means of reducing government outlays. He was also made an Officer of the Order of the British Empire (OBE) for his services to astronomy. This recognition was well received in Canberra where he was highly regarded as a one-time President of the Royal Society of Canberra and of the Public Service Professional Officers' Association. A committee in London, consisting of A. Fowler, E.A. Milne and H.H. Plaskett, was asked to suggest a candidate for the position of Director. In June 1939 it commended to the Department of the Interior the name of Dr Richard van der Reit Woolley, who came with testimonials from the Astronomer Royal, Sir Harold Spencer Jones, Professor Sir Arthur Eddington and Professor F.J.M. Stratton.[43] Woolley was just 33 years of age but possessed considerable academic ability and the qualities of leadership needed in an observatory director with responsibility for research students. He also made plain his desire to change the direction of the Observatory's work from solar astronomy and geophysical work to an exclusive concentration on stellar astronomy. The alternate candidate, should Woolley decline, was Dr David F. Martyn, although he was already being prepared for work in radiophysics.

Martyn, a Scottish physicist who came to Australia in 1930, was a highly competitive man and regarded Woolley, who was the same age, as something of a rival. Martyn had become known through his work on the upper atmosphere and in pioneering radiophysics. Had the selectors decided on Martyn, the work at Mount Stromlo would have remained unchanged and perhaps become entirely terrestrial in character. Another name mentioned at the time, the more mathematically minded Dr Subrahmanyan Chandrasekhar, might have resulted in the development of a theoretical astronomical institute. But, in any event, the candidate needed to have administrative and organisational skills to reverse the Observatory's recent fortunes. Many, like the British astronomer Sir William McCrea, did not rate the Commonwealth Solar Observatory highly during the 1930s. McCrea believed it was a 'small, isolated and neglected Observatory'.[44] The eventual appointment of Richard Woolley would, it was hoped, lead to profound changes at Mount Stromlo and a significantly higher international profile for Australian astronomy. As Woolley's tenure would show, the appointment of Director had a direct bearing on the course of Mount Stromlo's future.

Richard van der Reit Woolley was born on 24 April 1906 at Weymouth, England.[45] His father, Paymaster Rear Admiral Charles Woolley CMG, was a senior Royal Navy officer who had met his mother at the Cape Town naval base where her father, Frederic J. van der Reit, was Resident Magistrate and Civil Commissioner. He was raised at Alverstone near Portsmouth and attended All Hallows School at Honiton in Devon as a boarder, where he did well in Latin and mathematics but did not study science. When his father retired and decided to resettle in South Africa, they moved to Cape Town in early 1922 and the young Richard was sent to the University of South Africa. He had not yet turned sixteen but decided to enter an Arts course in Latin and study pure mathematics, as these had been the areas of his success at school. As he needed to matriculate, it seemed easiest for him to complete a course in physics to supplement the work he had already completed. From the outset, Woolley was attracted to physics but could not decide whether to major in it or mathematics. He decided to concentrate on both and graduated with first class honours in physics and mathematics during December 1924. A year later, he obtained an MSc when still aged nineteen. Woolley remained at the University of Cape Town for another two years, although he was yet to have any contact with astronomy. With support from his family and the University, he was able to pursue his studies at Caius College, Cambridge in 1926. When he participated in the 1927 Cambridge University expedition to Spitzbergen led by Gino Watkins, he came into contact with astronomers for the first time.

Returning to Cambridge in 1928, he resolved to study astronomy and was encouraged by Professor F.J.M. Stratton, then Senior Tutor at Caius and Deputy Director of the Solar Physics Laboratory. Stratton suggested to Woolley that he apply for a Commonwealth Fund Fellowship to study at the California Institute of Technology in Pasadena with the hope of working at the Mount Wilson Observatory in California for two years before returning to Cambridge to complete his doctorate. This was a chance for Woolley to meet most of the leading American astronomers of the day and saw him aligned with the 'solar faction' then in competition with a 'stellar faction'. On returning to Cambridge, Woolley was awarded the Isaac Newton Studentship that allowed him to complete his PhD research under Sir Arthur Eddington's supervision. At the same time, he married Gwyneth Meyler. When Harold Spencer Jones became Astronomer Royal in 1933, he offered Woolley the position of Chief Assistant at Greenwich. He was overjoyed and accepted without hesitation. However, the observing routine at Greenwich and the necessity of living in London led to Woolley and his new wife being less than satisfied with their domestic circumstances. When the post of John Couch Adams Astronomer and Chief Assistant at the University Observatory in Cambridge became vacant in 1937, Woolley sought the appointment despite being counselled by professional colleagues against such a move. By then he had settled on what would be his main research interest: the astrophysics of the outer layers of a star. His biographer, Sir William McCrea, felt that his appointment as Director of the Commonwealth Solar Observatory in 1939 was what 'Woolley had been in need of all his life . . . for responsibility to be thrust upon him'.

The Woolleys arrived at Mount Stromlo on 4 December 1939. The Minister for the Interior, Senator H.S. Foll, welcomed Woolley formally at a function held in the Observatory Library. The Minister expressed the Commonwealth government's pleasure that the new Director had 'left the academic portals of Cambridge to come to this fair land of ours to direct the affairs of our national Solar Observatory'. It was an auspicious day. Woolley began at Mount Stromlo committed to his objective of changing the emphasis of the Observatory from solar and geophysical work to stellar astronomy.

The third recommendation of the advisory committee was the formation of an Observatory Advisory Board. It consisted of scientists and administrators and met for the first time on 28 May 1940. The leading figures were Professors Kerr Grant (Adelaide), Oscar Vonwiller (Sydney) and Thomas Laby (Melbourne), who was elected Chairman. In his inaugural report to the Board, Woolley reflected on the work that had been done over the previous fifteen years and that he proposed for the new decade. In his view, the Observatory had enormous unexplored potential in the area of astrophysics. A program of spectroscopic work was devised for the still troublesome 'Reynolds' telescope while the 'Oddie' telescope would be used for the determination of colour temperatures. The photoelectric photometry work would be expanded along with the time service. New initiatives, such as the figuring of astronomical mirrors and the setting up of an aluminising plant, were forecast. A slit spectrograph would be acquired for radial velocity work. The Commonwealth government had also authorised an expedition to observe the total solar eclipse in South Africa on 1 October 1940, with apparatus prepared for the study of two corona problems. The first required a spectrum of the outer corona and the second photographs of the corona through polaroid screens and filters. Woolley and Allen were to observe the corona while Higgs would conduct ionospheric observations. Although a new program of research activity had been unveiled, another international crisis was developing that would affect much of the proposed work.

On 1 September 1939, Germany invaded Poland after a series of expansionary operations that had placed Austria, the Sudetenland and then all of Czechoslovakia under German control. Meanwhile, Mussolini's Fascist government in Italy had already disclosed a common cause with the Nazi regime in Germany with the militarist government in Japan revealing its imperial ambitions by asserting that it would be 'independent' rather than neutral in the event of war. As a consequence of continued German aggression and its refusal to leave Poland, Britain and France declared war on Germany. Because Australia was part of the British Empire, Prime Minister Robert Menzies made a national statement at 1.15 p.m. on 3 September 1939: 'It is my melancholy duty to inform you officially that, in consequence of a persistence by Germany in her invasion of Poland, Great Britain has declared war upon her and that, as a result, Australia is also at war'. The Union of South Africa declared war on 6 September and Canada four days later. By then, the German army had overrun Poland and by the end of the month had negotiated a partition of Poland with the Soviet Union that had on 17 September also invaded Poland. When the last large Polish force surrendered on 5 October, the partition of the country was complete. After attacking Denmark and Norway, the German army launched a massive assault on Belgium, Holland and France on 10 May 1940. Striking through the Ardennes, German armour separated the British Expeditionary Force from the French army. Six weeks later, France capitulated and the British army was withdrawn from Dunkirk across the English Channel. Shortly after the German army entered Paris (14 June 1940), Italy entered the war on Germany's side.

Two weeks after declaring war, the United Australia Party government announced its intention to recruit 20,000 men for the 2nd Australian Imperial Force (AIF). These men would form a division and auxiliary units for service either in Australia or overseas. The government also decided to call up the militia in two drafts to receive one month's intensive training. On 15 November 1939, the government proposed that conscription be introduced for military service within Australia before announcing that the newly created 6th Division would go to Europe on 10 January 1940, together with selected units of the Royal Australian Air Force (RAAF) and the Royal Australian Navy (RAN). By the end of March 1940, approximately 100,000 Australians had volunteered for overseas service in the war against Germany.

About one in every six eligible Australian men had signed up. But there was a need for those with specialist skills employed in occupations critical to the war effort to remain in their present employment. In an effort to limit the number of specialist personnel joining the Armed Forces, men in various age groups whose trade or profession was listed in the *Schedule of Reserve Occupations* were unable to volunteer unless they were specifically employed in their area of expertise by one of the forces. In 1939, the schedule included physics graduates. The country was now totally committed to prosecuting the war against Germany and preparing for possible hostilities against Japan. It came as no surprise when the staff and facilities of the Commonwealth Solar Observatory were called upon to assist.

chapter four

War on the mountain, 1940–1945

By mid-1940, with Britain and the Dominions standing alone against the combined force of Germany, Austria and Italy, it was apparent that supplies of military equipment from Britain could no longer be guaranteed. Australia would need to rely on local production for many items that had previously been obtained from Britain. This required an immediate reorganisation of Australian industry and a range of new challenges for local manufacturers. The newly formed Department of Munitions was tasked with ensuring that the Armed Forces had whatever equipment and logistic support they required to prosecute the war against Germany and its allies. The Manager of General Motors Holden (Australia), Mr Laurence (later Sir Laurence) Hartnett, was asked by Prime Minister Menzies to become Director of Ordnance Production within the new Department. As an engineer, he had oversight of a number of war-related industries. Hartnett quickly learned that, although there were efficient small arms factories, they were unable to build larger guns requiring telescopic sights. When it became apparent that optical sights for the 2-pounder anti-tank gun then being produced in Australia would not be available from the usual supplier in Britain, an immediate need for local production was created. Hartnett later commented:

> I began to feel a bit less confident when I realised an optical instrument industry just didn't exist in Australia. Apart from spectacles made from imported glass, hardly a single optical instrument had ever been made in Australia. The manufacture of optical glass was

> a highly specialised science, and the techniques involved were little known . . . Unless we could pull this one out of the hat, the [anti-tank guns] would be a waste of time . . . sights had to be used for everything but point-blank, open-sight shooting.[1]

Hartnett shared his concerns and desire for expert advice with the Assistant Director General of Munitions, Keith Brodbribb, who apparently replied: 'You want physicists? We've got half a dozen of them, roaring around, dying to do something to help, but no one's been able to use them'. The Department of Munitions then convened a meeting of physicists to advise on the possibility of producing small military lenses locally. There was an energetic response and a local optical industry was launched.

Professor Thomas Laby, President of the Australian Branch of the Institute of Physics, had earlier (August 1939) written to Prime Minister Menzies detailing the need for a consultative committee of Australian physicists to be established to advise the government on war-related scientific matters. In reply, Laby was told that 'foreseen requirements were at present well provided for', but that action would be taken should circumstances change. Laby was angered by this obvious rebuff of his good intention. On 26 June 1940, the gathering of physicists proposed by Laby occurred in response to Hartnett's immediate problem with respect to the 2-pounder anti-tank gun. The Optical Munitions Panel was created. Hartnett invited Laby to serve as Chairman, with another Melbourne University staff member, J.S. Rogers, appointed to the position of Secretary. Panel members included Lieutenant Colonel G.H. Adams (Assistant Director of Artillery, Army Headquarters), Frank S. Daley (Department of Defence, Ordnance Department, Maribyrnong), N.A. Esserman (National Standards Laboratory), Professor Sir Kerr Grant (University of Adelaide), Laurence Hartnett (Department of Munitions), Associate Professor Eric O. Hercus (University of Melbourne), Assistant Superintendent E.L. Sayce (Munitions Supply Laboratories), Professor Oscar Vonwiller (University of Sydney) and Dr Richard Woolley as Director of the Commonwealth Solar Observatory. Although Woolley had planned to observe the total solar eclipse in South Africa on 1 October 1940, he arranged for Cla Allen to go in his place and remained in Canberra to attend this meeting. This effectively marked the end of the research program Woolley had initiated on his arrival six months earlier.

By this time, France had fallen and the evacuation of the British Army from Dunkirk had been completed. It was now obvious that the war would continue for some years and involve enormous national effort. The Observatory and its staff were needed to contribute to Germany's defeat. But concern had already been expressed that Australia might be deprived of Woolley's services for the duration of the war with *Smith's Weekly* asking 'Will Britain commandeer van der Reit Woolley?'. Woolley was young and known to be from a family enjoying a proud military heritage. There was a suggestion that Woolley be commissioned as a Lieutenant in the Survey Corps or appointed Acting Commonwealth Meteorologist in order to secure his presence in Australia and to prevent the Observatory from dissolving without a Director. In the end, these inducements were not needed. There was plenty of valuable war-related work for Woolley to do in Australia and he had already suggested several ways in which he and the Mount Stromlo staff could contribute to prosecuting the war, including offering their expertise with optical instruments.

The Optical Munitions Panel met for the first time in July 1940 and on another five occasions before the end of the year. Thereafter, it met every two months. Most of the business was conducted in the Department of Natural Philosophy at Melbourne University where Laby was based, but several meetings were held at Mount Stromlo. It was apparent to all that these gatherings also constituted a unique opportunity for physicists to meet regularly and to develop a professional network that would strengthen their discipline after the war.[2] But the Panel's chairman, the formidable Professor Laby, soon clashed with the young Dr Woolley who felt the Panel was a group that suffered from an awkward administrative structure with a lack of clear distinction between its scientific, advisory and executive functions. Hartnett had earlier written to the Director General of Munitions, Essington Lewis, that Woolley was 'the highest qualified mathematician in Australia'. Woolley was later to argue (with hindsight) that the Panel did not have enough power to discharge its functions and that it could and should have been organised differently so as to improve its efficiency.[3] In many ways, conflict between Laby and Woolley was inevitable. Woolley described Laby, who was 26 years his senior, as:

> a very strong character and a very patriotic man; he burned with a great zeal to help Australia . . . He was, to say the least, not the soul of tact but the Panel was, I suppose, completely successful in accomplishing all that it had been set to do, including some supposedly impossible tasks. To these successes Laby's drive contributed materially.[4]

Despite differences of opinion and clashes with Laby, whom Woolley accused of duplicating and complicating his own work at Mount Stromlo, there was a clear commitment among the Panel members to working as a team. The Panel could not have functioned without goodwill and cooperation because each member realised that, at the time of the Panel's formation:

> there was no single centralised research institution in the Commonwealth capable of advising the Department of Munitions on all the problems likely to arise in any attempt to make optical instruments. It is true, a beginning had been made at the Munitions Supply Laboratories to study the problems of manufacture of telescopes, but the task that lay before the country was much too complicated and extensive to be undertaken with the resources of those laboratories alone . . . The principal task of the Panel was to advise the Directorate of Ordnance Production on the scientific and technical problems likely to arise in making optical munitions and also to give assistance to firms and annexes selected or established for their manufacture.[5]

In planning its work, the Panel decided very quickly against the creation of a centralised research institution. Instead, the need for particular equipment was individualised with each of the Panel members and their institutions given several instruments to research and design, with their manufacture managed through contracts with local industrial firms. Work was distributed among twelve

The Commonwealth Solar Observatory workshop in the World War II era Photo: MSO

government agencies, including the Universities of Melbourne, Sydney, Adelaide and Western Australia, Sydney and Melbourne Technical Colleges, the Commonwealth Solar Observatory, and two private firms: the Australian Optical Company and the British Optical Company. Woolley agreed 'to set up his place [Mount Stromlo] and direct it solely to optical work'. Specifically, the Panel 'agreed to place all optical design under [Woolley's] supervision', with the Director working seventeen-hour days, including Sundays, on optical problems throughout the first year of the Panel's existence.

The Panel's very first task was to produce 'Sighting Telescope 24B' to be used on artillery pieces, or mounted in tanks or on anti-tank guns. This was the sight originally identified by Hartnett that subsequently led to the Panel's formation. The initial order was for 3,500 instruments. A specimen instrument had been obtained from the War Office in London but no detail was given on the kind of optical glass used or the radii of curvature and thickness of the lenses used. Woolley made crucial measurements of the specimen at Mount Stromlo for use in the production of a modified model known as 'Sighting Telescope No. 124 (Aust)'. Although identical in external appearance, this instrument was made at Mount Stromlo using spectacle lens glass rather than optical glass. It was submitted to the Panel for evaluation. Having been formally accepted in October 1940 as a prototype, the instrument was sent to the Army for field trials:

> Here it failed to satisfy requirements for use in tanks but was good enough to function as an efficient aiming instrument in an anti-tank gun . . . by February 1941 [the British Optical Company in Sydney and the Australian Optical Company in Melbourne] had produced their first sighting telescopes. In the opinion of the Panel the instruments were still far from perfection but their performance was sufficiently promising to encourage the hope that the contractors, with a little more experience, might make really satisfactory instruments.[6]

A problem encountered by the contractors was the absence of test plates for checking their lenses. Once these had also been made at Mount Stromlo, both companies began producing satisfactory gun-sights with more than 900 instruments passing inspection by April 1942. A further 600 sights had been produced by the end of the war, despite Mount Stromlo staff producing an improved design that never entered production. Within six months of embracing the new optical work, the Observatory had produced three prototype instruments, and designed six complete instruments in addition to the manufacture of several highly accurate optical gauges.[7] In producing this expanding array of instruments, it was obvious that specialised glass production was an immediate priority.

Woolley agreed to have the Observatory examine the design and production of military optical equipment using the only two types of glass then available in Australia: spectacle crown and flint. Spectacle glass was more readily obtainable than flint glass, which could only be made of very special iron-free sand and which had not been made previously in Australia. In the early part of 1941, the Observatory constructed test flats and spherical test surfaces, and went on to make lenses for the prototypes of its experimental designs. During 1941, considerable attention was given to special designs for dial sights and directors without prisms because optical glass was unavailable. Mirrors were used instead and proved to be much cheaper to produce. Despite the demonstrable benefits, Woolley lamented the 'general tendency [to] prefer an Imperial design to a local product even when the local product is substantially simpler'.[8] Although necessity might have been 'the mother of all invention', by mid-1941 Woolley accepted overblown Australian esteem for Imperial production and was obliged to conclude that the 'time for experiment and research into entirely novel designs seems to have passed'.[9]

By this time, Mount Stromlo had been contracted to make 23 telescopes for the depression range finder. The British suppliers were not confident that Australian manufacturers would be successful in the production of flint glass. However, Ernst Hartung, Professor of Chemistry at Melbourne University and an astronomer who owned a 12-inch telescope, located suitable sand and the first large-scale batch of optical glass was poured in conjunction with Australian Consolidated Industries (ACI) during September 1941. Full-scale production began in 1942. By the end of the year, locally manufactured glass had made mirror-substitute instruments unnecessary.

The extent of the contract work and the staff involved required a substantial enlargement in the Observatory's buildings. The Ordnance Production Directorate initially extended the original mechanical workshop from 600 to 1,200 square feet (55 to 110 square metres) and later built an optical shop, assembly shop, store and heat treatment room along the north side of the quadrangle. These buildings, some air-conditioned, provided the Observatory with over 1,500 square feet (138 square metres) of additional floor space and a range of machine tools provided under the wartime 'Lend-Lease' program that would be useful in its postwar technical work.[10] By the end of the war, Woolley considered Mount Stromlo to be as well equipped as the Royal Observatory at Greenwich—if not better. The Official War Historian commented that:

> the Observatory ranked with the Supply Laboratories and the University of Melbourne as a centre for pioneering optical techniques and for diffusing them to industry. Representatives of most of the firms that became engaged in optical manufacture visited Mount Stromlo at some time.[11]

Those involved in wartime optical munitions work included the physicists Noel Chamberlain, Ben Gascoigne, Walter Stibbs (Research Assistant), Jim Dooley (Research Student) and Cla Allen, who were supported by Jim Banham (the long-serving Mechanical Workshop Foreman), Francis Lord (Optical Technician) and Syd Elwin (Assistant Optical Technician). They were a formidable team, boasting vast education and an impressive array of experience.

Noel Chamberlain arrived at Mount Stromlo to undertake meteorological duties after working at the Watheroo Magnetic Observatory in Western Australia. However, he was quickly drafted into service as Woolley's assistant in the design, manufacture and inspection of optical instruments. Chamberlain found the practical work to his liking and would leave the Observatory after the war for employment with the Commonwealth Mineral Resources Survey. Chamberlain would be appointed Head Geophysicist when the Survey became the Bureau of Mineral Resources, Geology and Geophysics.

A Stromlo staff outing in early 1942. L to R: Noel Chamberlain, Hans Meyer, Walter Stibbs, Gus Krentler, Ted McCarthy, Georg Froelich, Ben Gascoigne, Kurt Gottlieb, Ernst Treye
Photo: MSO

Ben Gascoigne was born at Napier, New Zealand in 1915 and educated at Auckland University College. After completing undergraduate studies at the University of Auckland, he gained his doctorate in astronomical optics from the University of Bristol and received an offer of a Research Fellowship from Woolley who believed his 'experience in optical work [was] unique'. In Woolley's estimation, Gascoigne had been 'trained in a way that no one else in Australia has been qualified' and was expert in the testing of large astronomical mirrors.[12] After working briefly on optical rangefinders in New Zealand, he arrived in Australia in August 1941 for the specific purpose of contributing his expertise to the optical munitions work at Mount Stromlo. In addition to a range of general duties, he worked on the design of a sighting telescope for an anti-aircraft gun. Another of his wartime tasks was the establishment of a Time Service, of which more will be said in the next chapter. Following his marriage to Rosalie in 1943, Gascoigne remained at Mount Stromlo for the rest of his professional working life, focusing postwar on stellar evolution, the distance scale and faint star photometry.

Douglas Walter Noble Stibbs, Deas Thomson Scholar in the School of Physics at the University of Sydney, was beginning to establish his reputation as a mathematical physicist when the war commenced. He had arrived at Mount Stromlo the same day as Woolley as his first Vacation Student and later, as an Observatory Research Assistant, joined the optical munitions work and the development of an ionospheric prediction service. While still in his early twenties, he designed a folded optical system for a gun-sight that later went into production. In late 1941, Stibbs also designed a 'Sun compass' for use in desert warfare before accepting a lectureship at the New England University College 'teaching mathematics and theoretical physics on topics related to the war effort'. During the war he would work with the RAAF on direction-finding problems relevant to submarine detection in the Tasman Sea with AVRO 'Ansons' using the Adcock High Frequency Direction Finding System (known as HFDF) and would lecture to Air Force officers at the Point Cook RAAF Base on radio propagation and anomalous direction-finding bearings with DF systems. He came to know Woolley very well, as they occasionally travelled to Melbourne together. In July 1945, Stibbs returned to the staff at Mount Stromlo at Woolley's invitation to embark on a study of radiative transfer problems involving non-coherent scattering and observational work in photoelectric photometry. He also worked on magnetic stars and became widely known for his theory of magnetic variability. During a visit to Canberra, Professor H.S.W. Massey, Quain Professor of Physics at University College, London, suggested that Woolley and Stibbs

Woolley and Stibbs, authors of The Outer Layers of a Star Photo: Walter Stibbs

collaborate on a book that was published in 1953 by Clarendon Press, Oxford with the title *The Outer Layers of a Star*. Stibbs left Mount Stromlo in 1951 to become the Radcliffe Travelling Fellow in Astronomy at Oxford University. He obtained his doctorate at New College Oxford in 1954. In 1959 Stibbs was appointed Napier Professor of Astronomy at the University of St Andrews in Scotland, a post he would hold for 30 years. He returned to Mount Stromlo on his retirement.

Jim Dooley was a research student at Melbourne University and was offered to Woolley by Professor Laby to meet a shortage of physicists at Mount Stromlo. Dooley was the only Observatory staff member to serve in a military unit, although physics graduates were precluded from active service as members of a 'reserved occupation' in great demand. Drawn between his work as a physicist and the nation's need for combatants, Dooley shared his dilemma with Woolley who assisted his recruitment into an anti-aircraft unit deployed in New Guinea. When Dooley's status as a physics graduate came to official notice, he was sent back to Australia where he was employed in the development of radar equipment. He later returned to Mount Stromlo and resumed optical munitions work. Like Chamberlain, Dooley also joined the Commonwealth Mineral Resources Survey after the cessation of hostilities.

Francis Lord was a key member of the Mount Stromlo staff. Born in Opava, Czechoslovakia in 1916, Lord studied optics in Paris under the famous Professor Fabry and gained experience in production work in his homeland from 1936 to 1939. On obtaining approval to travel to Rumania to fulfil an optics order, Lord felt in fear of his life in Nazi Germany and fled to Britain. On arriving in Australia, he worked for the British Optical Company in Sydney where he encountered suspicion on the factory floor due to his nationality. His employment was terminated. When Woolley arranged for the screening of all refugees, including the 'Dunera boys', in the hope of identifying individuals with the skills he needed at Mount Stromlo (seven staff members in all were recruited by this means), he learned of Lord's expertise and experience and immediately offered him work at Mount Stromlo. Because he was an 'enemy alien', the Manager of the British Optical Company escorted Lord to Canberra where he was taken into Woolley's custody. No time was wasted exploiting Lord's practical knowledge. He advised on the creation of a sophisticated optical workshop and supervised experiments on a range of lenses and prisms. Lord also lectured staff on elements of optical computing. Within twelve months, Mount Stromlo was able to produce a range of instruments that had been unthinkable before Lord's arrival.

Syd Elwin, born in Sunderland, England in 1906, was an ideal partner for Lord. With experience as a lecturer in manual training at Sydney Teachers' College and with a keen amateur interest in astronomy, Elwin had taught himself a range of skills in optical glass working. In addition to working with Lord in the creation and operation of the workshop, Elwin's teaching background was used in the instruction of staff from a number of manufacturing firms that had received government optical munitions contracts. He also conducted courses in glass working techniques and applied optics theory.

Having assembled a highly specialised team, the Panel met for the first time at Mount Stromlo in February 1941. They were immediately impressed by the new facilities and the eagerness of the staff, particularly those with whom they spoke—namely, Allen, Lord and Elwin—while Woolley was lauded by the *Canberra Times* as 'one of the most outstanding representatives of the science of optics'.[13]

By 1942, the Observatory had literally been transformed into an optical munitions factory where 'bomb and gun sights and other optical equipment for the armed forces [were] manufactured. At its peak, the factory employed up to 70 persons who carried out both design and manufacture'.[14] While Mount Stromlo did not produce the same quantities as other institutions or factories, it was the only facility to have every stage of production, from design to final inspection, under the one roof. In addition to Miss J.M. Foley and Miss M.T. Ryan, who had been recruited for astronomical computing work in early 1940, women were also employed at the Observatory in a variety of technical roles. They formed part of the optical design team, although their duties changed with wartime demands on the Observatory.

The question of their pay was the first Federal Capital Territory case considered by the Women's Employment Board. Woolley explained that the women were appointed on the recommendation of the Principal of the Canberra Technical College, while Lord explained that men and women did the same work and, in his view, were 'interchangeable'. The Board awarded the female optical workers at Mount Stromlo a significant pay rise. Mrs Lillian Wilson was one such employee at the Observatory between 1941 and 1943. She recalled, in an interview with Ms Louise Taylor in 1998:

> My job was to clean eye pieces (predictors) which were used in the periscopes of submarines. There were 40 to 50 people working in the big workshop, and seven in my workshop—five men and two women. We had no idea what the people in the other part of the workshop did and we were not encouraged to talk about it. None of the people in the workshop lived on the mountain. A special bus used to pick us up on the corner of Franklin Street in Canberra and pick us up again from Mount Stromlo in the late afternoon. There were three 'aliens' working up there, and with Dr [*sic*] Rimmer's permission, they were allowed out on Fridays on the bus but had to come back that night. They could not stay out for the weekend.

When Woolley found himself in urgent need of wartime staff, he learned that a number of refugees and wartime internees had 'first class optical experience', like Francis Lord. He sought three men in particular, arguing strongly that they would 'more than double the number of experienced men in Australia for this particular class of work'. There was a great deal of wrangling about the men's abilities and loyalties. Eventually the matter was referred to the War Cabinet, which agreed to release them into Woolley's personal custody. They were readily and easily absorbed into the Observatory's workforce. The first two, Georg Froelich and Hans Meyer, started work at Mount Stromlo on 1 July 1941. Meyer had come from a family of medical practitioners. His future wife worked at Mount Stromlo and after the war, when he settled in Sydney, he established a successful optical repair and maintenance business and later specialised in the zoom lenses in television cameras. Another three were to join them within twelve months. Most of the European refugees working at the Observatory had fled from Nazi Germany and were employed at Mount Stromlo to exploit their special skills in optics and mechanics. Kurt Gottlieb was one such refugee. He was born in Austria into a cultured Jewish family that fled Nazi anti-Semitism in 1940. When his engineering qualifications were not recognised in Australia, he enrolled in a technical college draftsman's course where he was 'discovered' by Woolley. He served throughout the war as a mechanical designer and became a research engineer after the cessation of hostilities. Gottlieb married, and he lived and continued to work at Mount Stromlo until his retirement.[15] A. Nisbet, a former employee of Barr & Stroud Pty Ltd, assisted him, together with another refugee, Gus Krentler, who was a qualified draftsman.

Despite the enlarged community on the mountain, the Observatory seemed even more isolated during wartime. This was a consequence of petrol rationing—although Noel Chamberlain owned a car that was fuelled with pooled petrol rations—and the closure of the Observatory to visitors after September 1941 in the interests of security. In addition to the single men who lived in the Bachelors' Quarters, there were a number of married couples: the Allens, Elwins, Dooleys and, later, the Gascoignes. The Allens maintained a close family life. Cla Allen was attentive to his roses and fruit trees. He was always genial company, although very committed to his research. Rose Allen was a generous host to Observatory staff and visitors. After their marriage in 1943, Rosalie Gascoigne became a long-term resident on Mount Stromlo. When she first arrived:

> there was nothing for the women and no real focal point for community life. Going up to the Observatory building to get the mail was just about the main form of social contact. The women were an assorted group from different backgrounds and with different levels of education. There was undoubtedly social demarcation between the professional families and the workshop. And so there was a certain amount of social isolation. Any news could quickly turn into gossip as is common in any confined community.
>
> It was a real pioneering kind of life. Nearly all the houses were cold; they had been built on the south side of the hill. Ours was a very big, cold house, one of the first built. There was only one family with a sunny house. Sometimes on a cold day one would find it warmer to sit outside the house then to stay inside.
>
> There was a community refrigerator. You had to book your day. It really was subsistence living. The women were really only kept going by the magic of the men's work. There was really nothing in the way of routinely organised social activities. The bachelors came mainly from various countries in Europe and were often lonely. They played chess sometimes, as did some of the other men. In desperation, really, the women formed a sewing circle, if only to warm up the house in which it was held. I became hungry for every sort of human communication. It was really company starvation.[16]

Walter Stibbs recalls of 1940 that:

> There were very few social occasions on Mount Stromlo except for a Wood Chopping Competition held at the back of the Bachelors' Quarters which was won by Zuriel Nowland, the Observatory Groundsman. Woolley took part but, unfortunately, someone had accidentally selected a rather knotty log for him and he eventually gave up chopping after all the other contestants had finished. This probably deterred him from taking part again in any physical contest with the inhabitants of Mount Stromlo except for tennis in which his height, high-kicking left-handed service and good reaction time gave him a considerable advantage. He was seldom beaten at tennis; in fact, it was generally accepted that it was preferable to give him a hard game but not to beat him.

An up-hill bicycle race was also held for the first time in 1941 from the Cotter Road at the base of the mountain to the Observatory offices on Mount Stromlo. It was organised by Walter Stibbs and held during office hours, and was treated as something of a 'gala occasion' by the staff. The inaugural participants numbered about twelve and they included Walter Stibbs, Jim Dooley, Ben Gascoigne, Ted Holmes and Ernst Froelich (no relation to Georg). Holmes won the race by a mere 10 yards (9 metres) from Stibbs. The up-hill race of bikes and later runners, the former now known as the 'Tour de Stromlo Classic', is still held more than 60 years after the staff sought a way to relieve the boredom at the Observatory.

Woolley's professional leadership was a counterpart to his social influence which was extended considerably when Observatory House (the Director's Residence) was occupied by the large number of single men who had come to work on the mountain in response to the enlarged war effort. Despite their diverse social and professional backgrounds, Woolley tended to keep the staff together and organised a series of recreational events that laid the foundation for close working relationships. Woolley hosted bridge evenings, tennis matches and horse riding. He sought out Francis Lord for games of chess and encouraged those with musical abilities to prepare performances for the staff and their families.

While new and additional staff arrived, members of the existing staff were relocated for specialist service. In November 1940, Arthur Hogg departed for Maribyrnong and wartime service with the Chemical Defence Section of the Munitions Supply Laboratories. There he was:

> engaged on research and developmental work on physical aspects of protection against chemical agents. Perhaps the most notable of a number of contributions he made was the development of an ionisation penetrometer to measure the penetration of charged

> particles through respirator filters. He also worked on the development of wool-resin filter materials and took an active part in the early work which led to the DSL Dust Respirator which was used extensively by the Services. He was transferred in mid-1944 to the newly formed Secretariat of the Chemical Defence Board and acted for about eighteen months as secretary of the Physical Sub-Committee.[17]

Arthur Higgs travelled to Victoria West in South Africa with Allen to observe the total solar eclipse in October 1940. Victoria West, a small town approximately 500 kilometres from Cape Town, was selected because it was situated not far from the centre line of the eclipse track for the ionosphere and because the Australians were offered the use of a disused radio station with an oil-driven generator supplying AC power. The expedition was assisted financially by a grant from the Joint Permanent Eclipse Committee of the Royal Society and the Royal Astronomical Society and materially by the Radio Research Board, and sought to understand the changes that took place within the ionosphere during a solar eclipse and its effects within the regions responsible for long-distance radio transmission. Astronomical observations were undertaken by Allen and a program of ionospheric measurements carried out by Higgs. Their findings were published by Higgs in a paper entitled 'Ionospheric Measurements Made During the Total Solar Eclipse of 1940 October 1', completed on 22 October 1941.[18]

As they prepared for the eclipse, and with the war gathering momentum, Higgs wrote to Woolley from South Africa explaining that he would 'very much prefer to be doing something toward "the nation's war effort"—we get the same old catchphrase just as often over here! When I come back I shall be champing at the bit to jump into something useful as soon as possible'.[19] He was pleased to receive Woolley's reply, which informed him that the Observatory was 'going into questions of the design of optical instruments for war purposes'.[20] On his return from South Africa in November 1940, Higgs was appointed a Research Officer at the Radiophysics Laboratory and remained there until the end of the war when he became its Technical Secretary. As a scholar who took naturally to the study of radiophysics, he did not return to the Observatory and devoted the remainder of his working life to the CSIR and the Commonwealth Scientific & Industrial Research Organisation (CSIRO), as it later became.

During Higgs' absence in South Africa, Stibbs assumed responsibility for the ionospheric work and solved a long-standing theoretical problem on the true height distribution of electron density in the E-Region of the ionosphere by a numerical solution of the Abelion integral equation. This was later outlined in a dissertation entitled 'The ultra-violet emission from the Sun in relation to the number of quanta forming the atmospheric ionised layers', submitted in March 1943, for which he was awarded a MSc from the University of Sydney.

This was a testing time for Woolley and the Observatory, when self-interest and the national interest appeared to be in competition as the male staff considered combatant service. As Sherratt and Conde observe:

> War is a time of uncertainty and conflicting emotions. Just as there are many reasons why soldiers choose to enlist, so there were many factors at work in the minds of the Mount Stromlo scientists. They hoped to stay at the Observatory, but what if they were needed elsewhere? It made sense for them to use their scientific skills, but was it just an excuse, should they join up anyway? They wanted to contribute to the war effort, but perhaps there were career opportunities as well? Motivations are never simple. They were not cartoon scientists, blinkered by their own research, seeking to be left alone. They were citizens, spouses, confused and concerned.[21]

Woolley was himself frequently absent from Mount Stromlo and spent a great deal of time in Melbourne, especially after Japan's attack on Pearl Harbor on 7 December 1941 led to war in the Pacific. With Australia threatened by a combined onslaught of Japanese land, sea and air forces, the War Cabinet resolved on 27 January 1942:

> 1. That an Army Inventions Directorate be established consisting of the following Directors: a) a Chief Executive Officer who is to be nominated by the Directors in session and who will devote the whole of his time to the business of the Directorate; b) six other directors as recommended by the Department of the Army. 2. That the functions of the Directorate be to receive, develop and adjudicate upon inventions and suggestions forwarded to the Directorate, and to examine problems presented by the Army with a view to ascertaining whether a satisfactory solution can be developed. 3. That the Directorate function directly under the Minister for the Army [Francis Forde] have its own full-time technical and clerical staff and have at its disposal such funds as may be authorised by the Minister and provided by the Treasurer. 4. That the Directorate be responsible for setting up the appropriate organisation for dealing with Army inventions in each military district.

Woolley became the Chief Executive Officer in April 1942 with Hartnett agreeing to be Chairman. By this stage, Woolley's reputation had been well established. Laby had earlier described Woolley as 'one of the finest mathematicians in the country'—a slight exaggeration—while Hartnett thought he was 'an excellent organiser and a fast worker'.[22] The Directorate's staff dealt with 21,645 submissions throughout the course of the war. Ninety per cent of the submissions proved to be fruitless, although each was assessed by at least two technical officers and by a military consultant. A very small number—127 in all—were actually developed, with very few rejected out of hand. Woolley later commented on his time with the Directorate:

> The work brought me into contact with a number of powerful civil servants, whose abilities I admired. I learnt the rules of Public Service in-fighting. Somebody has to be kind to inventors, and somebody has to see that he is not too kind. Somebody has to advocate the exploitation of any idea that looks good, and somebody has to point out that it really isn't better than what the Army has already got. At least the [Directorate] kept itself off the floor of the House of Representatives.[23]

Woolley found himself called upon to perform a number of other varied tasks during the war. Some involved policy issues but most were practical. In 1941 he produced a small pamphlet for the Australian Army entitled *Navigation by the Stars*. Printed by the Victorian Railways Printing Works, it was an example of the many everyday benefits of astronomy and the ways in which the Observatory assisted the war effort. In the introduction, Woolley explained:

> In modern warfare, where movement by night tends to increase, a knowledge of the stars sufficient to enable a rough bearing to be taken has become of importance to soldiers. It is not necessary to be an astronomer or a navigator to acquire such knowledge. A soldier, with little practice, can find his bearings by means of the stars, and avoid any feeling of helplessness due to having no means of determining his direction. This pamphlet has, accordingly, been written to give a general knowledge of the sky and of the apparent movements of the constellations.[24]

Yet Woolley remained the Observatory's Director throughout the war and he was responsible for all that happened on Mount Stromlo. As the wartime work conducted at the Observatory was not without its problems, he frequently returned to provide leadership and give direction.

A report produced by the Ordnance Production Directorate in January 1943 was very critical of the generally low standard of production work at Mount Stromlo. It appeared as though communication between the various work sections had created problems that were identified in the quality control phase. Unless the various section heads at the Observatory could work harmoniously, and the quality of the work improve, the report recommended that Mount Stromlo should 'be relieved of all its production work'.[25] The staff responded to this threat energetically. Training standards were reviewed while attempts were made to ensure the workshops were clean and efficient in all aspects of production. Woolley and Cla Allen worked closely with the section heads to reverse the criticisms made in the assessment. By mid-1943, the same Directorate official reviewed the quality of optical instruments produced on Mount Stromlo and pronounced his satisfaction with the Observatory's work. An embarrassing crisis had been averted. In the second half of 1943, Mount Stromlo's production had increased from 300 to 500 optical prisms per month mainly to fill orders that had been placed by the United States Army. The Observatory would produce over 15,000 items for the Optical Munitions Panel with an estimated value of £750,000.[26]

While optical munitions work constituted the bulk of the Observatory's work, scientific research building on insights gained in the 1930s contributed practically and substantially to the war effort. The foremost example was the ionospheric prediction work undertaken by Cla Allen, probably the Observatory's leading and most productive scientist to that time. With the outbreak of war and astronomy all but abandoned at Mount Stromlo, and with no apparent end to the war in sight and the Army needing young men, Allen told Woolley that he had 'been having visions of joining up, or perhaps being sent somewhere else'.[27] Allen discovered on his return from South Africa that a great deal had changed at Mount Stromlo and he found the adjustment difficult. A personal reflection contained in his diary was most revealing: 'The only variety at work is made by the mistakes I make and their subsequent difficulty'.[28] Together with his assistant Ted McCarthy (a science graduate from the University of Western Australia), Allen would find respite in research into radio waves and the ionosphere. It was the latter work that directly affected communications and related to some of the data analysis-oriented work in which he was already engaged on Mount Stromlo.

The ionosphere is a region of the Earth's atmosphere extending from about 70–300 kilometres above the Earth. It has a complex, layered structure. The radio waves most effective for medium and long-distance communication (distances exceeding 160 kilometres) are in the High Frequency (HF) band (3–30 MHz) and travel from transmitter to receiver along an indirect path—a sky wave—reflected off the ionosphere. The most important of the electrically charged particles implicated in the reflection of radio waves by the ionosphere are the free electrons ejected from atoms by the Sun's radiation. The density of electrons in any region of the ionosphere depends on the intensity of the Sun's radiation and therefore varies with the time of day and season of the year. During darkness, ionisation ceases and many electrons disappear, often causing communications failures. Because the ionosphere varies from place to place, there was a need for local observations so more could be known about the characteristics of the ionosphere and the upper limit in the frequency of the radio waves—the critical frequency—that could be used for reliable communications.

With the outbreak of war in 1939 producing the need for reliable HF communications, the Allied governments decided that scientific groups in the United States, the United Kingdom and Australia should make predictions for the different theatres of operations. In Australia the task was given to the RRB. Ionospheric prediction work at Mount Stromlo originated with a visit from Pilot Officer (later Squadron Leader) A.L. Hall of the RAAF to determine whether the Observatory could provide a prediction service for the maximum useable frequencies for radio communication at different times of the day each month.

By this time, Mount Stromlo had already taken delivery of ionospheric sounding equipment. Dr Pulley built the first ionospheric recorder in Australia during 1935, with Mount Stromlo becoming the first ionospheric station to be

Ionospheric work at the Kite House Photo: Walter Stibbs

equipped with automatic recording in 1937. The recorder was located in the Kite House (since demolished), near the Director's residence, and operated by Arthur Higgs. This was the start of a long series of observations designed to expand understanding of the nature and function of the ionosphere. Woolley was rather diffident about this kind of work. He had already brought a sudden cessation to meteorological work on becoming Director. He felt it was not astronomy in the proper sense, while he never felt entirely comfortable with emerging technology.[29] But as this work directly affected the war effort, Woolley agreed.

After Hall's visit, Stibbs was tasked with the collection, reduction and prediction of maximum useable frequencies from the data. When data about the ionosphere were collated and analysed, predictions of the critical frequency could be calculated. By the end of 1944, the accuracy of ionospheric condition predictions in the Southwest Pacific had achieved a reliability of 90 per cent. The residual 10 per cent uncertainty was due largely to the fact that conditions in the ionosphere are substantially influenced by two major—but, at that time, unpredictable—disturbances on the surface of the Sun.[30] The first is related directly to the intensity of X-rays produced in a solar flare. This causes an immediate effect known as a short wave 'fade-out'—interruptions in radio transmissions when the E and F layers in the ionosphere are prevented from reflecting radio waves in the usual manner due to enhanced absorption in the lower ionosphere where the gas pressure is higher—and can lead to a complete blackout of radio communications across the globe. Such events are not uncommon. However, in wartime their prediction became a matter of urgency.

The second effect has a time delay and is related to the release of a large cloud of particles by the Sun. These clouds travel to the Earth at approximately 1,000 kilometres per second and can produce geomagnetic storms one to four days after their release. Although there was speculation about the relationship between solar activity and geomagnetic activity in the 1800s, the first (white light) solar flare was observed by R.C. Carrington on 1 September 1859 and attributed as the cause of an enormous disturbance in the terrestrial magnetic field detected by a magnetometer at Kew Observatory. By this stage, scientists knew that solar activity affected predictions in normal conditions but they were unsure about predicting or estimating their probable effects.

The Australian Radio Propagation Committee, established by the Commonwealth government in 1942, asked Allen to investigate these phenomena. Although not experienced in radio work, he was becoming an authority on the Sun and was already working on solar–terrestrial relations and magnetic storms, being particularly concerned with the effect of *M*-regions on the Sun producing recurrent magnetic storms with the synodic period of 27 days. The Committee had presented Allen with a problem that was far from straightforward. It required high-quality data and a creative analytical approach. Although the average number and area of sunspots increased and decreased in cycles of approximately eleven years, individual sunspots appeared in a rather sporadic manner. The difficulty of prediction was magnified by the erratic and almost unique influence on the ionosphere of any particular spot.

Using a solar telescope and a spectrohelioscope, Allen maintained close observation on abnormally large, persistent spots and prominences that could cause severe magnetic storms. When an active sunspot approaches the Sun's central meridian, magnetic storms are imminent. The Sun rotates once in 27 days, and sunspots and other sources of magnetic disturbance tend to recur every 27 days. In the period 1942–44, sunspots were the main cause of disturbance. By correlating daily data on flares and sunspots with information about the Earth's magnetic field, Allen was able to issue the armed forces with forecasts of the optimal times for radio transmissions. Thus the effects of the recurrent storms could be predicted with acceptable accuracy. These forecasts, according to the official war historian, were 'greatly appreciated by all Australian services and did much to enhance the reliability of communications'.[31] The Optical Munitions Panel separately commended Allen for his work in designing and producing optical mechanisms for wartime use.

Allen's work made operations in places such as New Guinea—where there had been no previous experience—more successful, with the prospect of nearly 100 per cent reliability in communications. Thereafter, the RRB provided predictions for average conditions and the Observatory provided data to the Radio Propagation Committee on disturbances. The other notable outcome of this work was that it served to encourage Allen to further his work in the new science of radioastronomy. The analytical aspects of his work in this area were summarised in a paper he published in the *Monthly Notices of the Royal Astronomical Society*.[32]

As a result of his work on radio disturbance predictions, Allen was also able to elucidate the character of the so-called *M*-regions. In addition to the geomagnetic activity that seemed to follow large solar flares, astronomers observed recurrent geomagnetic storms that were related to the period of solar rotation. Chapman had referred to unknown regions of the Sun that produced these recurrent disturbances as *M*-regions. Allen tried to connect these regions to white light coronal streamers that could only be observed during times of total solar eclipse on the Earth's surface. (Allen participated in five solar eclipse expeditions: Japan in 1936, South Africa in 1940, Sweden in 1954, Ceylon in 1955 and the Canary Islands in 1959.) He believed that these streamers may have been indicators of the trajectories of corpuscular streams that, when they reached the Earth, produced the recurrent geomagnetic activity.

He showed that, whereas major storms are associated with sunspots and flares, with a time delay of about a day, recurrent storms are due to the action of sunspot magnetic fields on particles with considerably longer flight times.[33] He concluded: 'The *M*-region is not a small area coming into activity by some fortuitous chance like a sunspot. It is more likely to be an emission coming continuously from practically the whole Sun's surface and constrained to move in streams by forces in the Sun's atmosphere'. This was the first suggestion of the nature of the solar wind, many years before the theory was fully developed. Bray and Loughhead, in *Sunspots*, made reference to his later research into the origins of the solar cycle.[34]

Allen was correct in concluding that it was the arrival at the Earth of solar corpuscular material that produced geomagnetic activity. However, he (and Van de Hulst, who came independently to the same conclusion) was incorrect in his designation of the white light coronal streamers as the source of these particles.[35] The search for the *M*-regions was to continue for many more decades, with the answer coming from extended, high-quality observations from Skylab in the late 1970s. These observations showed that it was the dark regions of the X-ray Sun, known as coronal holes, which were the source of high-speed corpuscular streams. This is the exact opposite to Allen's hypothesis, although his theorising was sound on the basis of known data. The bright regions, whether observed in white light or X-rays, are regions where there are strong magnetic fields that confine the motions of particles close to the Sun. It is in the dark regions, or coronal holes, where the solar magnetic field is open to the interplanetary medium, that the solar particles are able to escape. Before being able to see the coronal holes in X-rays, astronomers drew calendars with a 27–28 day periodicity as a means of predicting recurrent geomagnetic storms. But knowing when the disturbance started and ended was difficult. With X-ray images available from satellites, the guessing has been eliminated; the task is now simply observing for the presence or absence of coronal holes.

When the war ended, it was decided that the RRB should return to its main task—radio research—and that a section be formed in the Department of the Interior to provide the prediction service. The Ionospheric Prediction Service (IPS) was formed in September 1947 for this task and was attached initially to

Mount Stromlo.[36] The IPS and the RRB worked together for some time, with several RRB staff transferred to the IPS before it became an entity separate from Mount Stromlo and the RRB. By 1949, the IPS was functioning as the provider of prediction services to Australian communicators, using Allen's wartime work as the basis for its prediction of disturbances. It was clear that the war had left a positive legacy in this area of physics and astronomy, with many new areas of research prompted by wartime problems. The same could not be said for optical work.

The Optical Munitions Panel was rightly credited with many important achievements. When Laby's resignation owing to ill-health was accepted in 1944,[37] Professor Sir Kerr Grant of Adelaide University succeeded him. Woolley was one of Grant's admirers. He later commented that, from the outset, Kerr:

> impressed me profoundly with his love of physics—the real physics of things that you actually handle—and his interest in literature. I suppose that few post-war undergraduates know what it is to be taught, rather than lectured at, or what it is to be taught science by a man of very wide interests. I was very lucky in my own undergraduate career, but I would have liked to be taught by Kerr Grant.[38]

Despite the Panel being renamed the 'Scientific Instruments and Optical Panel' towards the end of the war, in an attempt to give the organisation an enduring peacetime role, it was disbanded in November 1945. At the Panel's final meeting, a formal resolution noted that 'the success of the work of the Panel has been in no small manner due to the zeal and unsparing efforts of Dr Laby'. In looking to the future, the Panel considered the future direction of optics research and manufacture in Australia. Two possibilities were apparent: first, the need for optical instruments for the Armed Forces and government departments could be met by local suppliers; and second, a commercial market for optical instruments needed by schools and universities, such as microscopes, binoculars and telescopes, could sustain a local optics industry. In the end, the capacity for local research and manufacture was lost after the war in the absence of government assistance for the development of an export industry and the aggressive tactics of overseas suppliers.[39] At Mount Stromlo, the practical expertise and hands-on experience that was gained during the war became very useful when astronomical work was resumed after 1945.

Despite the demands of the Armed Forces, some pure scientific work had continued during the war. In September 1942, astronomical activity was resumed on a very limited scale with the spectrohelioscope being used to estimate the activity of sunspot groups. Whenever the weather permitted, sunspot sketches were made. Earth current recordings began in November 1942 and continued throughout the war. But in 1944, when the Advisory Board met again for the first time since 1940, the word 'Solar' was dropped from the Observatory's title as attention was beginning to return to Woolley's program for stellar research.

With the Japanese all but defeated in the Pacific, Bill Rimmer collapsed and died suddenly in the main Observatory building on 2 July 1945. His wife Kathleen survived him and there were no children.[40] His service at Mount Stromlo was long and dedicated but his capacity for work had been much reduced after an accident while working with the 'Oddie' telescope. He had fallen and remained unconscious on the floor of the telescope dome all night until he was found in the morning. Among his injuries was a detached retina. Although Woolley had a high regard for his work in the 1930s, Rimmer did not play a leading role at the Observatory after 1939, while Mary Lee Woods worked on much of the data that he had collected by the time of his death. The passing of Bill Rimmer marked the end of an era in more ways than one.

chapter five

From the sun to the stars, 1946–1951

THE PRE-WAR STAFF OF TEN HAD BEEN ENLARGED to 70 during the war. With the cessation of hostilities, it had settled at a permanent level of around 30. Of that number, only four had served before the war. Yet a number of wartime technical services survived and were considered indispensable to the Observatory's future. They included an engineering design section, an instrument shop, an optical shop, electronics and photographic laboratories, and the mechanical shop. The design section was responsible for the preparation of designs and drawings for new pieces of equipment from small filter mounts to complete telescope domes. The optical shop was tasked with the manufacture of lenses and mirrors with apertures as large as 50 inches (127 centimetres) from the unprepared glass or quartz blanks using machines that ground and then polished the optical surfaces. The prevailing standard of precision in optical surfaces was one-tenth of the wavelength of visible light or approximately two-millionths of one inch. The photographic laboratory developed much of the photographic material produced at the telescopes in addition to conducting tests on the photographic plates used in spectrographs. The laboratory also produced charts and atlases of star clusters, nebulae and sections of the Milky Way and the Magellanic Clouds. The electronics laboratory constructed DC amplifiers needed for photoelectric work, control units for telescope drives, devices for the conversion of information into digital form, equipment used for counting, dividing and shaping pulses, and the instruments used for precise comparison of clocks for the Commonwealth Time Service.

The Commonwealth Time Service, established by Gascoigne at Mount Stromlo during the war to assist Army surveyors in the definition of longitude, was maintained long after hostilities ended to fix 'Australian Standard Time'. It was apparent during the early years of the war that existing radio time signals were too erratic for the determination of longitude. When the existing Service needed to be relocated in 1944 with the closure of the Melbourne Observatory, Mount Stromlo seemed the obvious destination. Gascoigne and Gottlieb acquired an old survey telescope on an altitude-azimuth mounting for the Observatory. It was fitted with new optics and rebuilt as a reversible transit instrument. Time was derived from two Shortt pendulum clocks. One had been in service at Mount Stromlo for many years, a second was transferred from the Melbourne Observatory. The Shortts were later replaced with quartz clocks and a number of accessories and batteries installed as a security against power failures. Since each transit observer had a 'personal equation' compounded of an image bisection error and a trailing error, an artificial star travelling at equatorial speed was used to determine the observer errors in order to obtain 'error free' transit times for analysis.

Once established, H.J.M. Abraham, formerly a staff member at the defunct Melbourne Observatory, was appointed in charge of the Service. Don Thomas ably assisted him. An RAAF officer, Thomas was captured by the Japanese in Java and survived the hardships of Changi prisoner-of-war camp. An electronics expert, he joined the staff immediately after demobilisation. To provide such a service, 'correct time' was fixed with respect to the stars and then kept by quartz crystal clocks produced at the Observatory, in which a specially ground piece of crystalline quartz was kept vibrating by an electronic circuit using first valves then, later, transistors. The output currents or pulses were used to drive the clock dials and to generate precise time signals. As astronomers were aware that the rotation of the Earth was irregular, precise observation of the stars was maintained at Mount Stromlo using a photographic zenith tube (PZT) that entered service in 1956. Measurement of the photographic plates from the PZT disclosed not only how fast or slow the Earth was rotating relative to the atomic clock, but also the slight movement of the north and south poles over the surface of the Earth. Time signals from the Observatory were transmitted three times daily from the Navy's short-wave radio station at Belconnen on a number of frequencies and were also relayed by telephone to the time signals provided by the Post Office in each state and territory.

When the Observatory was later transferred to the Australian National University (see Chapter 6), part of the agreement included an undertaking from the University to continue the work of the Time Service. The Observatory's 1958 annual report commented:

> It seems indeed fitting that this be the case. Not only is it a fine tradition that the largest observatory in the country should have under its wings the National Time Service, but the presence of the Service in a University observatory provides day-to-day contact for staff and students with the precision techniques of measurements of time and positions—an art that is only too easily ignored in an observatory where the primary emphasis is on astronomical studies and research on the Milky Way system and other galaxies. The traditional activities of the Time Service have continued and in addition several are being extended in connection with the International Geophysical Year. The installation and placing into operation of the PZT is taking much of the staff's time. With the arrival of three new staff members an extensive programme of training has been initiated.[1]

The Time Service, renamed the 'Section on Positional Astronomy' in 1961, would remain at Mount Stromlo until 1968 when it was incorporated into the Division of National Mapping in the Department of National Development and Energy.

In May 1946, Woolley noted in a report on astrophysical studies that 'a program of observation of masses and surface temperatures of southern stars . . . is being developed on Mount Stromlo. This will be made to fit in with programs undertaken by various observatories in South Africa'.[2] Woolley remained impressed by the 'exceptionally clear sky' and the generally favourable conditions still prevailing on the mountain. But the existing telescopes were limited in their capacity to undertake this work. The Sun telescope was plainly nearing the end of its useful service, although there was still good spectroscopic work for which it would be used. The 10- and 15-inch mirrors (i.e., the coelastat and secondary mirror) had been aluminised at the University of Melbourne during the war. In 1949, sunspot sketches were made on 286 days and the magnetic polarity of sunspots observed on 74 days. The last task performed by the telescope was to photograph the transit of the planet Mercury across the face of the sun on 6 May 1957. Regular contributions of data obtained from the Hale spectroheliograph to the *Quarterly Bulletin on Solar Activity* were resumed. Other solar research work was initiated including an investigation of the electron density of the corona and an estimate of the amount of ultra-violet radiation emitted from it.

A series of double-star measures was made with the 'Oddie' telescope and the results published in *Memoire Number* 9. A determination of the stellar magnitude of the Sun was undertaken and variable star investigations were carried out, initially by Stibbs with a photocell on the 'Reynolds' reflector and subsequently with a 1P21 electron multiplier photometer installed at the 'Oddie' telescope. Stibbs used the 1P21 for his work on the variable magnetic star HD125248 which led to his 'oblique rotator' model. He was able to explain variable stars' properties in terms of a dipole magnetic field offset from the star's rotation axis, as is the case with the Earth's.[3]

Many of the pre-war problems with the 'Reynolds' were resolved by May 1946, with Woolley remarking that the telescope will 'be used to make a photoelectric and spectroscopic analysis of stars'.[4] An electron multiplier photometer with slitless spectrograph designed by Allen and Gottlieb, together with slow-motion drives, was added to the 'Reynolds' and operated from August 1947 to determine stellar photometric gradients and also monochromatic magnitudes by photoelectric means.[5] The program was designed as a southward extension of the Greenwich relative gradient program and closely imitated the Greenwich work in both instrumental equipment and general observing procedures. A nebula spectrograph was also constructed and spectra of nebulae and of the Milky Way obtained. Monochromatic magnitudes and colours of faint stars were also determined from very small dispersion spectra produced by a wire grating in front of the reflector. Radial velocity measurements of stars, including the Hyades, were made from spectra obtained with the spectrograph. After the telescope's wiring was improved and the observation platform reconditioned in 1950, the slitless spectrograph was removed and replaced with a refrigerated 1P21 multiplier from the Lick Observatory, which was used over eighteen months by Gascoigne and Kron. This equipment placed Stromlo in a better position than any other observatory in the world to undertake faint photometry of southern stars (see next chapter). Stars in the Magellanic Clouds (the nearest galaxies) down to magnitude 15.5 were observed. Two-colour light curves were obtained for nine Cepheid variable stars in the Small Magellanic Cloud in addition to magnitudes for 22 globular clusters that were then used in determining the distance of that galaxy. By this time, problems with the telescope's drive system and slow-motion controls were finally improved with the installation of an electronic oscillator, while the mirrors were aluminium coated.

While these activities were valuable in themselves, decisions were needed about the overall focus and essence of the Observatory's work. The Astronomer Royal, Sir Harold Spencer Jones, was invited to attend the 1947 meeting of the Observatory's Advisory Board. This meeting was particularly significant as there was detailed discussion of the Observatory's future research activities and the emerging need for a large telescope for astrophysical work. In a report prepared

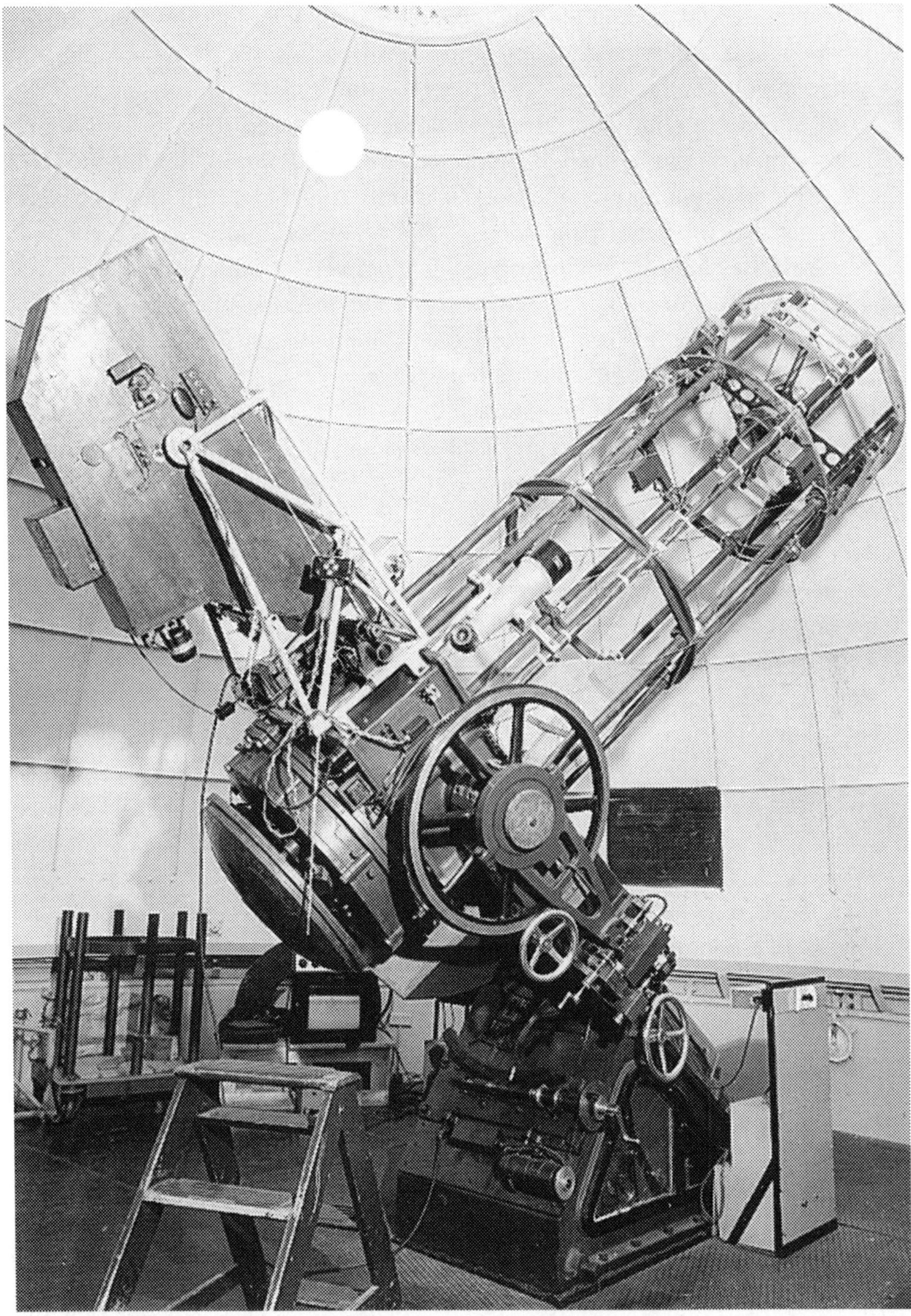

The 'Reynolds' Telescope with Cassegrain spectrograph in the 1950s Photo: MSO

for the Minister of the Interior in March 1947, Sir Harold strongly urged the acquisition of a major modern telescope for the Observatory to complement programs being conducted elsewhere in the world.[6] As a consequence of the meeting, the Commonwealth government was persuaded to install a 74-inch reflector similar to the 74-inch 'Radcliffe' telescope in Pretoria, South Africa and the 72-inch telescope in Victoria, Canada at a total cost of £100,000. A contract for the construction and installation of the telescope and a steel dome was signed with Sir Howard Grubb, Parsons & Company of Newcastle-upon-Tyne. Making use of an invitation to attend the 1948 meeting of the International Astronomical Union in Zurich, Woolley and Allen visited the contractors and settled a number of specific construction issues.

It was also clear that the limitations of the Observatory's existing equipment required new directions in research. Allen recognised that the solar telescope was no longer able to further his research and shifted his interest to hydrogen emission regions along the Milky Way.

Arthur Hogg returned in 1946 to a substantially changed institution with its focus now firmly set on stellar astronomy. Having considered the possibility of remaining at Maribyrnong, the 43-year-old Hogg had to think hard about joining the astronomers because he:

> had to leave two fields in which he had acquired real authority, for a new subject with new methods, new concepts, and almost a new philosophy. In the event he never moved as easily in astronomy as he had in atmospheric electricity or cosmic rays. He could not have been helped by the heavy administrative load he carried under Woolley—not, it must be added, that he found this uncongenial; he enjoyed responsibility, and a formal and orderly man himself, was well at home within the formal and orderly framework of the Public Service. All in all it is surprising that he published as much and achieved as high a reputation as he did; but he was too good a physicist not to seek out significant problems and to make important contributions to them.[7]

Hogg was persuaded to abandon his work on cosmic rays and to invest his energies elsewhere. He built a photoelectric photometer and began observing stars known from spectroscopic observations to be in binary systems. His hope was that some would prove to be eclipsing variables as well. When an observer's line of sight happens to lie in the plane of the binary orbit, the resulting stellar eclipses provide information about the size and shape of the orbit, and about the radii and masses of the component stars themselves. The observer needs to make both photometric and spectroscopic measurements to achieve this. Although a laborious procedure, it was one of the few means available for learning about the fundamental stellar properties of mass and radius. A total of 2,433 photometric observations were made with the 'Farnham' on 57 observing nights in 1947. Of particular interest were the stars σ Scorpii and μ Centauri, while the most significant was the eclipsing system ζ Phoenicis, first observed by Hogg. This work enabled Hogg to extend the mass scale upwards to stars about six times as heavy as the Sun.[8]

During the following year, with the aid of electronic slow-motion controls fitted to the 'Farnham' telescope, 4,675 photometric observations of stars were made on 126 observing nights. In 1949, a further 400 photometric observations were made on 30 nights. These results produced a complete light-curve for ζ Phoenicis and allowed a determination of its orbit to be calculated. The results of this research were prepared for publication, while a 4-inch spectrograph was mounted on the telescope to be calibrated on lamps in the adjacent Sun telescope dome. The latter innovation did not produce data of any great value. In 1951, the 'Farnham' was removed from its mount and packed for storage. The 'Reynolds' was used as an 'objective prism' in the Observatory's relative magnitude program, described above.

Another great challenge for the Observatory was relating its work to a quite remarkable contemporary development in Sydney—the birth of radio astronomy. As astronomers can now use the entirety of the electromagnetic spectrum to collect information about the Universe, it is easy to forget that from ancient times until merely 60 years ago, knowledge of the heavens was limited to that tiny window of the spectrum that we call 'light'. The spread of wavelengths that we can see with our eyes spans less than a factor of two from 0.39 to 0.70 of a micron. By contrast, the total spread of wavelengths now used by astronomers spans a factor of some one thousand million, million, million from the highest energy gamma-rays detected to the lowest energy radio waves.

Visible light has another property essential for making stellar observations in earlier centuries: it can penetrate Earth's atmosphere and reach human beings and ground-based telescopes. This is also the case for much of the radio domain but, because we cannot see radio emissions, it was not until the mid-twentieth century that people discovered they could build telescopes of an entirely different kind, which revealed a new, radically different and complementary view of the Universe. The radio 'window' of atmospheric transmission is actually considerably larger than the optical, spanning a factor of about 1,000,000 in wavelength.

In a sense, radio astronomy came about unintentionally. Radio scientists did not set out to use the equipment they were developing to view the Universe; they were far too preoccupied with such practical matters as the transmission and reception of radio signals for long-distance terrestrial communication. Instead, the Universe made its presence felt in a quite uninvited way. In 1932, Karl Jansky was investigating the source of some troublesome background hiss for the Bell Telephone Laboratories when he realised that the origin of part of this 'noise' was extra-terrestrial.[9] It was coming from the direction of the constellation of Sagittarius, now known to house the nucleus of our Milky Way galaxy. Seven years later, another American radio engineer, Grote Reber, built the first dish-shaped radio telescope, considerably improving on Jansky's work.[10] Then came World War II, and radio scientists were so immersed in urgent military tasks that they gave little further thought to this newly discovered cosmic phenomenon.

The greatest wartime effort was devoted to the development of radar, and the Allied countries tackled this in quite a coordinated way. The United Kingdom, the United States and Australia each set up a single research establishment and shared the classified output of all three. The Australian group was formed within the CSIR and named the Radiophysics Laboratory. It occupied a wing of the National Standards Laboratory in the grounds of the University of Sydney. The original Chief was David Martyn, who was succeeded by Frederick White in 1942. In early 1945, White was appointed to the CSIR Executive, and a year later Edward Gordon ('Taffy') Bowen became Chief. Bowen had the distinction of working successively at all three of the major Allied radar establishments during the course of the war.

The outstanding wartime achievements of the Radiophysics Laboratory have been recounted elsewhere and make heroic reading.[11] What matters more from the point of view of this history is what happened at the war's end. In Britain and America, many of the scientists and engineers involved in the wartime radar development returned to their pre-war occupations. By way of contrast, in Australia the Radiophysics Laboratory, with all of its equipment and most of its 300-strong staff, was retained intact. Not only that, the head of CSIR, Sir David Rivett, took a decision that his organisation would relinquish its involvement in classified research, and concentrate on work of economic benefit to Australia, along with some fundamental research. He gave Taffy Bowen an entirely free hand in determining the postwar role of the Radiophysics Laboratory. Bowen made a long list of possible research areas for his staff to pursue—intending, as time passed, to concentrate resources into those areas which were proving the most promising. It was not long before the work of the Radiophysics Laboratory focused on two fields: cloud physics and 'cosmic noise', as it was known before the name 'radio astronomy' was coined. Bowen headed the former work himself, while Joseph Lade Pawsey was the undisputed leader of the latter. Joe Pawsey's genius had inspired many of the group's wartime radar successes, and he continued to exert a commanding influence as they launched themselves into exploring this entirely new branch of astronomy. Indeed, Pawsey has been credited with coining the field's new name.[12] He was eventually appointed Assistant Chief of Radiophysics in 1952.

The worldwide genesis of radio astronomy in the immediate postwar years has been well documented.[13] Although other groups were active at the time, notably at Cambridge and Manchester, the Radiophysics Laboratory in Sydney was unquestionably the most successful. This was a novel experience for Australian scientists; participating in world-class research in those days generally entailed moving abroad. The excitement of working in such an era of discovery was intense, and evident to anyone visiting the Laboratory. The staff worked long and hard, at the oddest of hours and in the strangest of places. Their pioneering endeavours consumed their whole lives. When they weren't working, they even played together. About the only non-astronomical topic of conversation discussed in the tearoom was the exploits of the Radiophysics Cricket Team. Paul Wild has likened working at Radiophysics in those years to what it must have been like to be part of Rutherford's team at the Cavendish Laboratory in the pioneering days of nuclear physics.[14]

A listing of some of the astronomical successes of the Radiophysics Laboratory in the immediate postwar years reveals the extent of their achievements:

- In 1945 Joe Pawsey's group detected million-degree radiation from the Sun.[15]
- In 1946 the same group showed that sunspots give rise to an enormous amount of the solar emission.[16] They used an ingenious technique known as 'sea interferometry' to localise the position of a radio source. The source was observed with an antenna on a clifftop overlooking the sea at Dover Heights. Combining the output from the direct line of sight to the source with that reflected off the water gives an effective telescope size (for source positioning purposes) of twice the height of the cliff.
- In 1947 John Bolton's group used a sea interferometer to observe Cygnus A, the third brightest radio source in the sky; they showed it to be of very small angular extent, less than 8 minutes of arc.[17] Many believed this work was worthy of the Nobel Prize. By year's end they had discovered another five such sources. The following year they took their equipment to New Zealand, and managed to obtain sufficiently precise positions for four of them to look for their optical counterparts.[18] The results were a complete surprise. None

was a normal star. One was the Crab Nebula, the supernova remnant of a star that was seen to explode by Chinese astrologers in 1054 AD. The other two were rather peculiar galaxies, lying far beyond our Milky Way home. The fourth, Cygnus A, remained unidentified for several years until Walter Baade and Rudolph Minkowski used the 200-inch reflector on Mount Palomar, and an even more precise position provided by the Cambridge radio astronomers, to find its optical counterpart.[19] This proved to be the greatest surprise of all—a galaxy so distant it was only just observable with the world's largest optical telescope.

- In 1948 Jack Piddington and Harry Minnett observed the Moon, and were able to infer its surface structure, which was confirmed years later during the Apollo missions.[20]
- During 1949–52 Paul Wild's group discovered several types of solar radio 'bursts', events involving vast plasma ejections from the Sun's surface out into its corona.[21]
- In 1951 'Chris' Christiansen and Jim Hindman confirmed the newly detected emission line of the hydrogen atom (at 21-centimetres wavelength), and presented evidence for a spiral arm structure in our Milky Way.[22]
- Also in 1951 Christiansen built a 32-dish 'grating interferometer' at Potts Hill, making observations that laid the foundation for the Earth-rotational aperture-synthesis technique. This technique underlies the operation of virtually all of the world's most sophisticated radio telescopes today.[23]
- In 1953 Frank Kerr built a 36-foot dish at Potts Hill, and commenced a 21-centimetre hydrogen line survey of the whole of the southern Milky Way.[24]
- In 1953 Bernard Mills invented yet another radically different type of radio telescope, the Mills Cross, which he subsequently used to make an all-sky survey, finding 600 new sources in the first year alone.[25] His analysis of their strengths provided an important contribution to a cosmological debate that was raging at the time.[26]
- In 1953 Dick McGee and Bolton used a new telescope at Dover Heights to locate the nucleus of our Milky Way Galaxy.[27]

Such a frenetic pace of new discovery occurs only rarely in the annals of scientific research.

The pace was *so* frenetic that it took the astronomical world some time to appreciate just what was happening in Sydney. The observational techniques being used there, the scientific results being obtained and even the technical jargon spoken, were all so foreign to the practitioners of traditional optical astronomy that this was perhaps inevitable. Furthermore, at the start, few astronomers were personally acquainted with these radio scientists, since they had come from entirely different backgrounds. As Taffy Bowen recalled, at the end of the war the Radiophysics Laboratory had 'not a single astronomer on the staff, nor, for that matter, anyone who had done a university course in astronomy'.[28]

The work of the Laboratory did receive appropriate recognition from other radio scientists, however. In 1952 the International Union of Radio Science (URSI) held its General Assembly in Sydney (at the invitation of David Martyn, then President of its Radio Astronomy Commission). This was a notable honour because it was the first occasion on which a general assembly of any international scientific union had been held outside Europe or North America.[29] Nevertheless, the Radiophysics researchers found the response of Australian optical astronomers to their work disappointing. There were three exceptions and each was linked to the Commonwealth Observatory.

The first was the work of Cla Allen and Colin Gum on the Sun and the Milky Way. Allen had developed a friendship with Joe Pawsey, who allowed him to borrow a 200 MHz receiver from the Radiophysics Laboratory in 1946 in order to monitor the solar 'noise' and also to survey the radio emission from the southern part of the galaxy.[30] Allen was the first optical astronomer to undertake radio observations on his own initiative.

The second was the assistance afforded to David Martyn while he was the RRB's Senior Principal Research Officer at Mount Stromlo. At the request of Sir John Madsen, Chairman of the RRB, office accommodation at the Observatory was made available for Martyn, who was seconded to Mount Stromlo from 1944 to 1956 to research solar–terrestrial relationships. While at Mount Stromlo, he proposed a theory of atmospheric tides that won him election as a Fellow of the Royal Society. Martyn was also a Founding Fellow of the Australian Academy of Science. His secondment to Mount Stromlo strengthened the extant connection between the Board's research work and the Observatory, and also led to fruitful collaboration between the two organisations in the field of solar and terrestrial relationships. Towards the end of 1945, the Radiophysics Laboratory was able to demonstrate that the solar radio noise at 200 Mc/s was closely correlated with measurements of sunspot activity at the Commonwealth Observatory. By April 1946, regular measurements of solar radio noise were taking place at Mount Stromlo and this continued for a number of years. Theoretical studies of the physical condition of the chromosphere followed.

The third exception was Woolley. He became personally involved in radio astronomy through his theoretical research to determine the optimum physical conditions in the ionised plasma regions surrounding hot stars if they were to account for the observed Milky Way radio emission. His paper on this work was one of the earliest Australian radio astronomical publications.[31] Nevertheless, in these formative years of radio astronomy, the degree of interaction between Mount Stromlo and the Radiophysics Laboratory was not as great as might have been hoped.

The relationship was not helped by two quite unfortunate incidents, both involving Woolley. The first relates to an address he gave in Perth, during the 1947 meeting of the Australian and New Zealand Association for the Advancement of Science (ANZAAS); he was then the President of Section A.[32] He delivered a paper entitled 'Opportunities for Astrophysical Work in Australia' and failed to make any mention of the newly developing radio field! Noting this significant omission, during the question period that followed Ron Bracewell asked Woolley where he thought radio astronomy would be in ten years' time. 'Forgotten' was his response! To the radiophysics researchers, eager to establish their credentials in the world of traditional astronomy, this was an intolerable slight, and one wonders why Woolley inflicted it. While he was certainly conservative in some of his attitudes (he was one of the last astronomers to accept the use of computers to control large optical telescopes), his remark on this occasion can hardly have been due to any real antagonism towards, or even lack of interest in, the new field. After all, radio equipment was in operation at his own observatory at the time, and he had made an important personal contribution.

It seems far more likely that his remark stemmed from his rather idiosyncratic sense of humour! He was much addicted to the one-line put-down, and radio astronomers were not the only people to suffer in this respect. On another famous occasion, when arriving in Britain as the new Astronomer Royal and a year or so before the flight of *Sputnik 1*, he was asked by a reporter for his views on space research. 'Utter bilge,' he replied.[33] Nor were colleagues working in his own area of optical astronomy in his own observatory immune from such withering witticisms. When Gerard De Vaucouleurs' research on southern galaxies led him to the realisation that many of Earth's nearest neighbours form a supergalactic plane, he wrote up his work and, brimming with excitement, burst into Woolley's office to tell him of this spectacular result. Then, pausing, he asked where the Director thought he should publish it. '*Punch*,' Woolley replied.[34] It actually appeared in the *Astronomical Journal*.[35]

Whether an intended witticism or a considered reply, Woolley's remark was most unfortunate. It did nothing to further the integration of the radio astronomers into the discipline as a whole—indeed, the fact that the players in those early exciting years at the Radiophysics Laboratory, and its historians, have continued to record the incident even when writing many decades after the event bears witness to the depth of hurt that the perceived slight occasioned.[36] It is interesting, too, that 'A complete list of Australian publications on radio astronomy during the years 1946–52' compiled by one of these radio pioneers for a review paper fails to include Woolley's early contribution.[37]

Woolley's perceived attitude towards radio astronomy was also cited in support of the accusation that he was a conservative. But he has been defended as having taken nothing other than a reasonable view at the time:

> For, in the early days, the only star giving detectable radio emission was the Sun and this was so feeble that there was little chance of observing a similar source any further away; while the only Galaxy as such giving detectable emission was the Milky Way Galaxy and this was so weak that there was almost no chance of observing any other such source.

The Commonwealth Observatory Board of Visitors, 1950. L to R: S. Chapman, O.U. Vonwiller, K.E. Bullen, J.P.V. Madsen, E.J. Hartung, K. Grant, R. v.d. R. Woolley
Photo: Walter Stibbs

> And if, as was then thought, naturally occurring radio radiation was bound to be mostly temperature radiation, this situation was precisely what any astronomer would predict for himself. Of course it was interesting that the Earth's atmosphere should be transparent in these wavelengths, and of course one should see what they could tell about the Sun and Galaxy. But it would be a waste of resources to do any more about it. I am not sure whether Woolley said it in so many words, but I do feel sure that this was his instinctive and wholly reasonable reaction. It would have been somewhat on a par with his own later reaction to space travel, and Rutherford's famous reaction to any possible 'use' of nuclear energy.[38]

A second, rather more substantial, occurrence took place in 1951. On 31 January the Observatory's Board of Visitors held its tenth meeting. The membership read like a mini-*Who's Who* of university heads of department in the physical sciences. It also included William MacLaren, Secretary of the Commonwealth Department of the Interior, with responsibility for the Commonwealth Observatory, Sir John Madsen, Chairman of the Radio Research Board, and Mark Oliphant, Director of the ANU's Research School of Physical Sciences.[39] The Board carried the following resolution:

> that this Board re-affirm its previous recommendation urging the establishment at the Observatory of a section dealing with radio-astronomy, and considers that this would be an essential part of Observatory development.[40]

Twelve days later, Woolley wrote to Fred White at the CSIRO Executive, informing him of plans not only to establish a major radio astronomy group in Canberra, but also to erect, as soon as possible, a high-resolution radio telescope in the paddocks at the foot of Mount Stromlo.[41] All hell broke loose! White summoned Taffy Bowen to Melbourne to find out what was going on. After a meeting of the CSIRO Executive, he wrote to Woolley, guardedly welcoming any collaboration but making it quite clear that he considered radio astronomy to be the province of the Radiophysics Laboratory. The correspondence did not end there; it went on to embroil Bowen, David Martyn, then working at Mount Stromlo, Oliphant and McLaren. The issue eventually subsided when a meeting between the Observatory's Board of Visitors and the CSIRO Executive held in May 1952 agreed to support the status quo, with optical astronomy being regarded as the province of the Commonwealth Observatory, and radio astronomy that of CSIRO's Radiophysics Laboratory.

Historians still debate whether Woolley was serious about wishing to start radio research in Canberra in such an ambitious way. While he had made contributions in the field, radio astronomy was far from Woolley's major research interest. Nor was he short of developmental work to occupy his attention at the time; his campaign to acquire a 74-inch reflector for Mount Stromlo was well underway.

Perhaps most interesting of all is the fact that Woolley himself seems to have taken little further part in the copious correspondence that followed his initial letter. It would seem equally possible that he contrived the whole affair as a Machiavellian means of ridding himself of an unwelcome directive from his Board of Visitors. He may well have calculated (rightly) that by writing a single letter, addressed to a sufficiently high-level official within CSIRO, and brazenly announcing plans which must surely have been read as an attempt to usurp the pre-eminent position in the world of radio astronomy that the Radiophysics Laboratory had won for itself, he would provoke such an outcry that his Board members would have to resile from their resolution of a fortnight before, and allow him to get on with building his 74-inch reflector without interference. The same cannot be said for Mark Oliphant. He seems to have had a genuine ambition to establish a national radio astronomy centre in Canberra and, several months after the matter was put to rest by the May 1952 truce, he was still in heated correspondence with Bowen, arguing that the building of a major radio telescope, then in early contemplation, was too prestigious and costly a project to be undertaken appropriately by *a division* within the CSIRO.[42] In his view, it ought to be built as a national facility.

This whole episode, whatever its real motivation, was another unfortunate hindrance to a smooth integration of the new radio field into mainstream astronomy, and to the establishment of harmonious relations between researchers in the two areas. It is interesting to note that the Radiophysics Laboratory has since been careful to employ the word 'national' in its nomenclature at every appropriate opportunity. When the Parkes 64-metre dish was commissioned in 1961, it was styled 'The Australian National Radio Astronomy Observatory'.[43] Subsequent to the building of the even more ambitious Australia Telescope in 1986, CSIRO's radio astronomers, as a whole, adopted a new divisional title for their entire enterprise: the 'Australia Telescope National Facility'. It might also be remarked that, decades after the event, there was still a certain wariness within the Australian astronomical community at any suggestion that Mount Stromlo should assume a major managerial role in respect of national facilities. It goes without saying, of course, that time has proved the Radiophysics Laboratory to be perfectly capable of operating world-class radio telescopes. There is no impediment to such facilities being managed by entities having 'divisional' status within a larger institutional structure, given appropriate funding and staffing resources. Such a group inevitably grows to meet the challenge.

Woolley's apparently ambivalent attitude toward radio astronomy should not, however, overshadow the Observatory's contribution to its emergence as an exciting new development within astronomy.

By 1950, Woolley had achieved his goal of changing the focus of Mount Stromlo from solar to stellar research. The word 'Solar' had been officially dropped from the Observatory's name, while Woolley had proposed that the Director bear the new title 'Commonwealth Astronomer'. There were, however, a number of solar research projects that were yet to be concluded, written up and published. The last was *Memoire No. 11*, entitled 'Solar Radiation Observations on Mount Stromlo 1927–1939' and published in October 1950. The report was attributed to Rimmer and Allen, although the latter had gathered and analysed much of the data that had been collected and partly collated by Rimmer before his sudden death. In addition to making public a large amount of data collected over many years, the *Memoire* showed the relation of Sun and sky radiation with cloudiness and seasons and noted that 'radiation from the sky alone approximately equals half the loss by Rayleigh and dust scattering from direct [solar] radiation'. This report also marked the end of Allen's long and valuable work at Mount Stromlo before he accepted an overseas appointment. His departure was felt keenly at the Observatory. Allen was not only a first-class astronomer, he was also highly productive. After the war, Allen published what became a well-known and widely read text, *Astrophysical Quantities*. It had a worldwide audience and has gone through several editions, partly because he accepted a lower author's royalty so as to make the book cheaper and thereby more accessible to students.[44] He resigned as Principal Research Assistant in October 1951 when he accepted the newly created Perren Chair of Astronomy at the University of London. Following his retirement in September 1972, he returned to Canberra and became an Honorary Research Fellow at Mount Stromlo Observatory. He died in December 1987.

Cla Allen was an exceptionally gifted man who exemplified the capacity of some scientists to cooperate and to bring the insights of one discipline to bear on the problems of another. He did not depend on others to dictate or even suggest his research interests—he was a creative and lateral thinker. This was nowhere more apparent than in the substantial contribution he made to communications in Australia, especially in the field of solar ionospheric physics. There are three areas in which history shows his work to have had a lasting impact.

The first was in correlations between sunspot numbers and overall solar activity with various ionospheric parameters. The second was the identification of the corona as the source of corpuscular streams that give rise to recurrent geomagnetic activity. (Allen misidentified the actual coronal structure, but his identification of the white coronal streams was a good working hypothesis at the time.) His third achievement was the technique of using white light coronal scattering to deduce the coronal electron density as a function of altitude.

For all of that, for several reasons Allen's ionospheric work has never received the attention that it might have attracted in other circumstances. First, wartime secrecy restrictions meant that his vital work was not widely publicised, as it would have offered a distinct advantage to the Japanese. The second reason is the possibility that Australian astronomical research was just beginning to receive the recognition it would later command. Dr M.A. Ellison, Principal Scientific Officer at the Royal Observatory in Edinburgh and author of *The Sun and its Influence,*[45] acknowledged Allen's contributions to both solar and ionospheric research although, across the Atlantic, the 1947 reprint of Harlan True Stetson's *Sunspots in Action* makes no mention of him. This was a sign of the growing sense of community between British and Australian astronomers. Third, like many Australian astronomers of this essentially pioneering period, Cla Allen was a relative unknown. Like most of his colleagues, he was not a good self-publicist of either his abilities or his achievements. His communications work was technically advanced in a country far from adept at understanding, recognising or celebrating significant progress. An obituary published by the Royal Astronomical Society in June 1990 included the remark that:

> It is characteristic of C.W. Allen that he described his important conclusions as 'rather tentative'. He was always aware of scientific limitations both in his own work and in the research of others, a characteristic which tended to leave him undervalued among his contemporaries.[46]

By the time of Allen's departure, after contributions spanning nearly a quarter of a century, the Observatory was gaining a reputation for stellar research and for the opportunities it was able to provide for budding young astronomers. It was this capacity that Woolley hoped to exploit during the 1950s.

chapter six

A fruitful partnership, 1951–1956

After struggling with the limitations of the 'Oddie', 'Farnham', 'Reynolds' and Sun telescopes, it was clear that the Observatory needed a new suite of instruments. By default and design, four new telescopes entered service during the mid-1950s. The first was acquired by good fortune and commissioned through skill.

At the request of the Victorian state authorities, the Commonwealth government assumed control of the Melbourne Observatory on 8 March 1944. This marked the end of a protracted series of discussions on astronomy in Australia which had commenced in 1906 when the Commonwealth took control of the meteorological services. An examination of the work then being performed at Melbourne led to the conclusion that the Commonwealth would not be justified in maintaining a separate observatory in the city. Consequently, it was suggested that the Great Melbourne Telescope, which had been standing idle for several years, be transferred to Mount Stromlo. Mount Stromlo would also assume responsibility for the National Time Service with the assistance of a newly installed transit instrument, and continue the magnetic work being done at the Toolangi Magnetic Observatory.

Woolley offered to purchase the Great Melbourne Telescope for its scrap value.[1] It was acquired very cheaply and a 50-inch mirror made to replace the original 48-inch mirror that had been made of speculum metal (an alloy of approximately 68 per cent copper and 32 per cent tin). Speculum metal had been used because the process of silvering glass was an emerging technique which the design committee considered unreliable. Regrettably, this decision led to the telescope being an almost complete failure from the outset. The reflectivity of the polished alloy is around 60 per cent and inevitable tarnishing reduced this even further. This meant that the metal surface needed constant repolishing, although this never succeeded in restoring the proper value to the mirror. Another source of trouble was damage caused to the mirror's surface when varnish applied to protect the surfaces during the instrument's sea voyage to Australia from Ireland was removed. As the telescope was not enclosed within a dome, and given that the focal length was 160 feet (49 metres), even the slightest breeze caused unmanageable vibration in the instrument and made observing pointless. Quite apart from the varied technical explanations for its inadequate performance, Gascoigne offers some mitigation on behalf of the much-maligned mirror. Based on an exhaustive examination of diaries and logbooks, he contends that the telescope:

> failed not through any fault of the staff but because its design meant it could be used only for making pencil-and-eye sketches of nebulae: photography and spectroscopy were precluded. The speculum mirror was a handicap but not a fatal one; a silver-on-glass mirror would have made no essential difference. Despite its faults, the telescope significantly influenced the design of the 36-inch Crossley (or Common) reflector, and through that, of later larger reflectors.[2]

Plainly, according to Gascoigne, it was unreasonable to demand an instrument perform in a manner for which it had simply not been designed.

Woolley achieved his second, and more important, instrumental acquisition by means of a well-orchestrated campaign in which he made excellent use of the administrative and lobbying skills he had acquired during the war years.[3] Soon after the war's end, he launched his bid for a 74-inch reflector by arranging a supportive visit from the Astronomer Royal, Sir Harold Spencer Jones.[4] His campaign was successful, and the Governor-General, Sir William Slim, performed the opening ceremony on 8 November 1955.[5] The 74-inch would remain the largest optical telescope in the country, and equal largest in the Southern Hemisphere, for nearly two decades.

When the instrument was first brought into use, there was a great deal of astigmatism in the images that substantially reduced their usefulness. There was some difficulty in determining whether the problem was with the primary mirror, the mirror supports, or the secondary mirror. Although none of the staff had experience with this aspect of large telescopes, Arthur Hogg was given the

The Great Melbourne Telescope being installed at Melbourne Observatory in 1869
Photo: MSO

problem to solve. The trouble was traced eventually to the main mirror and it was returned to the manufacturer for refiguring. Gascoigne remarks:

> We learnt a great deal with it and a great deal from it . . . it was through experience gained with the 74-inch that Australian optical astronomy came of age, in the sense that it was brought face-to-face with urgent contemporary problems.[6]

As Dr Gwen Simonow had predicted at the end of her 1950 history of the Observatory, with the construction of the 74-inch reflector and the adaptation of the Great Melbourne Telescope for use at Mount Stromlo, 'the opportunities for astrophysical work will be greatly increased and the character of the institution will be settled'.[7]

The Yale–Columbia 26-inch refractor at Mount Stromlo Photo: MSO

The third telescope commissioned in 1955 was the 26-inch refractor of the Yale and Columbia Observatories which had, at Woolley's invitation, been relocated from Johannesburg, where it had been used for astrometric work since 1926. Woolley also invited the University of Uppsala to locate its new 20/26-inch Schmidt telescope at Mount Stromlo. The dome was completed in 1955, although the telescope itself was not brought into service until 1957. Both of these telescopes were operated as 'foreign stations', with resident observers from the two guest nations spending various periods at Mount Stromlo.

Even as Woolley was busily equipping Stromlo with this enviable suite of telescopes, the Observatory's technical horizons were widened in another way by the year-long visit, during 1951, of two Lick Observatory astronomers, Gerry Kron and Olin Eggen.[8] They were both experts in the rapidly developing field of photoelectric photometry, and they brought with them a photometer that Kron had built. Their stay marked a turning point in the Observatory's approach to photometric research.

One staff member who eagerly embraced this emerging technology was Ben Gascoigne. He had just embarked on a project to explore the implications of a recent discovery by the Mount Wilson astronomer Walter Baade that stars contained in galaxies can be sorted into two 'population types', each having its own characteristics.[9] This recognition resulted, *inter alia*, in a doubling of the distance scale of the Universe, and the publication of Baade's work caused a sensation. Gascoigne realised that the Magellanic Clouds provided Southern Hemisphere observers with a golden opportunity to investigate such matters. Clearly visible in the southern sky on a moonless summer night, these two small galaxies lie about 170,000 light years away from us, just seven times as far as the centre of our own Milky Way. Their masses are, respectively, only a tenth and a hundredth that of the Galaxy. They are so close, in fact, that they can best be thought of as satellites of our own, and they serve as unique stepping stones to the Universe at large.

Gascoigne planned to study the Cepheid variables in the Clouds. The intrinsic brightness of these objects is directly related to their period of pulsation, making them excellent 'standard candles' for the measurement of distance. Gascoigne and Kron began an observing program on the Cloud Cepheids in 1951, which Gascoigne would continue for more than a decade.[10] He established an excellent Cepheid calibration which fully confirmed the distance-scale implications of Baade's work, and which has stood the test of time remarkably well.[11]

As the workload and work rate at the Observatory increased, additional staff accommodation was required. Woolley wrote to the Department of the Interior (the Department had maintained this name since 12 April 1932 after three prior name changes) on 16 May 1949 stating that 'houses are required very urgently for the Observatory staff who, at present, have to travel up and down Stromlo, frequently at all hours of the night and often at 2 or 3am'.[12] Seven additional houses were built in a cluster to impart a sense of community among the married staff members, and the aged wooden Bachelors' Quarters were replaced. But these new buildings were suddenly threatened with destruction on 5 February 1952.

During a hot summer, a major bushfire started from a lightning strike near the junction of the Molonglo and Murrumbidgee Rivers to the northwest of the Observatory. It burned through the pine forests on the southern and eastern slopes of the mountain and headed rapidly towards the buildings. As Woolley recalls:

> We saw the actual flash at morning tea time, and saw the fire blowing towards the river and not coming our way until about noon, when the wind changed. At one [p.m.] we were told that there was still no danger but we sent for a bus and evacuated all the women except four who chose not to go (my wife, Mrs Hogg, Miss Davson and Miss Beech [later Mrs Clare Wehner]).[13]

Doris Hogg recalled:

> It was calm at first but then with a whoosh the workshop caught alight when a spark from the encroaching fire caught on some leaves in the guttering. We were all stamping out odd sparks on the lawns surrounding the Observatory . . . A wood heap close to the Observatory burst into flames and we used towels and our feet to try and help the men with their hoses. It wasn't long before the booster pumps for water gave out as the electricity failed. We felt we were all doomed as the fire roared around us but by the grace of God in the middle of a downward sweep towards the Cotter Road a light rain fell and the race of the fire halted.[14]

After the fire destroyed the electrical supply system and cut off the water pumps that supplied water to the hoses, the staff hastily formed a bucket brigade. Twenty-eight men and the four women remained behind to fight the fire that had spread across a 3-kilometre front:

> At two [p.m.] it was on us. At a time like that anything can happen. We had fire hoses which had electric pumps to give the water a head. It is no joke operating a full-sized fire

> hose solo. Unfortunately the forest blazed and the power line was cut so that after that there was no water. We banged the grass with bags. A man called Snowy suddenly appeared from nowhere. (It was said that he swam about in the town water supply until the fire had mostly passed.) We lost the workshop but saved everything else, chiefly on account of the magnificent spirit of the Australian staff.

In addition to the workshop located on the western wing of the main Observatory building, the records of the Great Melbourne Telescope, two storage buildings and machine tools valued at £50,000 were destroyed. The damage could so easily have been much worse. The greatest destruction seemed to have been caused by exploding oxygen and acetylene cylinders that brought the roof of the workshop down. The fire seriously affected research work at the Observatory as it would take six years for the destroyed facilities to be replaced.[15] But the incident was not without its humorous side, as Woolley records:

> When the fire had passed, up came a barrel of beer, and sandwiches, with the compliments of the Hotel Canberra. Finally there came the Fire Brigade who drank the beer. Then the Minister for the Interior [Wilfred Kent Hughes] who started giving the fire fighters instructions.[16]

Jim Banham at the burnt-out workshop after the 1952 fire Photo: MSO

The other positive outcome from the fire was the removal of a number of burnt trees that allowed a stock fence with netting to be built in 1953 to prevent further damage to the site by rabbits. But the close relationship that existed between Mount Stromlo and the Department of the Interior at the time of the fire was under some stress as Woolley sought a new administrative home for the Observatory—part of his new vision for its future. Not surprisingly, that new home was within a university.

It is perhaps surprising that the teaching of astronomy at Australian universities took so long to become established. In 1892, H.C. Russell proposed, without success, the creation of the 'Leake School of Practical Astronomy' at the University of Tasmania.[17] Between 1900 and 1930 there were occasional lectures at Sydney University. In 1906, T.F. Furber addressed engineering students on the topics of geodesy and astronomy. Between 1914 and 1926, the New South Wales Government Astronomer, William Cooke, who also held the position of Professor of Astronomy, lectured second-year science students on general astronomy. Having demonstrated its capacity to contribute to national life during the Second World War, Woolley was adamant that his discipline's cause would be advanced by its participation in the Australian National University (ANU) after its foundation in 1946.

Although the University grew out of the turmoil of the Second World War, suggestions that there should be such an institution can be traced as far back as the 1870s; they were never far from the surface in the years leading up to Federation and the eventual establishment of a national capital. The Commonwealth government's international design competition for the future city specified that a site be provided for a university. The location chosen by Walter Burley Griffin in his 1912 winning entry is, in fact, the one on which the University now stands. Nevertheless, the germination of these early seeds took place only when a number of influential people began to turn their minds to how Australia might handle the transition to a peacetime society, and even to a new social order, when the war was eventually won.[18]

In early 1943 the Commonwealth Treasurer, Joseph Benedict Chifley, appointed Dr H.C. 'Nugget' Coombs to head a Department of Post-War Reconstruction. More than any other person, Coombs is credited with responsibility for the creation of the ANU.[19] Over the next couple of years, the realisation grew that, although education was the responsibility of the states, there were certain matters 'of national importance' within the educational task which would require a Commonwealth involvement if the government's vision for a new postwar social order was to become a reality. To this end, the government approved recommendations in mid-1945 for the establishment of an institution to be called 'The Australian National University'.[20] Later that same year, Nugget Coombs and John Johnstone Dedman, who had recently taken over as Minister for Post-War Reconstruction, had a detailed proposal ready to take to Cabinet.[21]

The disciplines proposed for the new institution were determined partly by the enthusiasms of the early planners, and partly by Australia's geographical location and particular national needs. Medical research, matters relating to government and the social sciences, and studies of the Pacific area were all advocated at an early stage. The inclusion of the physical sciences seems to have been something of an afterthought, and might not have happened at all had not Coombs been sharing accommodation, at the time, with the Commonwealth Astronomer. According to Coombs, it was Woolley who persuaded him that no research university would be complete without an involvement in the natural sciences, and the awesome way in which the war in the Pacific was terminated at Hiroshima and Nagasaki left no doubt in anyone's mind that physics had suddenly assumed a stature 'of national importance'.[22] Even so, the inclusion of physics (or, more particularly, nuclear physics) was advocated only tentatively in the Cabinet submission of late 1945. Yet part of the proposal it remained, and a school of physical sciences emerged as one of the four founding schools specified for the University when Dedman steered his Bill through Parliament and into law on 1 August 1946.

Whereas some believed that research in the natural sciences and technology should be the province of the CSIR, Woolley argued passionately for their inclusion within the ANU. As a personal friend of both Nugget Coombs and R.I. Downing, Economic Adviser to Prime Minister Chifley, Woolley was well placed to influence government policy at a very high level—and this he did. However, he met another hurdle when Professor Mark (later Sir Mark) Oliphant, the newly appointed Director of the Research School of Physical Sciences who had been recruited from Birmingham, proposed to devote the school's resources towards his own research field, experimental nuclear physics.[23] Woolley and Oliphant had known each other from their days at Cambridge. Woolley then proceeded to acquaint Oliphant with extant research in astronomy, astrophysics and geophysics at Mount Stromlo. Specifically, he suggested that researchers at Mount Stromlo should be able to submit their work for the award of degrees. Both the quality of the staff and the breadth and depth of their published work ought to warrant a place within the national university. While these disciplines might be appropriately represented within the school, the proposal to incorporate the Observatory into the University was rejected by Oliphant on the grounds of the Observatory's continuing statutory responsibilities being incompatible with the structure and organisation of the school. The proposal received the same fate when put to a group of consultant scientists, including Sir John Madsen, F.W.G. White, G.H. Briggs and David Martyn (who would have been directly affected).[24]

A compromise solution outlined by Oliphant in October 1949 incorporated the disciplines being practised at Mount Stromlo without making the Observatory a university liability or a complication in its operations. As Director of the Commonwealth Observatory, Woolley would be appointed Honorary Professor of Astronomy with a right to appoint a Research Fellow to his staff and to supervise research students.[25] In effect, the ANU would have a *de facto* Department of Astronomy, and this gave Woolley plenty of room to manoeuvre. The two Research Fellows already employed at Mount Stromlo became doctoral candidates when the University accepted its first students in 1951. The first PhD degree awarded at the ANU was to Antoni Przybylski on 18 February 1954 for a thesis entitled 'The Maximum Effect of Convection in Stellar Atmospheres on the Observed Properties of Stellar Spectra'. Gerard de Vaucouleurs received the DSc on 15 March 1957 for a submission entitled '*Manuel de Photographie Scientifique*'.

De Vaucouleurs used the 'Reynolds' telescope from October 1952 to January 1955, with the assistance of an ANU Fellowship, to conduct a photographic survey of relatively bright southern galaxies in a study similar to that made with its 'twin', the 30-inch telescope in Helwan, Egypt, where galaxies north of –45° declination had been surveyed for several decades.[26] This survey was the first to provide systematic coverage of the extreme southern portion of the sky. Working with the instrument for this purpose led de Vaucouleurs to speculate that Reynolds might have given the telescope 'in the hope of promoting extra-galactic studies in the Southern Hemisphere'. In the course of his observations, de Vaucouleurs produced 250 one-hour exposures of galaxies brighter than 14th magnitude. He concluded that his survey filled 'the last major gap' in the nearer southern galaxies and complemented the work of astronomers at a number of observatories including Mount Wilson, Lick and Helwan. He considered that it also supported Harlow Shapley's theory that galaxies evolve from irregular to regular types and from spirals into ellipticals. An important outcome of de Vaucouleurs' research was his identification of a flattened Local Supercluster of galaxies, some 100 million light years in diameter and centred on the Virgo Cluster.[27] This would later earn him the 1980 Herschel Medal of the Royal Astronomical Society.[28] De Vaucouleurs' work made a major contribution to the study of the southern skies and was a very clear demonstration of the quality of research being undertaken at Mount Stromlo.

Woolley moved quickly to secure honorary University appointments for his leading staff in 1951. Just how proactive he was in taking advantage of Canberra's new academic opportunities may be seen from the fact that Oliphant himself wasn't able to move into his school's new buildings until 1952–53. Woolley also delivered a series of lectures on the subject matter of the manuscript for the book he had co-authored with Walter Stibbs, *The Outer Layers of a Star*.[29] Published in 1953, this became a well-known text in the field of stellar atmospheres.

Woolley made some of the earliest research student appointments in the University as well. The first of these was Colin Gum, a science graduate from Adelaide University who arrived in January 1951. His thesis, entitled 'A Study of Diffuse Southern H-alpha Nebulae', was submitted in June 1954. It was the first by a student recruited into the 'Department of Astronomy' by the usual means.[30] He received his doctorate in 1955. Gum used an improvised spectrograph camera to discover 44 hydrogen emission regions in the southern Milky Way; one of his discoveries was the great emission nebula in the constellation of Vela that now bears his name. He became a Research Officer in the CSIRO Radiophysics Division in 1956 and remained there until 1959 when he was appointed Senior Lecturer in Physics at the University of Sydney. Gum, who suffered periodically from mental illness, died in a skiing accident at Zermatt in 1960. Also notable among Woolley's student recruits was Alex Rodgers, who came as a physics graduate from what is now the University of New England in March 1954. His thesis was 'A Photometric Study of the Southern Coalsack'. Virtually all of Rodgers' notable career was spent at Stromlo and culminated in his eventual appointment as the sixth Director, the first Mount Stromlo graduate to hold the post.

By 1953, Woolley had developed a lecture and instruction program within 'his department' that graduate students were expected to complete within two years. The only complication was the participation of Observatory staff who were not recognised by the University. But, as Woolley pointed out and the Board of Graduate Studies agreed, as the Observatory staff were not to be remunerated for their contributions and were not in any event examiners, it should not be a matter of concern to the University. Where Woolley did find himself out of step with the ethos of the University was in the relationship between research and teaching. In an address on the physical sciences to a seminar marking the jubilee of the Commonwealth of Australia in 1951, he did not hesitate to declare that the primary duty of universities was to teach.

They were educational institutions with a duty to their students as well as to research. He repudiated the distinction that would separate teaching from research and vice versa. He was of the opinion that institutes set up specifically for research 'should be associated with a university and should take appropriate teaching activities'.[31]

Woolley's approach to the Observatory and its broad charter was, in part, a reflection of his experience in South Africa where the observatories had been constrained by government control (in this case, the British Admiralty), limited to the delivery of practical services, remote from the universities and a culture of intellectual freedom and had not attracted the best minds. It came also, in part, from his time at Cambridge where Sir Arthur Eddington had attracted many to the discipline and created a strong link between 'those who did' and 'those who taught'. Again he cited the South African experience:

> Very little contact has ever been established between the numerous Observatories in South Africa and the South African Universities. The Observatories have always had inefficient junior staff, and have always languished except under exceptionally vigorous direction: they have too often degenerated into remote-controlled appendages of their Northern parent authorities. On the other hand, the South African University physics departments, seldom in any way distinguished, have missed the opportunity of deriving real inspiration from real work going on in their midst . . . It is greatly to be desired that this mistake should be avoided in Australia, and that Australian astrophysical work should be well integrated with the Australian University structure.[32]

He consistently argued that working astronomers ought also to be teachers to ensure those entering the profession were acquainted with new research and best practice. There was a strongly nationalist element in his thinking as well:

> At the moment only about half of the Mount Stromlo graduate staff is Australian trained: it is easier to attract an established European scientist to the base grade professional appointment than it is to secure a good Australian graduate.[33]

If Australia wanted to remain an acknowledged 'player' within the international astronomical community, it needed to encourage young Australian graduates to join the profession. The majority of research fellows at Mount Stromlo were from overseas.

Cognisant of opposition within the ANU from those not sharing his views, Woolley recognised that he needed to move gradually towards achieving his objective of integrating the Observatory into the University. His position and standing were strengthened when he was elected a Fellow of the Royal Society (FRS) and made an Officer of the Order of the British Empire (OBE) in 1953. Woolley then wrote to the University Registrar and suggested that responsibility for astrophysical work at Mount Stromlo might be transferred from the Department of the Interior to the University.[34] The ANU Council considered the matter and endorsed the transfer in principle on the condition that it did not involve any additional expense on the part of the University.[35] Having gained the University's support, Woolley approached the Secretary of the Department of the Interior, Bill MacLaren, and recommended the wholesale transfer of the Observatory to the ANU.[36] Woolley recalled:

> The Department did not want to lose the Observatory although Bill Maclaren as usual backed me when I asked for the transfer despite his personal interest in science. The staff did not want to stop being public servants and thought the Observatory would lose financially. Oliphant, of course, helped vigorously from the University side.[37]

However, the Public Service Board and the Treasury, encouraged by Hogg's strong personal antagonism to the proposal, both opposed the transfer while the Minister for the Interior, Wilfred Kent Hughes, rejected the proposal outright despite MacLaren's personal support. He preferred collaboration rather than co-option. This appeared to be the end of the matter. Woolley certainly thought so, as there were no subsequent negotiations with either the University or the Department.

Believing that they would live in Australia for the remainder of their lives (they had no children), the Woolleys purchased a 'retirement' property on the New South Wales south coast. However—and with some reluctance—Woolley allowed his name to be considered as a candidate for the position of Astronomer Royal when Sir Harold Spencer Jones retired in 1955:

> His wife, although more thoroughly British than himself, was even more reluctant to leave Australia. Indeed, I have been told that he made a last minute signal to the Admiralty in London asking for his name to be withdrawn, but that this only arrived after the Admiralty had submitted his name to the Queen.[38]

This very late change of heart, something he shared with a number of staff members including Gascoigne, may have reflected some personal regret at being denied the possibility of reaping any of the observational fruits of all his planning and organisation for the acquisition of the 74-inch telescope. The future was sealed when he was officially offered the appointment and duly accepted.

With their departure from Australia set for the end of 1955, Woolley finally delivered his inaugural professorial lecture on 28 July 1955. It was also his last public address in Canberra. Entitled 'The Longest Tyranny' and borrowing from the poet Dryden, Woolley addressed the long-running 'subjugation of thought about the physical world to the philosophical views of Aristotle, with their defiance of observed physical and astronomical phenomena'.[39] He also managed to outline his ideas on teaching astronomy to undergraduates although the founders of the ANU had, despite Woolley's advocacy, decided against having undergraduate students:

> As a practical scheme I advocate a general course [of astronomy] in the first year (with no more mathematics than the Leaving Certificate prerequisite) for those who wish to offer only one science and also to whet the appetites of potential specialists [in astronomy]. For advanced undergraduate work I advocate an optional course in astronomy and astrophysics to be taken as part of a degree course in mathematics and physics. It would perhaps be a bad thing to type a student as an astronomer so early as his third or fourth year.

He concluded by thanking the ANU for 'finding a place for astronomy' and was delighted when, in 1960, undergraduate students were accepted.

Woolley had also taken an active part in the development of University House as chairman of the House Committee, as one of its Fellows and, from March 1954, as Deputy Master before the first Master, Professor Trendall from Sydney University, arrived in Canberra. Woolley and his wife were given a flat at University House and took a leading part in its social life. Woolley was also instrumental in the foundation of the Australian Academy of Science. When it was established in 1954, Sir Mark Oliphant was elected President with David Martyn serving as the Secretary. Woolley was one of the 23 Foundation Fellows and served briefly as Treasurer. He also served a term as President of the Australian and New Zealand Association for the Advancement of Science.

With Woolley's departure imminent, the University Council became concerned about the highly successful relationship that had been fostered under Woolley's directorship between the Observatory and the University. As Susan Davies observes:

> Suddenly the Observatory became very important to the University. Its Department of Astronomy was among its most successful ventures. It had an impressive record in research;

> its students were the first to graduate; it brought repute to the institution, and the University had no guarantee that the next Commonwealth Astronomer would be suitable or even willing to continue the arrangement entered into by Woolley. The University administration mobilised its very considerable resources to save its Department of Astronomy.[40]

The negotiations were greatly assisted by one member of the University Council, Allen S. Brown, who was also Secretary of the Prime Minister's Department. While it is most likely that Brown brought the matter to the attention of the Prime Minister, Woolley has suggested that it was the University Chancellor, Sir Leslie Melville, who raised the matter with Menzies.[41] On 7 September 1955, the Prime Minister met with Woolley and the University Vice-Chancellor at Parliament House. Brown was also present. After devoting 'an hour to the topic', in which he expressed his sympathy for the arguments presented by both the Observatory and the University, Menzies took advice from the Treasury and decided that the transfer would take place.[42] In November 1955, the Prime Minister informed the Vice-Chancellor that he was at liberty to appoint a Professor of Astronomy who would also serve as Director of the Observatory.[43] Woolley had succeeded after ten years of negotiation and diplomacy. The requisite legislation passed through the federal Parliament, receiving royal assent on 7 November 1956.[44]

Woolley and his wife Gwyneth left Mount Stromlo on 7 December 1955. Many things were said in praise of his sixteen years as Director, although his influence on Mount Stromlo was far from over. No sooner had the 74-inch telescope entered service (Woolley was able to participate in some of the final tests before his departure a month later) than he started thinking about the Commonwealth's acquisition of a 200-inch telescope to be used by both British and Australian astronomers as part of a long-term partnership. There was also the enduring contribution of his former students and the existing staff, all of whom had been encouraged and enriched by his mind and his creative energies. Woolley did his best scientific work in the first few years after the war. In 1946 alone, he published six papers dealing with subjects as varied as stellar energy generation, the mechanism of ionospheric ionisation and the future prospects for astronomy in postwar Japan. Over the following few years, he produced memorable papers on the solar corona and galactic noise, and his co-authored book with Walter Stibbs, *The Outer Layers of a Star*, was published by the prestigious Clarendon Press in Oxford. He also prepared articles touching on the administration of astronomy and continued to probe possibilities for joint and cooperative research in various continents. McCrea makes particular mention of his work ethic:

> [He] called both his father's and mother's families 'professional with some contact with university circles'. Once, after we had been seeing an American colleague of his at work on a project requiring the marshalling of voluminous information, he did nevertheless remark: 'You see how professional an American can be; we are all amateurs.' I knew what he meant, and it struck home. But personally I should make an exception of Woolley himself as regards the conduct of his science, even if not always his administration.[45]

As for his approach to scholarship, he 'was in fact one for *doing* science not reading about it. He told me once he never read the *Astrophysical Journal* (the leading academic journal) rather he relied on Stibbs and me to keep him posted'.[46] Gascoigne's assessment, made at some distance from the events, is worth noting:

> During his sixteen years in Australia he had set Mount Stromlo firmly on a course of stellar astronomy . . . increased many-fold its telescope power and its technical resources; watched over the growth of some vigorous and productive lines of research, and seen some fine papers published . . . True there were deficiencies: the Observatory was not yet in the first division, not in the sense that Radiophysics was; and the state of some subjects like spectroscopy left something to be desired. But these objectives were within reach, and were in fact to be achieved within a decade.[47]

But Gascoigne felt the incorporation of the Observatory into the ANU 'was Woolley's last contribution to the Observatory, and I have always maintained it was his best'.[48] Perhaps the most concise description of his directorship was a headline in the Melbourne *Herald* newspaper when his appointment as Astronomer Royal was announced. It simply read: 'This man gets things done'. And Mount Stromlo, the ANU, the Australian people and astronomy were the beneficiaries.

Woolley commenced his duties as the eleventh Astronomer Royal in 1956 and held the post until 1971. He was knighted in 1963 and elected as President of the Royal Astronomical Society for 1963–65. He was awarded the Gold Medal of that Society in 1971. After being involved in the development of the South African Astronomical Observatory at its new site at Sutherland, in 1972 he agreed to become its Foundation Director. He delivered his official opening speech at the new observatory in fluent Afrikaans before the Prime Minister of South Africa and the British Minister for Science and Education, Margaret Thatcher. He retired in 1976 and returned to England. His wife died suddenly in 1979 and he remarried later in the year, returning to South Africa in 1980. His second wife died in 1985 and he again remarried in the same year. Richard Woolley died on Christmas Eve 1986 at the age of 80.

Although the selection of a successor at Mount Stromlo was not his responsibility, at the invitation of the University Woolley produced what he described as a 'Little Essay', subtitled 'Notes on Directorship of Mount Stromlo'. These notes would be used as a background paper for Electoral Committees well into the 1970s. Woolley contended that the qualities needed in a new Director included:

> a) extreme intellectual eminence; b) acquaintance with, and perhaps mastery of, contemporary developments in equipment; c) ability to direct staff, getting the best out of them and leading them in research where desirable; d) ability to cooperate with his own governing body; and e) ability to cooperate with outside institutions.[49]

While accepting that it was 'unlikely that any person will excel in all these directions', he thought it important that the new Director should be able to 'cooperate with Radiophysics' and be able to work with astronomers from overseas, especially those based in Britain. He also made mention of one other qualification that he possessed in great abundance. This was a 'willingness to delegate matters to senior members of a supporting staff; not in the sense of taking no further share in any activity, but in the sense of letting the staff member do much of the work and then backing him up'. His final suggestions were quite specific. It would be 'a gamble' to select someone younger than 45 (Woolley was 33 at the time of his appointment), and he felt that 'everyone would like to see an Australian appointed—but not, of course, if he were not the best man available for the job'.

After Woolley's departure, Arthur Hogg was appointed Acting Commonwealth Astronomer pending the selection and appointment of a new Director.[50] As Gascoigne recalls:

> There followed a difficult fifteen months . . . [as] Hogg was strongly opposed to [the Observatory's transfer to the ANU], and apprehensive about the future. The simultaneous transfer to the University and change of Director did, in fact, affect his position a great deal; he had little to do with the students, who soon came to play an important part in the research activities of the Observatory, and he did not achieve as close a personal relationship with [Woolley's successor].[51]

Despite his anxiety about the Observatory's incorporation into the University, Hogg was appointed Reader in Astronomy at the ANU. Gascoigne was appointed a Reader as well. There was also a series of international visitors to the Observatory including Dr S. Gaposchkin, who arrived in mid-1956 as a visitor from Harvard University and returned to the United States in June 1957.

In early 1956, there was no obvious successor to Woolley. Although he did not anoint a successor, he would have resolutely opposed giving the Directorship to one potential candidate: the Dutch-American astronomer Bart J. Bok. As Gascoigne recalls:

> They were men of widely different character and temperament who detested each other. Once or twice I was left with the two of them, making uneasy conversation to each in turn while they declined to exchange a word.[52]

Of course, the appointment was not Woolley's to make, and a number of possible candidates were to be considered from across the world.

In January 1957, as a result of the *Mount Stromlo Observatory Act* 1956, arrangements were made for the transfer of the Observatory from the Department of the Interior to the Research School of Physical Sciences at the Australian National University, a more appropriate base for a pure research organisation. This administrative change also brought a new name to the site. On 3 January 1957, the establishment officially became known as Mount Stromlo Observatory.[53] The continuing statutory responsibilities of the Observatory were thereafter administered by a number of Commonwealth government departments, including the Department of the Interior (1957–72); the Department of Science (1972–78, 1984–86); the Department of Science and the Environment (1978–80); and the Department of Science and Technology (1980–84). The Act was repealed in 1986 by the *Statute Law (Miscellaneous Provisions) Act* 1986 and the Observatory has remained under the direct control of the ANU since that time. The first important decision it would make in conjunction with the Observatory staff was the appointment of a new Director.

chapter seven

Harvard in the south: Bart Bok at Mount Stromlo, 1957–1966

As THE TIME FOR WOOLLEY'S DEPARTURE approached, the senior officers of the Australian National University consulted widely and took a great deal of advice on the most suitable candidate to become Mount Stromlo's third Director. This was, after all, the first time that responsibility for filling the post had rested with them. Walter Baade and Allan Sandage were two of the names under consideration.[1] Woolley himself would probably have favoured Olin Eggen, who had made two substantial visits to Stromlo as a guest observer during the 1950s, and whose research interests dovetailed well with Woolley's vision of Mount Stromlo's future.[2] Certainly when Eggen was appointed as the Observatory's *fourth* Director a decade later, the Astronomer Royal was warmly enthusiastic.[3]

One factor which had a considerable impact on the University's choice was the pre-eminent position of the CSIRO's Radiophysics Lab in world radio astronomy. Although relationships between Radiophysics and the Observatory had sometimes been difficult (as noted in Chapter 5), the University was keen to seek the advice of the Sydney researchers. In fact, it was Joe Pawsey who suggested the name of Bart Bok.[4] He saw virtue in the appointment of someone with experience in *both* branches of the subject and, as we shall see, Bok's Harvard background fitted the bill. Such arguments were persuasive, and Bok was duly appointed from 7 March 1957.

Bok's arrival at Mount Stromlo marked a new beginning both for the institution and for the man. The transfer of the Observatory to the ANU had taken effect two months earlier (on 3 January 1957), normalising its erstwhile 'honorary' status within the University. Thus it was as 'Professor and Head of the Department of Astronomy within the ANU's Research School of Physical Sciences' that Bok came to Canberra. His appointment also carried the title 'Director of the Mount Stromlo Observatory', as the old Commonwealth Observatory had now been officially renamed.

The new appointment also marked a significant change for Bok himself.[5] He came to Canberra after a career at Harvard College Observatory spanning some 27 years. Born in Holland in 1906, he had moved to the United States in 1929 (though still enrolled as a doctoral student at the University of Groningen) to work at Harvard under the directorship of Harlow Shapley. Bok had first read about Shapley in astronomy magazines while he was still at school and developed an admiration for him that lasted throughout his life.[6]

Shapley's fame rested upon Milky Way research that he carried out at Mount Wilson Observatory during 1917–19. At the time it was believed that our Galaxy was a disc-shaped system of stars, about 20,000 light years in diameter, with our Sun at or near its centre. This was called the 'Kapteyn Universe' after the Dutch astronomer Jacobus Kapteyn who, in 1904, initiated a monumental international collaboration to map the structure of the Milky Way by detailed observation and statistical analysis of the stars lying in 206 'Selected Areas' distributed across the sky.

Shapley took a radically different approach. In a series of a dozen papers, he reported an epochal study of star *clusters* in our Galaxy.[7] In particular, he explored the *globular* clusters—giant agglomerations containing as many as a million stars each. It was well known that the distribution of these objects was extremely non-uniform on the sky, with nearly all of them lying within just one hemisphere, but there was little understanding as to why this should be so. Shapley derived the *spatial* distribution of the nearly 70 then known globulars. He observed the clusters' apparent diameters, and the magnitude of their brightest stars, to determine their distances, using a handful of clusters containing Cepheid variable stars to anchor his measures. A few years earlier, Henrietta Leavitt of the Harvard College Observatory had found that the pulsation period of Cepheids is related to their intrinsic brightness, making them ideal distance indicators.[8]

Shapley's work showed that the globular clusters occupy a volume of space which is vastly more distended than the Kapteyn Universe, though their distribution is still symmetrical about the *plane* of the Milky Way. Our Sun lies far removed from the centre of the system, calculated by Shapley to be about 65,000 light years distant, in the direction of the southern constellation Sagittarius. Shapley has been called the Copernicus of the twentieth century—what Copernicus did to dispel our illusions about the Earth being at the centre of the Solar System, Shapley repeated in respect of the Galaxy at large.

Kapteyn's star counts yielded a false picture of the distribution of matter within our Galaxy simply because vast clouds of dust coexist with the stars *in the plane of the system*; these obscure the stars lying at any great distance. Since he was looking *above* and *below* the Milky Way plane, Shapley's study of the globular clusters was less affected by this obscuration, revealing just how far from the centre of the Galaxy our Sun really lies. Even so, the dust affected the cluster observations far more than Shapley realised, causing him to over-estimate the distance to the Galactic centre. Today's belief is that it lies a mere 25,000 light years distant!

Having shown that the volume of space occupied by our Galactic system is vastly greater than had been thought previously, Shapley proposed that most of the bright, hot stars that we see about us in the sky must form a subordinate clustering within this overall system. Similar groupings undoubtedly occur in other parts of the Milky Way as well. Combining his observations with the results of several other workers, Shapley concluded that this 'Local Cluster' is also a disc-shaped system—though only 3000 light years in diameter—with its plane tilted about 12° to that of the Galaxy as a whole. Our Sun is noticeably off-set from even the local centre, which lies about 300 light years away in the direction of the southern constellation Carina.

Shapley's great advance provided the inspiration for Bok's own research career. The southern Milky Way became for him a region of intense interest, beauty and delight. His lifelong quest was to determine the Galaxy's size and shape in greater and greater detail, and to understand how its constituent parts relate to the whole. Together with his wife Priscilla, Bok later wrote a definitive textbook titled *The Milky Way*, which saw five editions between 1941 and 1981.[9] Even Bok's doctoral thesis, on the stars and emission nebulae in the Carina region, was shaped more by Shapley at Harvard than by his official adviser, Pieter van Rhijn, at Groningen.[10] The beautiful Carina region became one of Bok's firm favourites.

Bok's career at Harvard was terminated by events which had little to do with astronomy.[11] In the early 1950s, Harlow Shapley became embroiled in the anti-communist witch-hunts of the McCarthy era. He was questioned by the House Committee on Un-American Activities regarding some quite innocuous associations, such as his involvement with UNESCO. He survived, but not unscathed. In the general climate of xenophobia, even such an innocent activity as promoting international scientific collaboration was sometimes enough to result in an academic being shunned by many colleagues, lest they too should come under suspicion. Bok was *not* such a colleague. He openly supported Shapley in this time of persecution, and his own life at Harvard became increasingly difficult as a result.

Shapley retired from the directorship of the Harvard College Observatory in September 1952.[12] For some time he had regarded Bok as his natural successor and, until the McCarthy affair, Bok's appointment seemed assured. In the event, however, the post went to another senior staff member, the solar astronomer Donald H. Menzel.

This was a bitter blow for Bok, and others were soon to follow. The next concerned Harvard's South African outpost, the Boyden Station. Observations made from its favourable southern latitude were important for Bok's Milky Way research; he had, in fact, spent eighteen months there in 1950–51 happily observing the Carina region. Now, financial pressures forced the University to transfer Boyden to an international consortium of six institutions, greatly diminishing its accessibility to Harvard observers.

A final blow fell when the Smithsonian Astrophysical Observatory was invited to move to Harvard, under the directorship of another senior astronomer there, Fred Whipple, a specialist in cometary research. Bok felt that, in addition to distancing themselves from Shapley for political reasons, the senior Harvard staff were now seeking to wind down his whole field of Milky Way research as well. Bart and Priscilla concluded that Harvard was no longer for them. On 10 May 1956, Bok announced his resignation with effect from the following January.[13] Fortuitously, he had recently received approaches regarding directorships at some other institutions, one of them the Australian National University.

☆

Bok's response to losing direct access to the Boyden Station had been to transfer his energies into equipping Harvard for the rapidly developing discipline of radio astronomy.[14] He supervised the construction of a 25-foot dish, commissioned at the Agassiz Station in 1953—the first such in America—and then of a 60-foot dish completed in 1956. Much of the funding for these was obtained through Bok's personal efforts. Although he had informed Harvard's President of his decision to leave in late 1955, Bok waited until a fortnight after the dedication of the 60-foot dish before making a public announcement.

This move into radio astronomy was a calculated one on Bok's part. The new window had already shown itself capable of detecting neutral hydrogen gas, the most abundant constituent of the Universe. A spectral line emitted by the hydrogen atom at a wavelength of 21 centimetres had been predicted in 1944 and observed in 1951.[15] Since this radio 'light' can pass uninhibited through the absorbing dust clouds of our Galaxy, it was immediately apparent that here was a new and powerful means of mapping the structure of the Milky Way. In fact, even as Bok was equipping Harvard with radio dishes, 21-centimetre observations were being accumulated in Sydney by Radiophysics scientists, most notably Frank Kerr, and at Leiden by Dutch astronomers, most notably Jan Oort, a colleague of Bok's from his student days. The Dutch work was being done with German radio equipment acquired by Holland as war reparations. The first complete map of hydrogen in our Galaxy was published by Oort, Kerr and Westerhaut in 1958.[16]

Since mapping the Milky Way was one of Bok's cherished goals, he was intensely interested in this research, and he embarked upon his Directorship at Mount Stromlo with radio collaborations well in mind. Soon after arriving, he cited his reasons for accepting the Stromlo post (at barely half his previous salary). They included the opportunity of working 'in close collaboration with the radio astronomers in Sydney'.[17] The coming together of the two disciplines in Australia in the late 1950s and early 1960s—through such mechanisms as research collaboration, exchange of telescope time, cross-appointments between Mount Stromlo and Radiophysics, and conferences like the 1963 joint Symposium on *The Galaxy and the Magellanic Clouds*—has been of lasting benefit to the subject as a whole.[18] Tragically, the personal interaction between Bok and Pawsey was brought to a halt by Pawsey's untimely death in 1962, but it was largely as a result of their shared vision that the idea of radio and optical astronomers working in close collaboration became well entrenched in the ethos of Australian astronomy.

☆

There were other ways in which Bok's approach to his new job was a continuation of his Harvard career. Thwarted in his hopes of succeeding Shapley as director of that venerable institution, he set about shaping Mount Stromlo Observatory in the image of what might have been—a Harvard in the South. His enthusiasm for the Milky Way and the Magellanic Clouds, so integral to Shapley's research career and to his own, remained undiminished. There were even hints of the Shapley *style* in his approach to life in the Director's chair. Shapley's use of the Harvard Residence for social occasions was legendary, and Bok followed suit at Mount Stromlo.

Extrovert by nature, Bok relished the opportunity of embarking on a public education campaign that would have left most people exhausted. He was a tireless traveller and lecturer, often speaking several times a month at high schools and to groups such as Rotary, as well as taking every opportunity to make contact with the decision-makers of the nation. In his first year at Mount Stromlo, Bok augmented the traditional daytime guided tours and the small weekly viewing nights by starting a series of monthly Visitors' Nights. These were held between first quarter and full moon, and continued throughout his directorship. Most telescopes were opened for inspection, following which about 150 visitors were served coffee, frequently at the Director's Residence. Few observatory staff avoided this regular monthly commitment. Bok launched these open nights in a way calculated to derive maximum public relations benefit. The 1957 *Annual Report* records that they were 'by invitation only so far . . . Among the groups invited have been the staff and Council of the University, Members of Parliament and of various Government Departments and members of the Diplomatic Corps, all with their families'.[19]

If the politicians would not come to Bok, he went to them. Just months after his arrival, he used the news-breaking flight of *Sputnik 1* to engineer an invitation to address a gathering of Senators and Members of the House of Representatives during their dinner adjournment, an invitation which was subsequently repeated several times. Bok always took two people with him—the first was one of the Mount Stromlo graduate students, and the second was the ANU's Vice-Chancellor. The student's task, ostensibly, was to show the slides, and the Vice-Chancellor's to say a few well-chosen words at the end; but there was never any doubt as to who was the ringmaster and who were the supporting acts when Bok went to Parliament House.

Bok also made the most of any unexpected opportunities. Early one morning he was driving down from the mountain when he saw the Stromlo milkman standing beside his broken-down van at the side of the road. Stopping to see if there was anything he could do, he found that the milkman's greatest concern was that he had not yet made the delivery to Government House. Before the milkman realised what was happening, Bok was loading the Governor-General's order into his car. One can be sure that, when he pulled up at the guardhouse, he announced his identity loud and clear, and that there was much rattling of bottles as the vice-regal milk was unloaded at the kitchen door of 'Yarralumla'

B.J. Bok, Director 1957–66
Photo: MSO

that morning. Certainly Bok loved to recount his 'milkman' story at every opportunity for years afterwards.

The results of Bok's public relations efforts were spectacular and long-lasting. In his first year at Mount Stromlo, there were 1,500 night-time and 3,500 daytime visitors.[20] By the end of his directorship, the annual total of people attending daytime tours was close to 40,000; the 1964 *Annual Report* states that 'we have had to request the local radio stations *not* to advertise Mount Stromlo Observatory as a tourist attraction'.[21] By that time, of course, the booking lists for the monthly Visitors' Nights were so long that they became completely impracticable; they were discontinued soon after Bok's departure.[22] But for many years afterwards inquiries were still received from people all over Australia wanting to come on 'one of those Observatory Visitors' Nights that Professor Bok used to run'.

Another lasting legacy of Bok's directorship is the Mount Stromlo Graduate Program. Once again we see hints here of the Harvard College Observatory ethos. By the time Shapley became Director there, the real cutting edge of telescope development in the United States had migrated to the West Coast observatories. Nevertheless, Harvard still managed to maintain its considerable reputation as a world-class institution, and not the least important factor in this was the excellent graduate program that Shapley ran.[23] Bok was an enthusiastic participant in this—working with students was one of his passions.

With Mount Stromlo's transfer to the University immediately prior to his arrival, all the pieces were in place for Bok to replicate a Harvard-style graduate school. Woolley had done all the groundwork leading to the transfer, as we have seen, and had even sown the seeds of a student program. Colin Gum and Alex Rodgers had both been recruited by Woolley as ANU students. Nevertheless, they had both completed their theses by the time Bok arrived in Canberra, so he found no graduate students 'on course'.

Bok immediately set about rectifying this situation. It was not an easy task since, in those days, astronomy was taught only incidentally—if at all—in Australia's state universities, and there was no ready-made pool of potential graduate scholars. Nevertheless, Bok implemented a number of highly successful strategies for recruiting PhD candidates. He established a summer Vacation Scholarship program in which promising physics and mathematics undergraduates from all over Australia came to Mount Stromlo for a couple of months, and were thoroughly immersed in the life of the Observatory. In the first long vacation after his arrival, Bok took eight such students, drawn from four universities.[24] This pattern continued for the remainder of his directorship. Indeed, it continues today. These summer scholarships were the means whereby many students had their interest in a research career in astronomy quickened; several of the present staff of the Observatory had their first introduction to Mount Stromlo as participants in Bok's summer program.

His second line of attack was to make frequent visits to the state universities. During these he would always try to arrange two things—a public lecture and an opportunity to talk with undergraduates approaching the end of their courses. In fact, he was poaching, but it is a mark of his warm and enthusiastic personality that, despite this, he was a welcome academic visitor almost everywhere he went. It should also be said that he participated energetically in some of the state universities' *own* extension and recruitment programs, such as the University of Sydney's Summer Science School for High School Students, and many of his contemporaries had nothing but praise for his efforts in promoting the study of astronomy, and of physics, in the Australian university system as a whole.

Bok's third stratagem was to cast his eyes further afield than Australia. In his second annual report he announced an aim of having 'six scholars from Australian universities, and possibly one or two from abroad'.[25] This was more than fulfilled: of the first dozen students he appointed, five were from overseas. One of the earliest was Natarajan 'Vis' Visvanathan who subsequently spent most of his research career at Mount Stromlo.

Bok's efforts in promoting graduate training at the Observatory were spectacularly successful. He left Mount Stromlo with fourteen scholars on course, and a total of nearly 30 had been attracted to the Observatory during his tenure as Director. Most of these are still active researchers today, or recently retired, and many have held very senior posts in the observatories of the world. Ron Ekers and Jasper Wall became the Directors of the Australia Telescope National Facility and the Royal Greenwich Observatory, respectively, and in 2003 Ekers assumed the Presidency of the International Astronomical Union, the ultimate distinction in astronomy.

The Graduate Program established by Bok has continued as a core observatory activity ever since. It now boasts some 150 graduates and currently enrolled students. It is noteworthy that the PhD graduation rate established by Bok in the first decade of the program's existence is about the same as has prevailed, on average, during its subsequent history. It was indeed a flying start.

★

By the late 1950s it was becoming evident that the site originally chosen for the national observatory was uncomfortably close to the growing city of Canberra. For the first few decades of its existence, Mount Stromlo had enjoyed an acceptable isolation from the lights of civilisation. Canberra extended no further than the suburb of Yarralumla, some 8 kilometres distant, and although the reservoir supplying the city also lay in the same direction, the 3-kilometre branch road up to the Observatory remained unsealed until 1959, thus discouraging its use by casual traffic. Indeed, life on the mountain had something of a rural village atmosphere. There were about 20 staff residences, and a 'Bachelors' Quarters' provided for single folk and occasional visitors. Most people made the trip into Canberra weekly rather than daily. Life in such a working-and-living-together community had some interesting social aspects. A monthly 'ladies night' persisted into the 1960s. Attendance by staff and student wives seemed to be expected, and some of the younger arrivals found the experience quaint to say the least! From an observational point of view, however, this remoteness from the light pollution of a still relatively small Canberra was a great advantage.

All this changed when the suburban development of the Woden Valley began in 1962. The Observatory was then confronted with a doubling of Canberra's population, half of whom were now living virtually on its doorstep, and further expansion was sure to follow. In addition, Baracchi's original 1910 assessment of Mount Stromlo's quality as an astronomical site had proved to be over-optimistic. Cloud cover percentages were unacceptably high, especially at night, and the atmospheric steadiness (referred to by astronomers as 'seeing') was frequently poor. Obviously the time had come to think about seeking a better site.

Thus the early years of Bok's directorship saw the start of an active site-testing program. This was not commenced in isolation since the Yale–Columbia group (already operating the 26-inch refractor on Mount Stromlo) was considering placing a 20-inch astrograph at a good Australian site. The US National Science Foundation funded Yale University for a site-testing program, and in July 1957 Isadore Epstein came to commence work. Bok was keen to link Stromlo into this program; he attended a site-testing conference held in Sydney following Epstein's arrival. Some joint expeditions were carried out during 1957—both the Boks accompanied Epstein on a first survey trip to Western Australia, in which three test sites were selected (Geraldton, Yalgoo and Meekatharra), and Ben Gascoigne and Alex Rodgers accompanied him on trips to northwestern New South Wales. Arthur Hogg was also active, inspecting sites in South Australia and Western Australia.[26] In August 1957, an expedition comprising Gascoigne, Epstein, the New South Wales Government Astronomer Harley Wood, and Richard Twiss (who was looking for an interferometer site) inspected northern New South Wales, passing through Coonabarabran and the nearby Siding Spring Mountain.[27]

These initial explorations had two different goals. The first, which was uppermost in the minds of the Mount Stromlo participants, was to locate a site for an ANU 'field station'. Although observing conditions at Mount Stromlo were far from ideal, the Observatory valued its newly established relationship with the ANU highly, and had no wish to abandon its existing location entirely, or even as the headquarters of its operations. Thus the first goal was to locate a site, preferably within New South Wales, which would have a significant weather advantage over Mount Stromlo, but which would be close enough to permit frequent visits and easy transport of equipment. The program's second aim was

Site testing near Mt Serle, SA Photo: MSO

to locate the best astronomical site *anywhere* in Australia; this was important not only for the Yale plans, but also for Australia's longer-term hopes of acquiring a telescope of a size comparable with the largest instruments of the American West Coast observatories. Ted Dunham Junior, a newly arrived staff member, also planned to build a 70-inch spectroscopic telescope in Australia.[28]

Perhaps because of these two rather distinct aims, the Yale and the Mount Stromlo site-testing operations did not proceed hand-in-hand for long. By the next year, it had been decided to operate independently, merely exchanging information and reports.[29] It was at this point that the Stromlo team adopted the following criteria for the location of their 'field station'. The site should:

1. have 50–100 per cent more clear nights than Mount Stromlo, especially during the September–March Magellanic Cloud season;
2. be as far south as possible, preferably below latitude –30°;
3. be on a relatively isolated hilltop or mountain range, at least 800 feet, and preferably more, above the surrounding plain;
4. be free from dust and bushfire smoke;
5. not be too isolated, preferably within 30 to 40 miles of a fair-sized town;
6. preferably be within six to ten hours' driving distance of Mount Stromlo, and in New South Wales (to foster collaboration with Sydney University and CSIRO Radiophysics). [30]

Using these criteria, Stromlo staff expeditions, under the guidance of Harley Wood, selected five sites in New South Wales for detailed investigation.[31]

In 1959 it was decided to occupy one site as a provisional field station, so as to have a control for the testing program, and to gain experience in remote operations. The location chosen was Mount Bingar, near Griffith in southwest New South Wales, which had topped the previous year's statistics for highest-quality nights, and which became a firm favourite with Bok. A 26-inch reflector was constructed from various pieces of existing equipment, and installed at Bingar in time for observations to commence on Christmas night.[32]

In 1960, Arthur Hogg was placed in charge of the site-testing program, which continued to run as two projects—with broad and narrow objectives. Monitoring of sites in both South and Western Australia continued. The telescope on Mount Bingar was scheduled regularly, and by the end of 1960 it was becoming apparent that it was rather too far south to escape the northern extremities of the storm centres forming the predominant weather pattern in the south of the continent. Nevertheless, Bingar remained in contention for the field station, together with three other sites in New South Wales: Siding Spring (near

Coonabarabran), Boona Mountain (near Condobolin), and Mount Kaputar (near Narrabri). The latter two were eliminated in 1961.[33]

The final decision between Bingar and Siding Spring was due to be taken in early 1962. In the event, it was anything but straightforward. Bok, who had spent a great deal of time observing at Mount Bingar, and who had developed a real affection for the Griffith district in general, advocated that site strongly. Hogg, on the other hand, argued for Siding Spring. The site-testing program had yielded a wealth of statistics but, as is frequently the case, these conveyed different messages to different people. The debate encompassed not only the sky cloudiness and the seeing measures, but also such factors as the *distribution* of cloudiness between the winter and the summer months, the latter being regarded by Bok as particularly important, being the 'Magellanic Cloud season'. Ultimately the ANU Vice-Chancellor, Leonard Huxley, convened a meeting of the staff members principally involved. Bok voted for Bingar, while Hogg, Gascoigne, Rodgers and Westerlund (with varying degrees of conviction) argued for Siding Spring.[34] On 12 May 1962, Huxley announced that Siding Spring was to become the Observatory's permanent field station. He later cited its establishment as one of the highlights of his Vice-Chancellorship.[35]

The opening of Siding Spring Observatory, 5 April 1965 Photo: MSO

At the time, it was not apparent just how important this decision would become in the development of Australian astronomy. It was expected that, although Siding Spring was to serve the ANU's field station needs, continued Australia-wide site testing would probably locate a better site for the eventual development of a very large reflector. In the event, it was decided to build an Anglo-Australian 150-inch reflector at Siding Spring, followed by a British 48-inch Schmidt camera. With the further construction there of the ANU's 2.3-metre Advanced Technology Telescope, and the transfer of the Uppsala Schmidt, Siding Spring Observatory has become one of the foremost international optical observatories in the world.

The initial suite of telescopes ordered for Siding Spring was, however, much more modest. Forty-inch and 16-inch reflectors made by Boller and Chivens arrived during the final years of Bok's directorship, and were devoted primarily to photometry and photography. A 24-inch instrument, with a rotatable tube to facilitate polarisation observations, was commissioned in 1966.

At Mount Stromlo itself, the early Bok years saw a need for consolidation following the frenetic burst of telescope acquisition that had marked the close of Woolley's directorship. This was particularly so since the two largest of the new

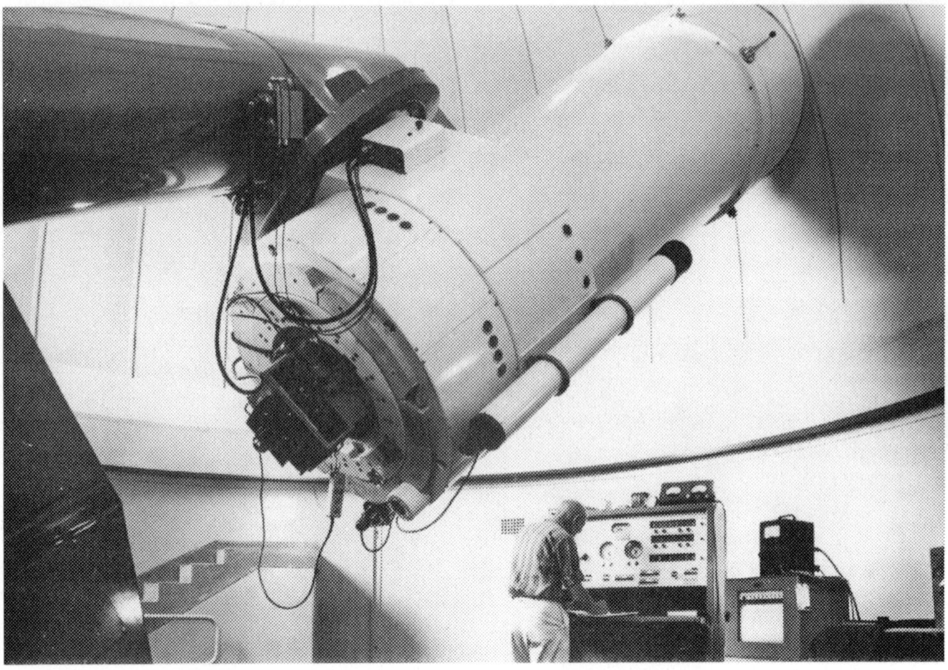

The 40-inch Telescope at Siding Spring Observatory Photo: MSO

The Great Melbourne Telescope at Mount Stromlo in the 1960s Photo: MSO

arrivals had unexpected teething difficulties which took several years to overcome. In the case of the 74-inch telescope, the primary mirror received from Grubb Parsons was found to have unacceptable astigmatism.[36] In 1959, the manufacturers replaced it temporarily by one of the same size which had been cracked while undergoing figuring for a new 74-inch telescope at Helwan Observatory in Egypt. The Stromlo mirror was sent back to Newcastle-upon-Tyne for refiguring and was eventually reinstalled in 1961.[37] In the case of the Great Melbourne Telescope, the original intention had been to convert it into a Schmidt camera and a new primary had been figured as a sphere.[38] It subsequently became apparent, however, that its mechanical design was inadequate for this purpose, so new conventional paraboloid–hyperboloid optics were ordered from Grubb Parsons.[39] These were finally commissioned in 1961 as well.[40]

There was also a need, at the time, to provide the new Mount Stromlo reflectors with an appropriate suite of auxiliary equipment. Telescopes themselves are, after all, mere collectors of light. It is the instruments used to analyse the light that provide us with a physical understanding of what is happening in stars, nebulae and galaxies. Moreover, the middle decades of the twentieth century saw something of a sea change in the nature of astronomical research—the rise of an intense interest in astro*physical*, as distinct from astro*nomical*, problems. Decades of studies concentrating on the positions and motions of astronomical entities suddenly gave way to a quest for an understanding of the fundamental processes at work *within* them. Questions relating to the chemical composition of the stars, and the ultimate source of their energy output, were more the stuff of this new era.

Thus, in the 1950s and 1960s, front-line research came to depend on fitting excellent spectrographs to the large reflectors then proliferating around the world.[41] These needed to cover a wide range of dispersions to match various tasks undertaken—from very high dispersion *coudé* instruments, for detailed abundance analyses of the atmospheres of bright stars and the study of interstellar clouds from their absorption lines, to low-resolution spectrographs for the analysis of faint emission nebulae, including those in external galaxies.[42]

The acquisition of the 74-inch telescope at Mount Stromlo went hand in hand with the building of a *coudé* spectrograph by one of the world's experts in such instrumentation, Ted Dunham Junior. He had visited Mount Stromlo in 1953 to discuss plans for the instrument and, in August 1957, took up a staff appointment, immediately commencing design work. The spectrograph he built has won world acclaim and enormously enhanced the telescope's capabilities.

The 74-inch reflector at Mount Stromlo Photo: Bob Cooper, ANU

Construction of the *coudé* was a considerable engineering challenge. The design called for a sizeable hole to be dug into solid rock, just outside the dome, and adjacent to the north pier of the 74-inch telescope. After many unsuccessful experiments with such ingenious ideas for shock-free rock breaking as drilling holes, filling them with water and utilising its expansion on freezing during the Mount Stromlo winter, the pit was finally excavated in a much more conventional fashion, using small, carefully controlled explosive charges. One of the best tales from this time relates how, after one of the blasting sessions, some of the wags from the workshop carted broken rock 200 metres from the *coudé* excavation to where they were working on the 50-inch telescope, scattered it about, and then called in the Observatory engineers, Kurt Gottlieb and Don Thomas, in order to enjoy the look of horror on their faces. One of the ring-leaders subsequently suffered apparent retribution in the form of a serious accident in which a rock fell on him, breaking his leg. Irrepressible as ever, the other wags promptly made, inscribed and attached a brass plaque to the offending boulder, which read: 'Clem Bright was laid by this stone, August 13, 1959'.[43]

The construction of the spectrograph itself was no less heroic and time consuming, with Dunham going to great lengths to ensure mechanical and optical stability. Although spectra were taken from 1960 onwards, this was achieved by a variety of makeshift (sometimes wooden) elements, which it took several more years to convert into their final form.[44] At the other end of the dispersion range, a nebula spectrograph for the 74-inch telescope, made by the Davidson Manufacturing Company, entered service in 1962.[45]

By the early 1960s, photoelectric techniques for the photometry of single objects were firmly established, but their use for spectroscopy was only just beginning. This was well before the development of solid state detector arrays, and the first assault on the problem of photoelectric spectrophotometry lay in developing spectrographs which scanned the spectrum past a single tube or an array of individual tubes. Mount Stromlo obtained an accelerated entry into this technology by virtue of a year-long visit by a leading practitioner, Lawrence Aller, who in 1960 brought with him the single-tube Michigan scanner.[46] The advantages of this new technique over traditional photographic spectroscopy were enormous, and it was not long before Alex Rodgers commenced construction of multi-channel instruments. A two-channel version was commissioned on the 50-inch telescope in 1964, followed by a 33-channel device which was brought into service in 1972.[47]

Another development during the early 1960s was polarisation work. The first

The 74-inch dome at twilight. The top of the coudé *room is silhouetted to its right*
Photo: Bob Cooper, ANU

such studies, on Magellanic Cloud stars, were undertaken by Visvanathan in 1962, using a polaroid inserted in the filter wheel of a standard photometer mounted at the 26-inch reflector at Mount Bingar.[48] A purpose-built, double-channel polarimeter, designed by Ben Gascoigne, was also being constructed at this time, and was commissioned at the 50-inch telescope in 1963.[49] This activity gave rise to a strong interest in polarisation studies as a means of investigating the properties of the interstellar medium and its associated magnetic field, which culminated in the commissioning at Siding Spring of the 24-inch rotatable reflector in 1966.[50]

Bok's directorship also saw the introduction of the digital computer into the Observatory. The first computer purchased anywhere in the ANU was an IBM 610 bought by Mount Stromlo in 1960.[51] This machine was programmed by means of punched paper tape, and had a plug board for wiring special functions; it was scarcely more than a glorified calculating engine. Everyone at Stromlo obtained a personal plug board in the hope of wiring it with a host of useful functions—and everyone discovered (quite independently) that it would hold the series expansions for the sine, cosine, exponential and logarithm functions, and no more! Upgrades followed, the first being to an IBM 1620 in 1964.[52] At first the computers were used solely for doing repetitive calculations in the reduction of observational data. No one, at that time, had any idea just how ubiquitous they would become in all aspects of observatory work, from instrument and telescope control, to mass data storage and retrieval, to large-scale theoretical calculations, to electronic and mechanical design work, to round-the-world communication, to keeping the budget, to preparing manuscripts online—and all in the space of just a few decades! Nevertheless, the Observatory, under Bok's directorship, embraced this new technology with enthusiasm and, as early as 1965, a committee, chaired by Ben Gascoigne, was established to oversee the integration of digital processors into the research work.

☆

Bok's own research, during his time at Mount Stromlo, continued to concentrate on the structure of the Milky Way system, and on studies of the Large and Small Clouds of Magellan. He observed frequently, but his many directorial duties prevented his output being as significant as it had been in his earlier career. His contribution lay rather in directing the research efforts of those around him, which he did with some vigour.

On his arrival, he soon adopted a policy that, since the Southern Hemisphere had a dearth of large telescopes, all the staff and students at Stromlo should undertake *observational* projects on *objects which can be best observed in the south*. Purely theoretical studies were forbidden and, since his residence overlooked the 74-inch dome, he could—and did—monitor the direction in which the telescope was pointing. Many an observer received a nocturnal telephone call from Bok demanding to know why they were looking to the north. Fortunately the stock reply, 'I'm just observing a standard comparison star', was usually effective—if not always truthful! Bok's policy was especially hard on researchers such as Antoni ('Bill') Przybylski, who had done some excellent theoretical work on stellar atmospheres under Woolley, but who made a far less distinguished observational astronomer.

Magellanic Cloud research at Mount Stromlo was especially noteworthy during the Bok years. In particular, Ben Gascoigne brought a couple of major projects to fruition. In one of these, already described in the previous chapter, he obtained a distance measure for each of the two Clouds, in partial collaboration with Gerry Kron.[53] His second major project related to globular clusters—the most populous star groupings found within galaxies, including in our own Milky Way and the Magellanic Clouds.[54] Careful measures of brightness and colour for many dozens of stars in any particular cluster are capable of yielding information as to how long ago the cluster was born (the interpretive tool is a plot called the 'colour-magnitude diagram'). All the globular clusters in our own Galaxy are immensely old, having been formed some ten thousand million years ago, not long after the Big Bang itself. In the Magellanic Clouds, by contrast, Gascoigne found that, while some of the globulars are equally ancient, there are others with ages only a tenth to a hundredth of this value. These 'blue globulars' have no known counterparts in the Galaxy, and this completely unexpected result must indicate that there is something startlingly different about the early and continuing history of cluster formation in these galaxy systems. Gascoigne's student, John Menzies, wrote his thesis on the colour-magnitude diagrams of four globulars within our own Milky Way.[55]

Also at this time, Mount Stromlo observations yielded the first measurements ever obtained for the chemical composition of objects lying beyond our own Milky Way. Emission nebulae are frequently the brightest entities in a galaxy, making them the easiest objects to investigate for chemical abundance. In 1959, Hugh Johnson, a visitor to the Observatory, derived the helium-to-hydrogen ratio in the giant 30 Doradus Nebula in the Large Magellanic Cloud, showing it to be quite similar to that in the gaseous nebulae of the Galaxy.[56] This work provided the first quantitative indication that the two most abundant of the chemical elements—hydrogen and helium—might have been formed by some common mechanism operating in every part of the Universe. It is now believed that they were indeed created, with much of their currently observed abundances, in the Big Bang itself. Johnson's work was soon extended by Lawrence Aller, during his year-long visit, and by his student Don Faulkner, to several other elements, and to many more nebulae.[57] Stellar abundance studies in the Clouds also followed, the first being by Bill Przybylski for the star HD 33579.[58] Thus a picture of the chemical composition of the Clouds was gradually built up.

Many other staff and students were involved in Magellanic Cloud research, including Bengt Westerlund, Lindsey Smith and Louise Webster.[59] They surveyed various types of stars and nebulae in the Large Cloud, reaching important conclusions concerning the star formation history and the kinematics of that system.

Bok's other passion, the southern Milky Way, also received a good deal of attention. Distance investigations for prominent stellar features, some using novel techniques, became thesis topics for several Mount Stromlo students, John Whiteoak, David Sher and John Graham among them.[60] Studies were also made of the gaseous interstellar material in the Milky Way. Alex Rodgers, Colin Campbell and John Whiteoak produced *An Atlas of H-α Emission in the Southern Milky Way*, mapping and cataloguing 182 emission regions.[61] These nebulae emit light because their hydrogen gas has been energised by ultraviolet radiation from luminous, hot, newly formed stars. Thus they are useful markers in delineating the bright spiral structure in a galaxy. The *RCW Atlas*, which has become a standard reference for such Milky Way nebulae, was produced using very modest telescope equipment—an 8-inch Schmidt camera attached to the 6-inch 'Farnham' refractor.

In spite of this Milky Way research, the problem of dealing with dust obscuration in the Galaxy remained as intractable as ever, and Bok's hopes of mapping the spiral structure by such optical studies were always doomed to have limited success. The real breakthrough came, as has been noted, at the hands of the radio astronomers with their 21-centimetre line surveys of the Galaxy's neutral hydrogen, an advance that Bok applauded, embraced and publicised enthusiastically.[62] While no Mount Stromlo staff member had been involved personally with radio observations since Cla Allen's departure in 1951, some of the earliest of Bok's student recruits chose radio astronomical thesis topics, and spent as much time working at CSIRO's Radiophysics Lab as they did at Stromlo. They included Jim Koehler, Marc Price and Ron Ekers.[63]

Apart from work on the Magellanic Clouds, little extragalactic research was undertaken by Mount Stromlo staff during Bok's tenure. This seems somewhat surprising in view of the advances made by de Vaucouleurs in the 1950s. Nevertheless, the extragalactic tradition was kept alive by one of the students, Bob Shobbrook, who put the Observatory's equipment to good use in a comprehensive study of some 50 southern galaxies lying in four different groups.[64]

The Observatory also benefited from a succession of international visitors. Hugh Johnson and Lawrence Aller were only two of many overseas astronomers who made lengthy visits to Canberra during these years. Bok's wide circle of colleagues and associates from his Harvard days brought Mount Stromlo into close contact with the mainstream of world astronomy as never before. Other visitors included Walter Baade, Karl Henize, Gerry Kron, Bernard Pagel, Wal Sargent, Bill Tifft and Frank Wood.

Of even greater consequence was a spate of new staff appointments, mostly from overseas. Ted Dunham Junior's has already been mentioned. Alex Rodgers became a Research Fellow soon after completing his thesis, the first of several Mount Stromlo graduates who made their subsequent careers at the Observatory. Bengt Westerlund, a Swedish observer running the Uppsala southern station, transferred to the ANU in 1961. In 1963 Leonard Searle arrived, greatly enhancing the astrophysical emphasis of the Observatory's research; and in 1965 Don Mathewson came to Mount Stromlo from the CSIRO's Radiophysics Lab. This appointment was one of the first to cross the radio–optical divide, further cementing the growing interaction. An early appointment in the opposite direction was that of John Whiteoak, one of Bok's first students, to the Radiophysics Lab. Two of the above, Mathewson and Rodgers, eventually became Directors at Mount Stromlo.

Many Stromlo staff and students enthusiastically embraced the new astrophysical paradigm in their research, undertaking spectral studies to derive the chemical make-up of various types of astronomical objects in the hope of illuminating the chemical history of our Galaxy, and learning about its initial formation. Thus Alex Rodgers and Roger Bell investigated the atmospheres of Cepheid variable stars; Roger Bell those of white dwarfs; John Danziger those of red variable stars nearing the end of their lives; Robin ('Harry') Hyland those of peculiar stars exhibiting magnetic properties; and Mike Bessell concentrated on a class of short-period variables (resolving a significant puzzle as to their evolutionary status).[65] One real chemical oddity—which came to light in an observing program designed for other purposes—is the unique 'holmium star' HD 101065, also known as 'Przybylski's Star' after Bill Przybylski, who discovered it and investigated its chemical composition.[66] Its spectrum, dominated by the lanthanide rare earths, caused consternation when first discovered, and even 40 years later it still evokes comments such as 'arguably one of the most chemically bizarre objects in the Galaxy'.[67]

Alex Rodgers and Leonard Searle made an intensive study of the peculiar object η Carinae, a star about one hundred times the mass of the Sun, embedded in a cloud of circumstellar gas which the Hubble Space Telescope has now revealed to be one of the most beautiful and fascinating of all the objects it has imaged.[68] The history of this object has been extraordinary. In the 1840s, it had a giant outburst, becoming, for a time, the second brightest star in the sky. It has since faded again and is currently about sixth magnitude. Rodgers and Searle's paper was seen as a classic example of the new style of astrophysical research, even though some of their detailed conclusions have required modification in the light of more recent work. This same object was also one of those studied by Visvanathan with his polarimeter.[69] He found that light from its core is being scattered by free electrons within a circumstellar shell.

By the end of Bok's tenure, Mount Stromlo was well positioned, in terms of both staff and equipment, to make a real contribution to the new astrophysical era then emerging—and it began to show. Though the remarkable rise of radio astronomy had dominated the Australian scene in the immediate postwar years, the gap between Sydney and 'the rest' was beginning to narrow. Ben Gascoigne recalls an exchange that occurred during the first coffee break of a joint Stromlo–Radiophysics symposium held towards the end of Bok's directorship: 'Leonard Searle . . . came up to me and said, "Ben, I think we're ahead on points". I thought so too, and it was a great day. We had caught up with them at last'.[70]

Mount Stromlo had indeed come of age. Bok even relaxed his earlier insistence that all his staff should be observers. This was perhaps inevitable since much more demanding theoretical studies were required to support the new astrophysical research work. Numerical modelling of stellar atmospheres became absolutely essential for the quantitative determination of chemical abundance, as did stellar interior modelling for the interpretation of the evolutionary development of stars, and of the galaxies containing them. In 1965, shortly before Bok's departure, another Stromlo graduate, Don Faulkner, was appointed to the staff, and he set about developing the Mount Stromlo Stellar Evolution Code—an *entirely* theoretical enterprise—thus making the break with Bok's former edict complete.

☆

One of the chief objectives of Bok's public relations efforts was to advance a campaign launched by Woolley in 1953, to acquire a large, multi-national telescope for Australia. Even before he had completed his 74-inch, Woolley proposed that an even larger British Commonwealth reflector (perhaps a 200-inch) should be built at Mount Stromlo. He discussed his ideas with Mark Oliphant, who in turn conveyed them to Sir John Cockcroft at Cambridge.[71] Nevertheless, nothing had eventuated before Woolley left Stromlo.

His successor's directorship could hardly have fallen at a more propitious time with regard to the general funding climate in Australia. The years of postwar austerity had given way to a period of sustained economic prosperity. Furthermore, in Robert Gordon Menzies, Australia had a Prime Minister who was well disposed towards scientific research in general, and university work in particular. He established the Murray Commission which reported on the Australian university system in 1957, and personally saw to it that its recommendations were implemented. He set up the Australian Universities Commission in 1959, which led to the most beneficial expansion of tertiary education Australia has ever seen. In the same year he established a formal research funding body, the Australian Research Grants Committee. So substantial was Menzies' patronage of scientific research that the youthful Australian Academy of Science (AAS) elected him to its Fellowship—a rarely bestowed honour for someone who is not a practising scientist. Sir Frederick White's biographical memoir of him in the Academy series makes stimulating reading.[72] Concerning Menzies' support for his own institution, White wrote: 'Menzies was Prime Minister during the period of the greatest expansion of the activities and facilities the CSIRO had ever experienced'. Bart Bok cultivated a relationship with Bob Menzies that was probably as close as the Prime Minister ever allowed.

The proposal for a British Commonwealth telescope raised complex issues, not least for the British themselves since they were considering several other possible collaborations. Notable among these was an invitation to join with five other European countries in setting up a European Southern Observatory (ESO) in Chile. Woolley, by then Astronomer Royal in Britain, continued to press for a British Commonwealth project. In 1962, on Woolley's advice, the British government finally decided against becoming an ESO member. (Britain would eventually join the consortium four decades later, in July 2002.) Even before this 1962 decision, however, the concept of a southern British Commonwealth Observatory had received support from several high-level science policy committees in Britain, and in 1959 the President of the Royal Society, Sir Cyril Hinshelwood, made a formal approach to the AAS to explore the possibility.[73]

The reception of the proposal in Australia was scarcely more straightforward. In this case, however, the debate was not about astronomical priorities—rather it took on interdisciplinary dimensions, with several biological Fellows of the Academy resenting the proposition that so much funding should be devoted to astronomy. In March 1960, Bart Bok and Sir Mark Oliphant appeared before the Academy's Council, following which the AAS gave the telescope proposal its scientific imprimatur, but withheld any other endorsement until a study of the financial implications could be undertaken. In response to a request for advice from the government in June 1961, the Academy begged for another two years to prioritise the dozen scientific proposals it had before it. Thus, both in Australia and in Britain, the large-telescope proposal entered a two-year hiatus.[74]

Bok became restive at this continuing delay, particularly after May 1962 when the ANU's Siding Spring decision seemed to have identified at least one acceptable Australian site for the construction of a large international telescope. In August of that year, he decided to hurry things along. Taking advantage of an official address as President of Section A of ANZAAS, he spoke publicly of his frustration at its Sydney congress.[75] His remarks were widely reported. The chief target of his criticism was the AAS, whom he accused of failing to consult optical astronomers in its assessment of the large telescope proposal, and of placing international scientific relations at risk by its tardy response to the Royal Society. The President of the Academy, Thomas Cherry, who was in the audience for Bok's address, was outraged. He sent Bok a stinging letter, telling him that he had offended 'the Establishment', outlived his usefulness in Australia and had best return to the United States. Bok was fearful that he might indeed have gone too far, thus ruining any hope of a large telescope. Not long afterwards, however, he was present at a formal ANU function to welcome the King of Thailand. As the official procession passed him, Bob Menzies unobtrusively poked him in the stomach and whispered: 'You are a bad, bad boy!'. Immediately Bok knew that all was *not* lost.[76]

Over the next three years, Woolley in Britain and Bok in Australia continued their promotional efforts. Several committees met—some national, some joint—to organise support, and to draft technical specifications. It was at a meeting convened by Woolley at Herstmonceux in October 1963 that a consensus emerged that the instrument should be of 150-inch diameter (rather than a 120-inch), and that it should have a conventional equatorial mount. The fact that it fell to Woolley and Bok to serve as the chief advocates in their respective countries may, in itself, have presented some difficulties. As has already been mentioned, Ben Gascoigne recalls that they 'were men of widely different character and temperament who detested each other'.[77]

By June 1965, the Australian Academy had satisfied itself that it could support the large telescope proposal, and an agreement was reached with the Royal Society to make a joint submission to both governments seeking funding on a 50/50 basis.[78] Given that this marked the end merely of the scientific campaign, and the political battle was still to be engaged, Bok was now resigned to the fact that a final victory was unlikely to be achieved during his time at Mount Stromlo. He had long intended to return to the United States on turning 60, to take up one last appointment on home soil before retiring.

One set-back at government level occurred when Senator John Gorton, the Minister responsible for Education and Science within the Prime Minister's Department, initiated an enquiry into whether a better overseas partner might be the United States rather than Britain.[79] This eventually came to nothing despite causing many months' delay.

Thus it was that, on 19 March 1966, Bart Bok concluded his directorship at Mount Stromlo a rather frustrated man. In his last Annual Report, he commented on what he considered to be his achievements and failures as Director.[80] His feelings regarding the delay in funding a 150-inch telescope were expressed so forcefully that the University persuaded him to provide an expurgated version for publication—the original is held in the Stromlo Library. Approval for the Anglo-Australian 150-inch was eventually forthcoming, as will be recounted in the next chapter; and, in spite of his personal disappointment regarding its timing, Bok's strenuous public relations efforts contributed in no small way to that successful outcome.

Bart and Priscilla Bok returned to the United States, as they had planned, with Bart taking up one final directorship—at the Steward Observatory of the University of Arizona. There he commissioned one final telescope, a 90-inch, built on Kitt Peak and completed in 1969.[81] Bok had accepted the Arizona post with one eye on his eventual retirement. The head office of the Kitt Peak National Observatory was in Tucson, and astronomers from all over the world were constantly passing through. It was an ideal location for someone wishing to maintain active contact with friends and colleagues. Bok stood down from the Steward directorship in 1970 so as to have more time for his research interests. In May 1972, Priscilla, who was ten years Bart's senior, suffered what was thought to be a mild stroke. In fact, it presaged a gradual deterioration of her mental capacity which led eventually to Bart becoming her devoted full-time carer. He retired from the University in 1974, and Priscilla died on 19 November 1975.

Priscilla's passing affected Bart greatly; for many months he was a complete recluse. By the middle of the following year, however, he felt able to attend a meeting of the American Astronomical Society, and he gradually resumed a schedule of observing, writing, lecturing and international travel which was as energetic as at any previous stage of his life. He died of a heart attack in his home at Tucson on 5 August 1983, aged 77 years. His larger-than-life personality attracted many tributes, including an entire commemorative issue of the journal *Mercury*.[82]

chapter eight

The rise and rise of astrophysics: the Eggen years, 1966–1977

WHEN OLIN JEUCK EGGEN TOOK UP HIS APPOINTMENT as Mount Stromlo's fourth Director on 1 July 1966, he was returning to familiar ground. He had previously made two substantial visits while a staff member at the Lick Observatory on Mount Hamilton, California. The first was during 1951–52, in company with Gerry Kron, also from Lick. They were a formidable duo, and brought the very latest equipment, techniques and expertise in photoelectric photometry to Mount Stromlo.[1] Eggen came back in September 1955 for another six-month stay.

Eggen's professional career took him to many famous institutions. In an invited autobiographical essay published when he was aged 73, he described his life as a 'perfect, if accidental, combination of time and place; Mount Hamilton in the 1940s, Herstmonceux Castle, Sussex, and Capetown in the 1950s, Pasadena, Mount Wilson, and Palomar in the 1960s, Australia in the 1970s, and Chile in the 1980s'.[2] He cherished the opportunities which had come his way to ride the crest of a wave from continent to continent as the mid-twentieth-century flowering of the new astrophysical paradigm made first one location and then another the most exciting place on Earth to work.

This paradigm posed an entirely new set of questions to tease the astronomer's curiosity. Previously, interest had centred mainly upon what was where in the Universe, and how things were moving. Now instrumental advances—especially in spectroscopy—and the advent of the digital computer made it possible to ask more penetrating questions concerning what was happening *inside* objects such as stars—what made them tick, in other words. In particular, astrophysicists began to investigate how things developed over time, and—most intriguing of all—how they originated in the first place.

In this respect, Olin Eggen came to Mount Stromlo with his reputation already made—the University appointed him 'by invitation'; the chair was not advertised.[3] In 1962 he had co-authored a paper which is now widely acclaimed as one of the 50-odd seminal astronomical contributions in twentieth-century astronomy: 'Evidence from the motions of old stars that the Galaxy collapsed' (Eggen, Lynden-Bell & Sandage).[4] This paper is so famous that it, and the concept it proposed, are usually referred to simply by the initials 'ELS'. Allan Sandage and Donald Lynden-Bell both spent periods of sabbatical leave at Mount Stromlo during Eggen's directorship.

To some extent, Eggen's work spanned the old paradigm and the new. His observational specialty was the measurement of the brightness of stars by photo-electric techniques, and he combined these observations with data on their motions to investigate their orbits within our Milky Way, a pursuit not unfamiliar to astronomers of the old school. What made the difference in Eggen's work, however, was that he then went on to ask new and interesting questions. It became apparent to him that the kinematical properties displayed by stars were linked to how long ago, within the Galaxy's history, they had been born. The oldest stars tend to travel in a large, nearly spherical volume of space, while those born more recently form a flat, rotating disc. Thus he realised that our Milky Way has not always been as we see it today. It appears to have originated in a type of giant collapse, with a shrinking proto-galactic gas cloud forming generation after generation of stars as time went by.

This was a revolutionary idea at the time, but everyone now accepts the fact that we live in an evolving Galaxy. The picture has become a good deal more complex, however, since we have also discovered that our Galaxy has swallowed other smaller ones along the way. We shall see in due course that this has probably contributed a fair amount of material to the old, spherical component—a process that continues even today. Thus Eggen's technique of using stellar motions to unravel the Galaxy's history is more important than ever.

Olin Eggen's directorship contrasted greatly with that of his predecessor. He was a man who devoted virtually all of his time to his research work, and who lived only for astrophysics. His lifestyle was disciplined—indeed, almost monastic. His working day started about noon and terminated shortly before dawn; thus his body clock required little adjustment between his observing runs at Siding Spring and his normal office routine at Mount Stromlo. He took

O.J. Eggen, Director 1966–77
Photo: MSO

a single meal each day, a mid-evening restaurant dinner, and when he found a venue to his liking he would patronise it habitually. He had no great liking for administration. He was conscientious enough in attending to the directorial tasks required of him, but he also saw advantage in his largely nocturnal working habits since they placed him half a day out of phase with those in the University who were addicted to the telephone and to meetings.

Eggen was rather enigmatic in personality. He was a large man with a somewhat gruff, confronting presence. He did not slip easily into conversations, and there was more than a hint of shyness in his approach to people, particularly those outside the work context. While he was comfortable enough with his staff, he seemed ill at ease when he encountered, sometimes unexpectedly, other members of their families on the Observatory premises. He was a bachelor, and had little use for the rather expansive Director's Residence that came with the job. He found it convenient to sleep there, since it was just two minutes' walk from his office, but he solved the problem of its upkeep by giving the run of the house to the Observatory gardener and his wife. He made the residence lawn available to all the Stromlo staff, and bought them a complete set of croquet equipment. For several years, a lunch-time club flourished—although Eggen himself never played. That was, after all, the hour of the day when he was just venturing forth to his office!

He was selective in the friendships he formed, but those relationships were usually close and long lasting. One such was with the University's Secretary, Ross Hohnen, third in the hierarchy of the University's senior officers, and one of the most competent and influential people in the Chancelry. Hohnen and Eggen became close colleagues, and comrades-in-arms when it came to defending what they saw as the University's interests in a major political battle which erupted within the astronomical community during Eggen's directorship (see pages 176–83).

Eggen was a dedicated observer; he spent a week each month at the Siding Spring Observatory, which he regarded as his home away from home. He stayed in a small Director's flat attached to the Lodge, and enjoyed life in an entirely astronomical environment almost completely free of administrative chores. He gathered prodigious amounts of photoelectric data with the 40-inch telescope, mostly designed to further our understanding of the evolutionary history of our Galaxy. The 40-inch was, in fact, Eggen's favourite instrument—just as it had been Bart Bok's.

In his first year as Director, Eggen elevated the profile of Siding Spring in world astronomy by adding its name to the official title of his institution, which thus became the 'Mount Stromlo and Siding Spring Observatories'.[5]

☆

One non-astronomical indulgence that Eggen enjoyed was ownership of a sports car. He brought with him to Australia a red and black Austin Healey, which soon became a familiar sight on the Canberra roads. It was capable of a somewhat greater turn of speed than the local police pursuit vehicles, as a patrolman discovered when he gave chase one weekend only to see his quarry promptly disappear over the horizon. The officer made fruitless pursuit all the way to the Cotter River, when it occurred to him that the object of his quest might well have turned off up the Mount Stromlo side-road. By the time he retraced his route to explore this possibility, Eggen had the Austin Healey safely garaged out of sight, and himself safely ensconced in his locked observatory contentedly reducing photoelectric observations in his office. Unfortunately, however, the policeman took it into his head to make a few inquiries of any mountain residents he found doing their weekend gardening, and one of the Observatory's engineers, showing more honesty than diplomacy, spilled the beans.

Eggen ruefully recounted the story of his near-triumph over the law for some time afterwards. He had, in fact, quite a fund of stories, but one was never quite sure where the boundary between fact and fiction lay. He asserted, for instance, that this same Austin Healey had three carburettors. This was simply not true, but the claim provided a generic name around the Observatory for some of his more outlandish tales—one of Olin's 'three-carburettor' stories.[6]

Eggen had rather an obscure sense of humour, but he could be extremely witty at times. He could also be quite acerbic. He once reviewed a conference volume by someone who had made a career of organising scientific meetings and then editing and publishing their proceedings. He made no attempt to disguise his disdain:

> The Editor has contrived to cite himself rather more frequently in the name index than his contributions to the conference would seem to warrant. One reference is to the conference photograph—in which he doesn't appear—perhaps he was the photographer! . . .
>
> Perhaps the best thing I can say about this volume is that I found my copy entirely self-destructing. Each leaf detached itself from the spine as I turned the page.[7]

★

The year in which Olin Eggen became Mount Stromlo's new Director also saw another new beginning—the creation of the Astronomical Society of Australia. For some time there had been suggestions that the Australian astronomical community was now of sufficient size and distinction to warrant the formation of a learned society. Ben Gascoigne and Bill Buscombe, both long-time Stromlo staff members, were among its early advocates.[8] In 1966 an *ad hoc* committee was formed to advance the proposal, and the inaugural meeting of the Society was duly held at the University of Sydney on 30 November. Ben Gascoigne was elected as a founding Vice-President. Eggen became the Society's third President, serving during 1971–73. While never one to relish administrative activity for its own sake, he saw to it that Mount Stromlo was supportive of the Society in its formative years. Three of the Society's first seven annual meetings were hosted in Canberra.

August 1973 marked another milestone in Australia's astronomical progress when the fifteenth General Assembly of the International Astronomical Union (IAU) was held in Sydney.[9] This was the first occasion that the IAU's triennial meeting had taken place outside Europe or the United States. In the event, something of a shadow was cast over this distinction when Poland was awarded a week-long 'Extraordinary General Assembly' in September of the same year. While ostensibly to commemorate the quincentenary of the birth of Copernicus, its announcement generated antipodean suspicions that the IAU Executive had bowed to pressure by providing an alternative meeting for the benefit of northern (especially European) astronomers not wishing to make the long journey to Sydney. Eggen responded by resigning his membership of the Union. This may sound like a significant protest, given that resignation from the IAU was something unheard of at the time.[10] In the close-knit astronomical world, however, many colleagues would have known of Eggen's somewhat ambivalent attitude towards membership of *anything*, making this grand gesture less of a sacrifice for him than it would have been for a more gregarious scientist.

Eggen's personal dedication to the research task set the tone for the whole Observatory. The scientific output for the period of his directorship bears witness to this. During the decade of Bok's directorship, publications in refereed international journals averaged eighteen per year. Under Eggen, this more than doubled to 42 per year.

Eggen himself published 99 papers during the eleven years he was at Mount Stromlo, most relating to the way that stars evolve, and how agglomerations of stars can tell us about the life history of our Milky Way galaxy as well. Such stellar clusterings are presumably born in a single star-formation event.[11] Some are quite evident simply from looking at the starry sky itself; these are called 'clusters'. Others, called 'associations', are not so apparent, but they can be discerned by isolating those stars sharing some particular physical characteristic. For instance, one might concentrate only on stars that are blue in colour, and thus very hot. In physical terms, this means that the stars concerned are quite young (having been born out of the interstellar gas just some millions of years ago!).

Even more difficult to isolate are 'moving groups'. The stars belonging to one of these entities can be located all over the sky, but their participation in a unique episode of star formation is betrayed by the fact that they all share a common motion within the Galaxy. Eggen made the study of such groups his specialty and, in fact, they are sometimes referred to as 'Eggen groups'.

A star's motion in the line of sight (its 'radial velocity') is reasonably easily derived from the Doppler wavelength shift of the dark lines in its spectrum.[12] To obtain the two velocity components in the plane of the sky, on the other hand, one must measure the minute change in its position over an interval of some decades. This *angular* motion (or 'proper motion') must then be converted into *linear* terms (kilometres per second) by finding the distance to the star—itself a complex process. When all this is done, however, one can plot a three-dimensional diagram representing the star's velocity components, in the same way as a star's position in space can also be represented by three coordinates. A bunching together in this 'velocity–space' diagram betrays a common origin for the stars concerned just as convincingly as does their bunching together in ordinary space.

By such painstaking methods, Eggen examined thousands upon thousands of stars, sorting them into their various groupings. He was then able to compare the position and motion of each group with other observed properties, such as the time that has elapsed since the group was born (the stellar age, in other words) and the chemical composition of the material out of which the stars were made.

This chemical make-up of the stars provides vital information concerning events in our Galaxy's early history. Most of the hydrogen and helium atoms that we see in the Universe are believed to have been made right at the start, during the Big Bang. Nearly all the other elements, however, have been manufactured inside stars, as products of the thermonuclear reactions that occur deep in their cores, providing all the energy that they radiate throughout their lives. Since hydrogen and helium are the two lightest elements, the manufactured ones—the carbon, oxygen, iron, uranium, etc.—are generally lumped together under the name 'heavy elements'.[13]

Some of the heavy element material manufactured deep inside stars is regurgitated into the interstellar gas when a star dies. Supernova explosions are the best examples of this, but other, gentler forms of mass loss can occur in lower-mass stars as well. Thus, over the long star-formation history of our Galaxy, the heavy element abundance in the interstellar medium has been gradually increasing, although it still amounts to only a few percentage points by mass. As new generations of stars form from the interstellar gas, their chemical composition reflects that of the gas at the time. Some stars have been found in which the heavy elements are 10,000 or even 100,000 times less abundant than they are in newly forming stars at the current epoch. By piecing together such abundance information for moving groups, one begins to obtain a picture of both the dynamical and the chemical evolution of our Galaxy itself.[14]

One unexpected result of Eggen's work was his identification of some moving groups in the Halo of the Milky Way—a region which, on the ELS picture, one might have expected to be rather well mixed. Eggen's discovery was greeted with some scepticism, and some of his claims are still in doubt, but studies employing even more sophisticated techniques have now verified that such 'Halo streams' do exist. They are the tidal debris of small galaxies which have been swallowed by our Milky Way during its past history—galaxies of similar size, perhaps, to the Magellanic Clouds. When Eggen, Lynden-Bell and Sandage wrote their seminal paper, the Milky Way was considered to be an entirely isolated galactic system, but now galactic mergers are much on the agenda as its life history is being progressively unravelled.

Approaching the topic in a somewhat different way, Alex Rodgers focused his attention on the region of the sky towards the galactic poles—well away from the plane of the Milky Way. He undertook a study of moderately hot stars (of spectral type A, with temperatures of 8,000 to 10,000 degrees Kelvin).[15] An A-star has a spectrum that permits reliable estimates to be made of the star's distance, its radial velocity and its heavy element abundance. Rodgers' sample contained some 60 stars, and he found that about half of them had abundances typical of the current interstellar gas, thus indicating that they are relatively young objects. In spite of this, some of the stars concerned are located at great heights in the Galaxy's Halo where, on the simple picture of its formation, one expects to find only very old stars.

At the time, Rodgers proposed that these rogue A-stars were actually born in the plane of the Milky Way and subsequently ejected into its Halo. The suggestion was that each had belonged to a binary star system, the other component of which had ended its life and exploded as a supernova. The A-star was thus set free from its binary orbit, and travelled off at high velocity in a type of slingshot mechanism.

Rodgers maintained a long-term interest in the study of these stars, eventually reaching the conclusion that they were born in a far more dramatic event—an intergalactic collision between our Milky Way and an external system of about the same size as the Large Magellanic Cloud about 2,000 million years ago.[16]

★

Eggen's hiring policies also contributed to Mount Stromlo's increased productivity. He chose his recruits carefully, gradually building up the strength of his research staff. The number of tenured astronomers grew only modestly, from eleven to thirteen—although, with some turnover following departures, a younger and more vigorous staff profile emerged. Six tenured appointments were made: Mike Bessell, Don Faulkner, Ken Freeman, Harry Hyland, Agris Kalnajs and John Norris. Of equal importance was the constant stream of young researchers passing through in non-tenured and post-doctoral appointments. There were always six to eight in post during Eggen's latter years. Four later became permanent staff members as well: Mike Dopita, Bruce Peterson, Vis Visvanathan and Peter Wood. Harry Hyland is now Deputy Vice-Chancellor at James Cook University, but the other nine Eggen appointees have made Stromlo home for their entire careers. An extremely stable tenured staff has been a marked characteristic of Stromlo over many decades. It is notable that six of Eggen's ten recruits obtained their PhD degrees in the Observatory's own Graduate Program, leading to a perception, at times, that Mount Stromlo has suffered from in-breeding.

Ken Freeman's appointment proved particularly outstanding. A West Australian, he did his doctoral studies at Cambridge, writing a thesis on the dynamics of barred spiral galaxies under Leon Mestel. Ben Gascoigne once heard Mestel remark that he had been basking in Ken's reflected glory ever since![17] Freeman was elected to fellowship of the Australian Academy of Science in 1981, and of the Royal Society of London in 1998, thus becoming the first Mount Stromlo researcher to win a Fellowship of the Royal Society (FRS) since the days of Woolley and Martyn.[18]

Freeman's main field of research relates to the kinematics and dynamics of stellar systems such as galaxies and clusters. In 1970 he published an influential paper investigating the physical implications of the form of the light distribution exhibited by spiral galaxies, as previously described by Gerard de Vaucouleurs.[19] He discovered a remarkable uniformity in the central surface brightness of the discs in these spirals. The interest in such studies lies not so much in the light distribution, but rather in what it tells us about the corresponding mass distribution within the systems. This provides another window into the evolution of galaxies and, ultimately, the Universe as a whole.[20] Freeman's paper also drew early attention to a 'dark matter' problem in spiral galaxies (dealt with in much greater detail in Chapter 10).

Freeman has been thesis adviser to many students, several of whom have themselves gone on to outstanding astronomical careers. One of his earliest, Garth Illingworth (now Professor of Astronomy and Astrophysics at the Lick Observatory), applied a novel Fourier technique to use the integrated spectra of globular clusters to study their stellar kinematics. Once again, the end point of the analysis was an estimate of the mass of each cluster. In his thesis, Illingworth studied ten globulars, obtaining mass estimates between 160,000 and 1.33 million times that of the Sun.[21] It was Agris Kalnajs who made the Fourier technique suggestion—one of many which have, over the years, earned him a reputation as an 'ideas man' at Mount Stromlo. The same idea occurred independently to a researcher at Columbia University, Sue Simkin, who later came to Stromlo as a Senior Research Fellow (1977–79).[22]

With heavy element enrichment processes playing such an important role in the evolution of galaxies, a great deal of effort was devoted, during the Eggen years, to chemical composition analyses of various types of stars in our Milky Way. With the advent of digital computers, it became possible to make sophisticated models for the atmospheres of stars which, in turn, yielded increasingly reliable estimates for their chemical abundances. Mike Bessell, Ken Freeman, John Norris and Alex Rodgers were all much involved in such work, and abundance analyses also formed a central theme for some excellent Stromlo theses produced at the time, notably those by Harvey Butcher and Jeremy Mould.[23] Mould would eventually become the seventh Director of Mount Stromlo, while Butcher currently heads the Netherlands Foundation for Research in Astronomy.

One intriguing discovery made at this time was that the stellar population of the massive globular cluster ω Centauri exhibits quite *a range* of heavy element enrichment.[24] Two possible explanations suggest themselves. The first is that this cluster is *so* large—over three million times the Sun's mass, and the biggest in our Galaxy—that its formation may have involved more than a single generation of star formation. Thus the heavy elements manufactured by the stars formed initially may have already enriched the remaining gas by the time the last generation of stars was born. The second possibility is that some mixing process is occurring within individual stars, bringing the heavy elements being made in their cores to the observable stellar surface layers—but only in some of them. The two theories were hotly debated at the time—so hotly, in fact, that John Norris (in favour of the second hypothesis) bet Ken Freeman a bottle of whisky that 'the abundance inhomogeneities in ω Centauri [will be shown to be] due to mixing'. It eventually became apparent, however, that *both* physical causes must be invoked in order to explain all the observations. Although everyone was partly correct, the very explicit way in which the wager had been worded meant that John Norris paid up!

On the theoretical front, three of Don Faulkner's students—John Robertson, Peter Wood and Bob Gingold—greatly enhanced the Mount Stromlo stellar structure program in order to model a much more extensive range of stellar physics. The group was particularly interested in understanding the complex, and calamitous, events that occur when stars are nearing the end of their lives. Our Sun is currently only halfway through the consumption of its nuclear fuel, but the time will come—in another 5,000 million years or so—when it will become a giant star, so large that the current orbit of the Earth will lie inside its surface. It will undergo repeated pulsations as it exhausts more and more of its nuclear fuel, and the last nuclear reactions draw ever closer to its surface. Finally it will eject up to half its mass as a shell of gas moving outwards into space at about 20 kilometres per second. Seen from a distance, such expanding shells provide glorious displays of emission; they are known as 'planetary nebulae'. The remaining solar core will then collapse until it is merely the size of the Earth, becoming one of a well-known class of stars—the white dwarfs. The computational modelling of such complex physical events produced two fine Mount Stromlo theses, by Wood and Gingold.[25]

John Robertson, on the other hand, used the enhanced computational suite to calculate the evolutionary properties of heavier stars found in cluster systems. Clusters are always a key tool in piecing together the life history of stars and their parent galaxies. In the last chapter we noted how Ben Gascoigne discovered a remarkable new type of cluster in the Magellanic Clouds. Robertson now extended these observations, and complemented them with theoretical modelling that provided an interpretation in terms of the evolutionary development of the Clouds.[26]

Gascoigne developed and refined the other aspect of his Magellanic Cloud research—the use of their Cepheid variable stars as stepping stones in establishing the distance scale of the Universe.[27] He confirmed that Cepheids in different galaxies obey a common relationship which links their intrinsic brightness to their measured pulsation period and their colour. Thus they make excellent 'standard candles' for deriving the distances to more remote galaxies.

It was at about this time that computational codes were being developed to model the pulsational properties of the Cepheid variables. Bob Stobie, a leading practitioner in such research, spent several years at Mount Stromlo, and he and Alex Rodgers were involved in trying to deduce the precise physical properties and evolutionary status of these stars.[28] There was, at first, a worrying inconsistency between the stellar masses implied by the new pulsational models and those implied by the older evolutionary results, and it was many years before the incorporation of higher quality physical data into the modelling programs finally yielded a satisfactory resolution.

Another two major computational weapons were added to Mount Stromlo's research armoury during the Eggen years. The arrival of Agris Kalnajs in 1973, and the growing capacity of the Observatory's computers, led to a continuing interest in 'n-body' calculations. These computer-intensive studies are used to gain insight into how a system consisting of many smaller entities—such as a galaxy of stars—will behave in response to the gravitational attraction exerted on each other by all these 'particles'. Over the years, Kalnajs has used his computer codes to explore the large-scale morphological features exhibited by galaxies, such as spiral structure and the prominent central 'bars' which characterise some of them.

A totally different weapon—a computer program for modelling shock waves in interstellar plasmas—was developed by Mike Dopita following his arrival in 1975. This versatile code can be used to investigate and interpret the physics of such diverse objects as supernova shells, T Tauri objects (bright emission regions associated with the birth of a hot star still shrouded in its gaseous placenta), or the nuclei of 'active' galaxies where vast jets of material are expelled at enormous energy.[29]

By the start of Eggen's term, the CSIRO's new 64-metre Parkes Radio Telescope had been in operation for over four years, and large numbers of discrete radio sources were being catalogued. An immediate challenge was to identify their optical counterparts. Of intense interest at the time were the quasi-stellar radio sources, or 'quasars'. In the early 1960s, the tentative hypothesis that these compact radio sources were objects of some unknown type within our own Galaxy had to be abandoned when two of them (3C 273 and 3C 48) proved to have optical spectra displaying large redshifts.[30] Suddenly it was realised that quasars must be extragalactic in nature, their large velocities reflecting the expansion of the Universe first discovered by Edwin Hubble in the 1920s.[31] The remoteness of these first quasars implied that they must be unbelievably bright, radiating 100 times the energy of the most luminous galaxies. Furthermore, the emission regions concerned proved to be quite small, scarcely 100 times the size of our Solar System. This is implied by the fact that the luminosity of 3C 273 was soon found to vary on a timescale of months.[32] Thus some previously unknown physical mechanism must be at work. These days most astronomers believe that it is the in-falling of vast amounts of matter into massive black holes that is responsible for such prodigious rates of energy release within such small volumes. Quasars were quickly acclaimed as one of the most fundamental astronomical discoveries ever made, capable of providing insight into physical processes occurring at the very earliest epochs in the history of our Universe.

John Bolton, of the CSIRO's Radiophysics Division, had an intimate association with the discovery of the first quasars, both during his six years at Caltech (1955–61) and when he returned to Australia as Director of the Parkes Telescope.[33] During the Eggen years he had a lively interest in making further identifications. By then the name 'quasi-stellar objects' (QSOs) was sometimes used instead of the more specific 'quasi-stellar radio sources' (quasars), since it had become apparent that by no means all of these objects emit in the radio domain. Bolton had many Californian contacts, and was able to access the large optical telescopes on Mount Wilson and Palomar in the QSO hunt. In addition, however, he established a collaborative arrangement with Bruce Peterson at Mount Stromlo, and they made use of the 74-inch telescope to extend the search to the southernmost regions of the sky. At its peak, such extragalactic research was allocated half the dark-of-moon time on the 74-inch.[34]

Peterson and Bolton examined some hundreds of Parkes sources exhibiting radio spectra characteristics of QSOs ('flat' or 'inverted' spectra). Their positions were photographed in blue and ultraviolet light using an image-tube camera on the 74-inch. An observed ultraviolet excess provides convincing evidence that an object near the radio-source position is indeed a QSO. Over a three-year period, a total of 129 QSOs were confirmed in this way, the vast majority being new identifications.[35] Leonard Searle, Bruce Peterson and John Bolton also made spectral observations at the 74-inch to obtain redshifts for many extragalactic objects, mostly QSOs.[36]

While the 74-inch QSO project was quite successful, the larger diameters of the Californian telescopes were a great advantage for this type of work, and Bolton keenly anticipated the commissioning of the 150-inch Anglo-Australian Telescope (AAT) at Siding Spring. The peak years of this 74-inch work (1970–73) coincided with those of the bitter dispute regarding the management of the AAT (see pages 176–81), and Bolton's relationships with Mount Stromlo became rather strained as a result. In particular, he felt that Eggen had gravely under-estimated the level of technical support that would be required to run the AAT, and he pressed this point home whenever any problems arose in his use of the existing Mount Stromlo telescopes.

At the high-energy end of the electromagnetic spectrum, optical identifications were also being sought for the X-ray sources being newly discovered in space missions, notably by the first X-ray satellite, *Uhuru* (Explorer 42).[37] Mike Bessel, Bruce Peterson, Nessio Vidal and Dayal Wickramasinghe were much involved in this endeavour, with others also contributing.[38] Modelling work was done to investigate the physical processes giving rise to this high-energy electromagnetic radiation. It gradually became apparent that gas was streaming on to a collapsed stellar remnant such as a neutron star—only about 10 kilometres in size—or even a still more compact black hole. The angular momentum of the gas prevents it falling directly on to the collapsed star. Instead, some sort of accretion disc builds up, which allows the gas to spiral down on to the collapsed object more gradually. This disc is, in fact, the actual source of the X-ray emission.[39]

It was particularly exciting, at the time, to study *binary* systems having an X-ray source as one of the companion stars. The stellar orbits in such systems can provide a mass estimate for the collapsed object. A result greater than a few times the mass of the Sun is strong evidence for the stellar remnant being an example of the long sought-after black hole.[40] The Stromlo team did indeed find an excellent candidate, and there is now no doubt that black holes can be formed when some massive stars end their lives.

The Eggen years saw a flourishing of infrared work at Mount Stromlo, led by Harry Hyland and his team. Hyland was also part of a collaboration which arranged in March 1977 for NASA's Kuiper Airborne Observatory (KAO) to make three flights out of Melbourne to observe large emission regions in the Magellanic Clouds, particularly the 30 Doradus Nebula.[41] The KAO was a C-141 jet aircraft specially equipped for making astronomical observations at around

12.5 kilometres' altitude.[42] The team also used the AAT to observe 30 Doradus at shorter infrared wavelengths. The combined results of these studies provided a comprehensive picture of this giant star-forming region, revealing both its stellar content and the properties of vast surrounding dust clouds heated by the radiation of the newly born stars. Hyland concluded that there had been two major bursts of star formation in 30 Doradus over the last ten million years.

Another intriguing discovery, at the time, was that natural masers can occur in the dense shells of gas and dust that surround some of the largest and coolest known stars. Such shells are assumed to have been lost from the stellar surface as the outer layers pulsate in the low-gravity conditions prevailing there. The maser activity was first detected by radio astronomers as bright emissions generated by the hydroxyl radical OH at a wavelength of 18 centimetres. Because of the enormous amounts of dust obscuring them, the stars concerned are best observed at infrared wavelengths, and both radio and infrared observations are required to discern how the mass is being lost from the star. Harry Hyland and his colleagues made infrared studies of several of these interesting objects.[43]

Don Mathewson initiated two notable collaborative programs during the Eggen years. In the first, Mathewson and his Research Officer, Vince Ford, used the 24-inch rotatable telescope at Siding Spring to make polarisation observations of 1,800 stars in our part of the Galaxy (mostly within 5,000 light years of the Sun).[44] Combining their measures with those of other workers, they were able to produce a map of stellar polarisation over the whole of the sky. Since polarisation is caused by the scattering of the starlight by dust particles aligned along the interstellar magnetic field lines, this map also depicts the magnetic field patterns in the nearby Milky Way. These proved to be quite complex, showing a helical structure.[45] This work also revealed a remarkable correspondence between the magnetic field-line patterns and the spurs and ridges observed in maps of the Galaxy made at radio wavelengths. Mathewson and Ford went on to make polarisation measures of some hundreds of stars in the Magellanic Clouds, finding evidence for a large-scale, regular magnetic field there as well.[46]

In the second project, Mathewson worked with Jim Clarke of Sydney University to carry out a combined radio–optical search for supernova remnants (SNRs) in the Magellanic Clouds. SNRs are vast shells of radiating material excited by shock waves emanating from the catastrophic explosions which end the lives of massive stars. A supernova is quite a rare occurrence in any particular galaxy. Our Milky Way is thought to experience only one or two events per century, and many are obscured from our view by the clouds of interstellar dust that lie around us in the plane of the Milky Way. Using the new Mills Cross at Molonglo and the

The 30 Doradus Nebula in the Large Magellanic Cloud

The Crab Nebula, the remnant of a star that exploded in 1054

The Eagle Nebula

The Great Nebula in Carina, Bart Bok's favourite

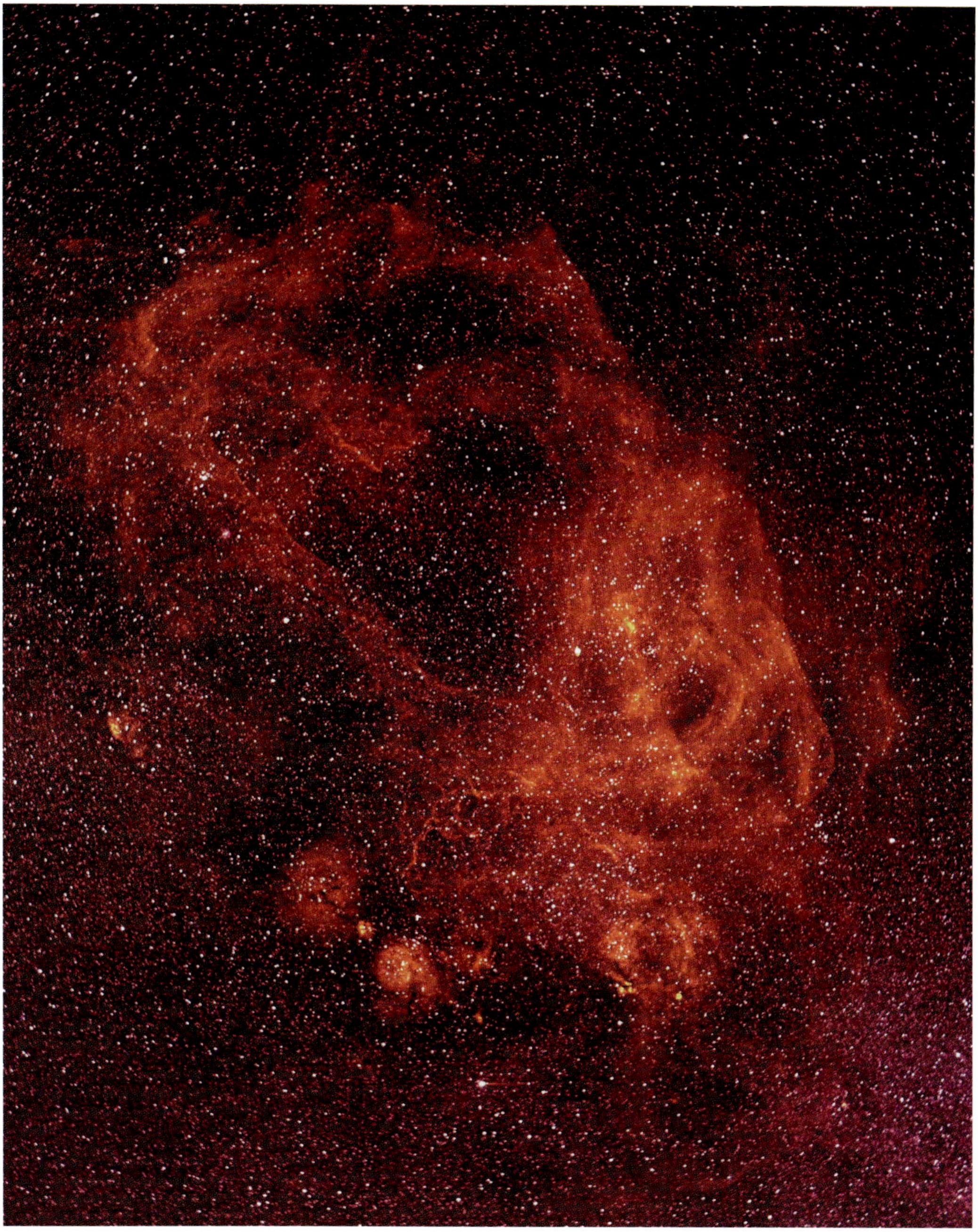

Glowing hydrogen gas in the Gum Nebula

The Helix Nebula; the luminous shell has been discarded by a dying star

The Horsehead Nebula, a dark dust cloud in Orion

The Orion Nebula, a vast stellar nursery

The Triffid Nebula

The Coalsack, a dark cloud near the Southern Cross. The Pointers (α and β Centauri) appear to the left, the Carina region to the right.

The centre of our Milky Way Galaxy is 27,000 light years away in the direction of the constellation Sagittarius

The Large and Small Magellanic Clouds

The Large Magellanic Cloud. Several million stars in this galaxy were monitored for lensing events in the MACHO Project

The Andromeda Galaxy, the closest major galaxy to our own Milky Way

The spiral galaxy NGC 6744

Comet Hale-Bopp, which appeared in 1996–97

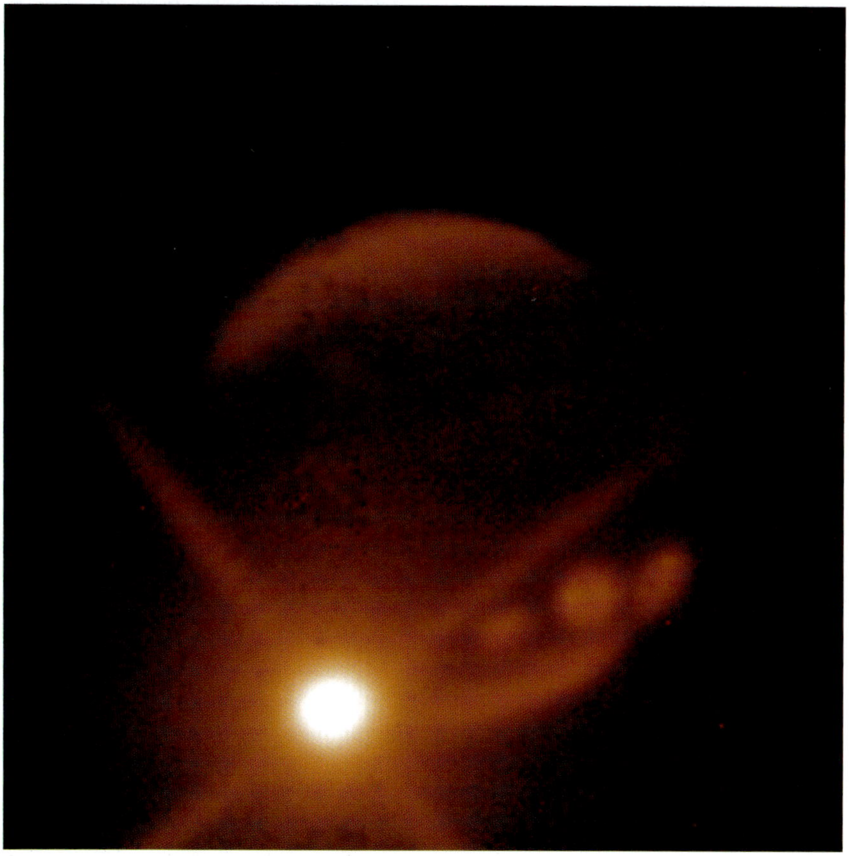

The impact of Comet Shoemaker-Levy on Jupiter, captured with the 2.3-metre Telescope's infrared camera (Photo: Peter McGregor)

Centaurus A, one of the first radio sources identified

40-inch reflector at Siding Spring, Mathewson and Clarke discovered nine remnants in the Large Cloud, and two in the Small.[47] One problem with investigating SNRs in our own Galaxy is that their distances are quite uncertain. The study of these examples in the Magellanic Clouds (which are at known distances) proved a very useful step in determining SNR distances in general.

A decade later, the Mount Stromlo work on Magellanic Cloud SNRs received new impetus when X-ray observations from space—notably by the *Einstein* X-ray Observatory—became routinely available.[48] By combining X-ray, optical and radio data, Don Mathewson, Mike Dopita, Ian Tuohy and Vince Ford (together with colleagues elsewhere) developed effective new techniques for recognising previously undiscovered remnants in the Clouds.[49] This led to the publication of catalogues listing 27 SNRs in the Large Cloud and eleven in the Small—more than double the number known previously.[50] As a result of their research, the team deduced that the Large Cloud experiences a supernova event about every 275 years on average, and the Small Cloud one every 800 years—roughly in proportion to the mass of stars they contain. Thus astronomers of the present generation are indeed fortunate to have witnessed the February 1987 explosion in the Large Cloud (see Chapter 10).

One of the remnants studied by Dopita, Mathewson and Ford—N86 in the Large Magellanic Cloud—became quite celebrated at the time. Its appearance through the telescope bears a striking resemblance to a gnarled caricature that had been created by the *Canberra Times* cartoonist, Pickering, to depict a former federal Attorney General, Lionel Murphy (by then a Justice of the High Court). Taking something of a liberty, the trio of Mount Stromlo researchers nicknamed N86 the 'Lionel-Murphy SNR' when they published their work in the *Astrophysical Journal*.[51] News of this naturally leaked out and, to their great relief, both the cartoonist and Justice Murphy were delighted. The *Canberra Times* seized the moment with a rare astronomical–political cartoon on the front page.[52] The judge ordered a large mounted photograph of the 'Lionel-Murphy SNR' to display in his chambers.

☆

By far the most ambitious instrumental development on the national astronomical scene during Olin Eggen's directorship was the building of the Anglo-Australian Telescope (AAT)—in which Eggen played a central, and at times tempestuous, role. Gascoigne, Proust and Robins have written a definitive history of the AAT's creation.[53]

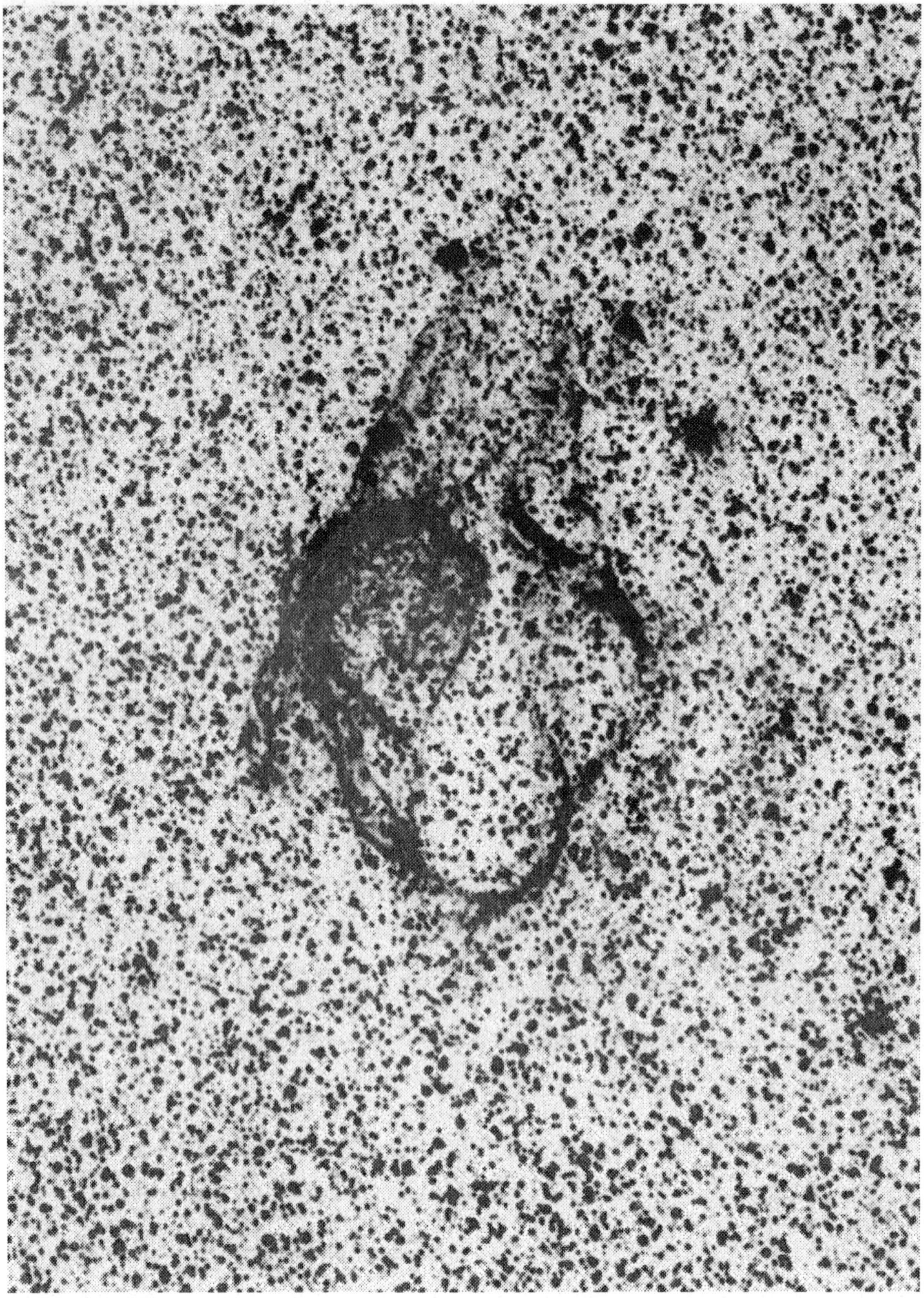

The Lionel-Murphy supernova remnant Image: Mike Dopita, Don Mathewson & Vince Ford

Although official government approval for the project was still proving elusive when Bart Bok left Mount Stromlo, it was achieved about a year later. On 14 April 1967, the Australian government signalled to the British government that it was prepared to join it as an equal partner in building and operating a large optical telescope in Australia, at an estimated capital cost of A$11 million.[54] Several conditions were stipulated: that the telescope be located at Siding Spring; facilities there should be shared with the ANU; a joint controlling body with equal national representation should be set up; the operational control should be in the hands of a person chosen by this controlling body; and observing time should be shared equally by the partner nations. The British accepted these Australian conditions within the month, although the drafting of a formal document took another two years. The final signing was on 25 September 1969.[55] Australian legislation was still to follow since the telescope was to be sited there; no corresponding British legislation was required. The *Anglo-Australian Telescope Agreement Act* 1970 received royal assent on 1 September 1970.[56]

Rather remarkably, the venture itself proceeded apace without waiting for this enabling legislation, or even for the signing of the formal bi-national agreement. Immediately after Australia indicated its willingness to participate, a Joint Policy Committee (JPC) was established, with the intention that it metamorphose into the continuing control authority once all the legalities were completed. The JPC would become the Anglo-Australian Telescope Board (AATB). The JPC first met in London in August 1967.[57] It comprised one government member and two scientific members from each nation—Ken Jones (of the Department of Education and Science), Taffy Bowen (Chief of CSIRO's Radiophysics Lab) and Olin Eggen represented Australia; Jim Hosie (Science Research Council), Herman Bondi (King's College London) and Richard Woolley (Astronomer Royal) represented the United Kingdom. Fred Hoyle (of Cambridge University) replaced Bondi after the first meeting.

After such a long gestation period, it is not surprising that the Australian and British astronomers were keen to proceed as rapidly as possible. At its first meeting, the JPC appointed a Project Manager, Michael Jeffrey. Tragically, Jeffrey died just over a year later.[58] Jeffrey took up his appointment in early 1968 and set up a Project Office in Canberra, which eventually grew to have a staff of about two dozen. Even earlier, however, a Technical Committee consisting of John Pope and Roderick Redman from Britain, and Ben Gascoigne and Herman Wehner (both Mount Stromlo staff members) from Australia had been appointed to oversee preliminary developments. Following the formal commencement of the project, Pope and Wehner were engaged as full-time senior engineers, while Redman and Gascoigne became 'astronomer-advisers' in the Project Office and, subsequently, to the Board.

Design work soon led to actual expenditure.[59] The primary mirror blank was bought in late 1969, and a contract for the mount was placed in August 1970—all before the enabling Act had received royal assent! Besides Eggen, Gascoigne and Wehner, several other Mount Stromlo staff members became involved from time to time in a variety of roles.[60] For instance, Alex Rodgers chaired an important working party, established in 1971, to provide technical input from astronomers on matters of the telescope pointing and instrumentation.[61]

It is a matter of history that all this activity had a resoundingly successful outcome. The Anglo-Australian 150-inch is one of about ten instruments of similar diameter built in the 1970s and 1980s.[62] It is widely acknowledged to have been the best and the most productive.

Between the post-funding euphoria and the acclaim born of ultimate success, however, a storm struck which clouded relationships among astronomers on both sides of the world for a considerable number of years. It was all about the control and management of the telescope.

Looking back on events, it seems quite astonishing that the AAT project was able to progress as far as it did before the enabling legislation was enacted. It is even more astonishing that the Act, when eventually passed, and the bi-national Agreement appended to it, addressed the issues of control and management only in the broadest terms. Most astonishing of all is that the little the Act and the Agreement did say was at variance with the initial joint recommendation of the Royal Society and the Australian Academy to their respective governments.

Their submissions specified, *inter alia*, that: 'A permanent staff consisting of a director, scientific and other personnel is to be employed'.[63] This was sensible advice given that an equally funded, bi-national telescope was to be built, and that there was no appropriate pre-existing Anglo-Australian entity which could take the new facility under its wing.

The Act, on the other hand, noted that the AAT was to be built at a site where there was already an observatory—operated by the Australian National University—and appeared to advocate that *no* independent management infrastructure should be established. Article 4 of the Agreement contains clauses such as: 'The arrangements for the provision by the University of facilities and services for the purposes of construction, operation, and maintenance of the telescope shall be such as are agreed upon between the Telescope Board, and the University'.[64] Taken as a whole, Article 4 could well be read as enjoining the

Telescope Board to deal with the University alone. Olin Eggen, the University's senior officers and Ken Jones read it in exactly that way, and successive Australian Ministers of Science were similarly persuaded. Even Richard Woolley, although a British Board member, saw this as a desirable management structure.

With two such diametrically opposed management concepts enshrined in the historical and legislative background to the telescope, it is not surprising that something of a fight should develop when, in March 1970, the JPC eventually gave its attention to the matter—and a battle there certainly was. It would become an extremely bitter one and last more than three years. A full and well-documented account of events occupies the whole of Chapter 8 in Gascoigne, Proust and Robins' history of the AAT.[65]

Little by little, the advocates for an 'independent' management structure won the day. A significant milestone in the debate occurred in April 1972, when the AATB convened an extraordinary meeting, on neutral ground in La Jolla, California, to decide its own attitude once and for all. Margaret Burbidge, who had succeeded Woolley as Director of the Royal Greenwich Observatory, and thus as a British Board member, proved to be a strong supporter of an independent structure, and was quite persuasive. Nevertheless, there was no unanimity and, in a tense meeting, the AATB resolved the matter by taking a vote, an expedient it had never had to adopt before. The proponents of an ANU-based management—Eggen and Jones—lost four votes to two.

Another crucial occasion occurred on 11 July 1972, when the head of the Commonwealth Department of Education and Science, Sir Hugh Ennor, convened a meeting of seventeen senior Australian astronomers to discuss the issue, following which he ascertained their views in confidence. All but three informed Ennor that they favoured the AAT being run by an independent staff. One, in fact, indicated that Eggen's vigorous espousal of the ANU case had done more to alarm than to persuade him.[66]

The ANU mounted one final rearguard action when the Australian government changed in late 1972. This delayed, but did not prevent, the inevitable. The new federal Minister for Science, Bill Morrison, initially inclined towards the ANU's position but was persuaded against it following another gathering in Canberra of the nation's senior astronomers. He finally signalled the Australian government's acceptance of the AATB's majority position in June 1973. Olin Eggen's stance on the matter remained unchanged. He responded to this final defeat by resigning his membership of the AATB on 23 August 1973.

Perhaps anticipating the inevitability of having to deal with an independent institution owning a major telescope located on the ANU's site at Siding Spring,

Eggen had, in early 1973, arranged for the ANU to set up an Observatory Services Unit (OSU).[67] The Mount Stromlo academics remained in the Department of Astronomy but all other Observatory staff were transferred to the OSU. Eggen himself was the only person common to both entities, being the effective head of both (although the Vice-Chancellor had ultimate responsibility for 'the conduct and management of the OSU'). This was Eggen's way of ensuring that any formal inter-institutional negotiations at Siding Spring became his personal responsibility, and perhaps of ensuring that the other Stromlo astronomers would be freed from any such involvement, so that they could continue scientific collaborations with colleagues at other institutions unhindered by recriminations resulting from any inter-institutional disputes that might develop.

The management battle resolved, the AAT Project went forward to completion, with the Prince of Wales performing the official opening on 16 October 1974. Two Mount Stromlo staff members headed the vital commissioning phases of the Project—Herman Wehner, who became the fourth and final Project Manager from July 1973, and Ben Gascoigne, who served as Commissioning Astronomer for eighteen months from January 1974 on full-time secondment.[68] The telescope became available for general use on 28 June 1975.[69]

The Anglo-Australian Telescope at Siding Spring Observatory Photo: MSO

The AAT management debate was not entirely a power struggle between the British and Australian astronomical communities. Such tensions certainly existed, but there was also a considerable blurring of national allegiances. Only six of the eleven astronomers holding fellowship of the Australian Academy of Science at the time of the storm were born in the antipodes.[70] The other five had come to Australia from Britain.[71] Indeed, there was a period (somewhat later, in 1975–77) when both of the Australian scientific members of the AATB (Street and Wild) were born in *Britain*, and one of the two on the British side (Massey) was born in *Australia*! Some expatriate British-born astronomers had become quite 'Australian' in outlook and interest, but others continued to see things through English eyes—both then and later. On the Australian side of the fence, there was a certain post-colonial wariness born of personal experience. It has been remarked of the then ANU Vice-Chancellor, Sir John Crawford, that he 'was very pro-Australia. He had lived through a period of Australian history when British views, especially in foreign policy, had influenced Australian policy to British advantage. He was first and foremost a politician'.[72]

Of comparable importance to the Australian–British dimension of the debate were some bitter tensions that existed within the Australian astronomical community itself. There were, at the time, three pre-eminent centres of research activity—Mount Stromlo in the optical domain, CSIRO's Radiophysics Division in the radio and Sydney University in both. Sydney had emerged as by far the strongest astronomical centre within the state universities. In the 1960s it had launched three major endeavours: the Molonglo Observatory where Bernie Mills built his new cross telescope; Chris Christiansen's synthesis telescope at Fleurs; and Hanbury Brown's optical intensity interferometer near Narrabri.[73] Both Mills and Christiansen had moved to Sydney University from Radiophysics, and Hanbury Brown had been recruited from England by the dynamic head of the University's School of Physics, Harry Messel.

All three sides of the Stromlo–Radiophysics–Sydney University triangle harboured lingering tensions born of past tussles over funding and/or scientific territory. In the case of Stromlo and Radiophysics, the unease generated by Woolley's 'radio astronomy' foray of 1951 (described in Chapter 5), while considerably assuaged during Bok's time, still lurked below the surface. In the case of Radiophysics and Sydney, Chris Christiansen and Bernie Mills had left the CSIRO largely as a result of its decision to fund a major dish telescope at Parkes and a radio-heliograph at Culgoora, at the expense of competing proposals for next-generation cross-type instruments. On the third side of the triangle, the fact that Mount Stromlo had, in 1957, become part of the ANU's

Institute of Advanced Studies cast something of a shadow over its relationship with Sydney, as with all state universities. When the ANU was born, many in the existing universities resented the creation of its Research Schools, seeing them as elitist centres with privileged access to research funding—indeed, many still do.

A significant sub-theme in the dispute involved access to observing time on the AAT. The ANU position was seen as a blatant attempt by Mount Stromlo to obtain privileged use of the facility and, as the debate proceeded, Eggen's intransigence resulted in this becoming as important a factor for the other Australian astronomers as it was for those in Britain. Indeed, the whole affair had one wry consequence within the Australian astronomical community—its members (except of course those at the ANU) achieved a notable degree of harmony and concord. Mount Stromlo became, for a time, the place that everyone else loved to hate.

Even after the telescope management issue was finally decided, the mutual mistrust between Mount Stromlo and other Australian centres persisted for some time. On 16 October 1973, Ennor convened yet another meeting of senior astronomers to discuss arrangements for the allocation of Australia's half of the telescope time. Some of the ANU attendees, fearful that an Australian Time Allocation Committee with an entirely local membership might be ill-disposed towards Stromlo, proposed some involvement of international experts from 'neutral' countries such as the United States. In the event, an entirely Australian composition was agreed, but with strict institutional affiliation limits—the ANU was to nominate three members to a committee of seven. John Bolton, of CSIRO, was the inaugural Chairman.

Another repercussion concerned the choice of a location for the AAT's management headquarters—now seen as being better sited at a major population centre rather than on or near Siding Spring itself. On the face of it, Canberra seemed an obvious choice. Even though the AAT staff were now to be completely independent of the ANU, they would surely have more in common, scientifically, with Mount Stromlo astronomers than with those at any other centre. Furthermore, the AAT's Project Office had been in Canberra. Such considerations did not prevail. In 1974, the AATB decided to locate the headquarters on a temporary basis at a site adjacent to the CSIRO's Radiophysics Division in Marsfield, Sydney. The whole operation, telescope and headquarters then became known as the Anglo-Australian Observatory (AAO).[74] The Marsfield location was made permanent by a Board decision of March 1976.[75]

The aftermath of the storm clouded relationships between Mount Stromlo and the AAO itself for many years, at least at the political level—although less so, fortunately, at the scientific. Although the ANU is home to the largest and most prestigious group of optical researchers in the country, no Stromlo astronomer was invited to serve on the AAT Board between Olin Eggen's resignation from it in August 1973 and Jeremy Mould's appointment in January 1994. ANU representation was achieved, instead, by the appointment as Board members of successive Directors of the University's Research School of Physical Sciences (and Engineering).

On a happier note, it is a matter of history that the organisational structure eventually adopted for the AAT has worked every bit as successfully as has the telescope itself—a further example of the principle that any designated entity can grow into the task of managing a major scientific facility, provided it is given adequate funding and resources to do so. A sufficiently beneficent environment will always attract sufficiently capable people.

As to scientific progress at Mount Stromlo, astronomers weathered the storm—and they certainly lost no time in putting the AAT to good use in addressing exciting astronomical opportunities. Their time-allocation fears proved unfounded: Stromlo applications were notably successful, winning over 70 nights per year, on average, during the decade 1980–89, far more than any other single institution in either country.[76] Their work on the AAT has also been notably productive. A decade into the AAT's operational life, the names of Stromlo staff and students were appearing on well over a quarter of all the scientific publications resulting from its use.[77]

Having control of, and responsibility for, a major national facility such as the AAT can be a mixed blessing. Providing the user support expected by guest observers always comes at a cost of staff time that might otherwise have been devoted to research. Indeed, one could well reflect that the pre-eminent position that Mount Stromlo enjoys in Australian astronomy today (see the final section of Chapter 11) is due in no small measure to its never having been saddled with the burdens of national facility operation. It has sometimes been remarked that not the least important of Olin Eggen's contributions as Director was in *losing* the battle for control of the AAT!

☆

Preoccupied as he was with AAT events, Eggen made no addition to the suite of ANU telescopes that he inherited, but he did initiate a campaign to acquire a

The 'Reynolds' Telescope after its 1970 refit
Photo: MSO

1.5-metre photometric instrument—for which there was a demonstrable need. The century-old Great Melbourne Telescope, recommissioned as a 50-inch in the early 1960s, was rapidly deteriorating and had to be decommissioned in 1973.[78] It had provided more than ten years of sterling service for photoelectric photometry and infrared observations, and its breakdown left a noticeable gap in the Observatory's resources. On more than one occasion during Eggen's latter years, the University made submissions to the Australian Universities Commission for funding to purchase a new 50-inch, but without success.[79]

Two of the existing telescopes were, however, substantially refurbished. A newly appointed engineer, John Hart, supervised a mechanical reconstruction of the 30-inch 'Reynolds' reflector during 1969–72, also re-equipping it for spectroscopic use.[80] In the late 1970s, improvements were also made to the 74-inch: these included a new top end and modifications to the primary mirror cell.[81]

Eggen was fully aware of the necessity of keeping Mount Stromlo's auxiliary equipment at the technological cutting edge, as is evident from the introduction in one of his early annual reports:

> This continuing emphasis on auxiliary instrumentation is one of the keys to the international reputation enjoyed by Mount Stromlo. By world standards our major telescope—the 74-inch reflector—is a small one. There are two similar telescopes in Canada, one in Japan, a British one in South Africa, another in France and several more. None of the observatories possessing these instruments [has] the scientific stature of Mount Stromlo. The reason is a simple one. They are insufficiently funded to build or purchase auxiliary equipment (spectrographs, photometers, spectrum scanners and the like).[82]

Eggen backed up this rhetoric in his management decisions. The Observatory's technical staff complement trebled during his directorship, and the astronomers were given almost free rein in developing new instruments, sometimes in collaboration with visitors invited to Mount Stromlo specifically to tap their technical expertise. Almost every annual report announced the commissioning of some new piece of equipment—an image-tube spectrograph for the 74-inch (developed by Kent Ford of the California Institute of Technology); a new spectrograph for the 74-inch (bought from Boller & Chivens); an *échelle* grating system for the *coudé* spectrograph (Mike Bessell and Harvey Butcher); infrared photometers for several wavelength domains (Harry Hyland); a 33-channel spectral scanner (Alex Rodgers, Herman Wehner and Peter Rudge); an electronic camera (Gerry Kron); a multi-mode Fabry-Perot interferometer for nebula work (Mike Dopita); and several more.[83] In retrospect, it might have been better had the rein extended by the Director been a little less free. Several of the instruments developed during this period were notably unsuccessful, and saw little service on the telescopes.

In his battles on behalf of his Observatory, and the ANU in general, Olin Eggen enjoyed quite remarkable support in the University's senior echelons—Crawford had been as single-minded in the whole AAT affair as had Eggen himself. Sir John's term as Vice-Chancellor concluded in March 1973, shortly before the matter was finally resolved. More staff changes were looming in the Chancelry—Eggen's other staunch ally, Ross Hohnen, retired in 1975—and there could be no assurance of Eggen forming an equally close alliance with the new regime. Indeed, there were early indications to the contrary, and Eggen began to give thought to his own tenure as Director.[84] He started looking for another position, and told some of his senior colleagues that they could not expect to have him around much longer. His directorship concluded on 30 September 1977. His departure was typical of his rather reclusive lifestyle—he left without fanfare; one day he was in his office as usual, the next he had left the country.

Eggen immediately took up an appointment as a staff astronomer at the Cerro Tololo Inter-American Observatory in Chile, where he worked productively for the last 21 years of his life. One of the terms of his contract was a month's annual leave with return travel to his place of recruitment. He invariably availed himself of this entitlement to revisit Mount Stromlo, and took a lively interest in all that was going on there. He seemed to retain a real affection for his erstwhile observatory. In fact, he wrote in 1993 that, while he regarded as 'irretrievable' the exciting ethos at most of the places where he had worked during their heyday, 'Australia still retains its allure'.[85] With the passing years, he mellowed and acquired a sociability for which he had had little time during the busy years of his directorship. He looked forward to his regular annual contacts with former Stromlo colleagues, and invariably took the opportunity to relive old times and old battles with Ross Hohnen.

Eggen's contract at CTIO specified no retirement date, and he never did retire. In his latter years he used to boast that his years of service-related pension entitlement now exceeded his annual salary, so his institution no longer had any incentive for persuading him to go! During his last few years he did move to a part-time appointment, but astronomy still remained his life. He suffered a heart attack in 1998 on arrival at Canberra Airport for his annual visit to Mount Stromlo. He died without regaining consciousness in Canberra Hospital on 2 October.

chapter nine

A time to build: the Mathewson era, 1977–1986

FOLLOWING OLIN EGGEN'S DEPARTURE ON 30 September 1977, Don Mathewson became the Acting Director at Mount Stromlo. Being asked to serve during an *interregnum* is not intended to convey any messages regarding the appointment to follow, but the University's choice on this occasion was an interesting one since Mathewson was one of two Professorial Fellows on the staff at the time, and the other, Alex Rodgers, had held professorial rank for five years to Mathewson's three.[1] In the event, Mathewson and Rodgers became, respectively, the fifth and the sixth Directors.

The University went about the task of making the substantive appointment in its usual careful way. In particular, the Electoral Committee had before it Sir Richard Woolley's 'little essay' entitled 'Notes on the Directorship of Mount Stromlo'.[2] In addition to listing the general qualities one would hope to find in the director of any major scientific institution, Woolley highlighted the specific skills that would be needed in dealing with the 'particular problems' then facing Mount Stromlo. Tactful handling of relationships with Radiophysics, the Anglo-Australian Observatory and with the Australian and British astronomical communities in general received specific mention.

Don Mathewson had graduated in Science from the University of Queensland. He joined CSIRO's Radiophysics Division in 1955, and spent a total of eight years there, in two spells, before coming to Mount Stromlo in 1966. He also spent time in Europe working at some of the leading radio observatories of the day. During 1958–60 he undertook his PhD studies at Manchester, using the recently commissioned 250-foot Jodrell Bank telescope to make observations of the Andromeda Galaxy, the Coma Cluster of galaxies, Cygnus X and the Cygnus Loop.[3] Somewhat later, in 1971, he took a year's study leave at Leiden, once again working with a new, ground-breaking instrument—the Westerbork Synthesis Radio Telescope.[4]

In the early 1950s, Chris Christiansen had pioneered the idea of using the Earth's rotation to sweep an array of radio antennae around each other so as to synthesise a giant effective dish.[5] Christiansen had verified the principle by producing a simple image of the quiet Sun using the CSIRO array at Potts Hill, although it took him six months' numerical labour to complete the Fourier calculations involved. Fifteen years later, by contrast, radio astronomers were dealing with these onerous computations by harnessing increasingly powerful digital processors to their telescopes, and general-purpose radio synthesis arrays came into operation at a number of overseas centres. Mathewson worked with the Dutch array of twelve 25-metre dishes, two of them moveable along high-precision tracks. The instruments' high spatial resolution meant that radio observations could be turned into detailed visual images for the first time.[6]

Mathewson and his Leiden colleagues used the new instrument to study a number of galaxies.[7] Their observations of the Whirlpool (M51) proved particularly interesting. Both the radio and the optical images of this galaxy show spiral patterns, but the two do not coincide. The radio emission is brightest in regions where the optical picture exhibits dark dust lanes. The radio 'light' is also polarised. This work provided convincing evidence that most of the radio emission in the galaxy is coming from high-velocity electrons moving through magnetic fields *associated with its spiral structure,* rather than from *within its supernova remnants*, the alternative conjecture at the time.[8]

Thus, by the late 1970s, Mathewson could boast an impressive career. He had acquired a familiarity with some of the most innovative instruments in the world. He was equally at home in the optical and the radio domains of the subject—perhaps as much so as any other astronomer at the time. Significantly, too, he had a close past association with CSIRO's Radiophysics Division, something of an asset at a time when inter-institutional relations between Mount Stromlo and the Division were at a low ebb.

There was a large field of applicants for the Directorship, and the selection process took about eighteen months. In the end, the Electoral Committee concluded it need look no further, and Mathewson was confirmed in the post. He took office as Mount Stromlo's fifth Director on 5 April 1979 for an initial period of seven years. This was the first occasion on which the Stromlo directorship had gone to an internal appointee, and the first when a fixed term was specified—the University's traditional practice of appointing departmental heads until retirement age had by then become a thing of the past.

Mathewson's approach to his own research was an effective one. He liked undertaking massive studies, some of which took him several years to complete. The observational techniques he used varied greatly from project to project, and his habit of interspersing radio and optical work continued throughout his career. His studies of the magnetic field structure in our Galaxy, and of supernova remnants in the Magellanic Clouds (described in the previous chapter), are good examples of the type of project he relished.

At the time he became Director, Mathewson's chief research interest lay in observing neutral hydrogen gas in and around the Large and Small Magellanic Clouds. As we have noted, these two small galaxies lie about 170,000 light years away from us, just seven times as far as does the centre of our own Milky Way. They are of much smaller mass than the Galaxy, and can best be thought of as satellite systems. They provide a useful stepping stone for investigations of the Universe at large, such as Ben Gascoigne's calibration work for the Cepheid distance scale (described in an earlier chapter).

Each of the three galaxy systems—the Milky Way and the two Clouds—feels the gravitational pull of the other two and, given their proximity, it is likely that they have suffered tidal distortions. Gerard de Vaucouleurs first suggested this in the early 1950s, when he used a twin Aero-Ektar camera in his 'backyard observatory' at Mount Stromlo to obtain wide-angle photographs of a large area of the sky in the neighbourhood of the Clouds. He detected some extended streams of luminous matter emanating from them, seemingly tidal in origin.[9]

The discovery of the 21-centimetre line of neutral hydrogen opened up the possibility of obtaining *radio* evidence of such tidal interactions as well. Since galaxies condense from hydrogen–helium gas clouds in the first place, the distribution of any gas remaining after star formation carries useful information about a galaxy's past history. Obtaining 21-centimetre observations of the Magellanic Clouds had been an early priority for the CSIRO's new 64-metre dish at Parkes, and the maps made by Jim Hindman, Frank Kerr, Dick McGee and Jan Milton, all of Radiophysics, soon revealed a bridge of gas linking the two Clouds.[10] This removed any possibility of considering them simply as isolated galaxy systems.

In 1973, Mathewson, his student Martha Cleary and John Murray of Radiophysics made a startling discovery. Mathewson had an idea that a 60°-long filament of neutral hydrogen reported by Northern Hemisphere observers might extend southwards and join up with the inter-Cloud bridge of gas.[11] To test this, he and his colleagues used the smaller (18-metre) dish at Parkes to obtain a 21-centimetre picture of a large area of the sky in the region.[12] His hunch proved correct. The team's observations revealed a tapering ribbon of hydrogen, which starts at the Clouds and stretches some 100° across the sky lying almost along a great circle. They named this giant filament the Magellanic Stream. Their survey also found some isolated patches of gas along the projection of the filament on the other side of the Clouds. Taken together, these hydrogen features extend over nearly 180°. Higher-resolution observations of the Stream itself, made with the 64-metre dish, showed that it too incorporates a string of six distinct cloudlets.[13]

The origin of the Magellanic Stream was keenly debated for many years. One theory is that it was indeed caused by the tidal stripping of material from the Clouds during a close encounter with the Galaxy. Tidal tails and bridges are apparent in some other interacting galaxy pairs, but the chief problem with this explanation is that gravitational forces operate equally well on both stars and gas. Thus the Stream might be expected to contain a stellar component accompanying its hydrogen, but careful searches have failed to find any associated stars.[14] Another possibility is that the gas of the Stream has been stripped from the Magellanic Clouds by the ram pressure of a very tenuous gaseous halo in our Galaxy. Mathewson favoured this kind of mechanism for a time, and three Mount Stromlo researchers—Gerhardt Meurer, Geoff Bicknell and Bob Gingold—computed a plausible model of the process.[15] Increasingly detailed models using both the tidal and the ram-pressure theories were still being published twenty years after the Stream's discovery.[16] Whatever the mechanism that caused the Stream, however, there can be no doubt that gravitational forces have been responsible for some of the other observational features in the region. The 'bridge' between the two Clouds, which does contain both stars and gas, is a prime example.[17]

Adding complexity to the picture, the whole main body of the Small Cloud is now believed to have been severely disrupted by tidal forces. Evidence for this first began to emerge with the earliest 21-centimetre observations. Since neutral hydrogen emits at a specific wavelength, the observed radio spectral profiles provide information on the line-of-sight motion of the gas.[18] In the case of the Small Cloud, the velocity picture turned out to be complex, with many of the 21-centimetre profiles showing multiple peaks betraying the existence of several distinct masses of hydrogen lying in the same direction. A double peak, for instance, could mean that one is observing an expanding shell of gas, the material on the near side appearing at a different velocity from that on the far side. Using the higher resolution obtainable with the Parkes 64-metre dish, Jim Hindman had shown, in the 1960s, just how pervasive this complexity was. His conclusion at the time was that the Small Cloud contained at least three such expanding shells of neutral hydrogen.[19] Subsequent studies showed that similarly complex velocity patterns are exhibited by other types of object in the Cloud as well—emission nebulae, hot young stars, planetary nebulae and Cepheid variables.

The Cepheids are particularly useful since, as previously noted, one can obtain an estimate of the distance of a Cepheid from the Earth by measuring its pulsation period and its brightness. In the late 1980s, Mathewson, Vince Ford and Vis Visvanathan made such measurements for 161 Cepheids in the Small Cloud.[20] The results demonstrated that the disruption the Cloud has suffered is much more severe than was previously thought. In fact, the Cloud has been stretched out, in one dimension, by a factor of five or more. Unfortunately, the dimension concerned is that lying along our line of sight, which makes it difficult for us to sort out the exact details of the Cloud's chaotic structure.

The clear bifurcation of most of the 21-centimetre line profiles suggests that the Small Cloud may have actually been torn apart, with the two fragments being superimposed from our point of view. If so, the more distant component (hiding behind what we see as the Small Cloud) already has a name—Mathewson called it the 'Mini-Magellanic Cloud'.[21] The Cloud's disruption must have been caused by something quite dramatic; it has been suggested that a near-collision between the Large and Small Clouds about 200 million years ago was responsible.[22] Mathewson considered it possible, in fact, that this same collision could have played a role in the creation of the Magellanic Stream.[23]

In the late 1990s—a quarter of a century after Mathewson's discovery of the Magellanic Stream—a new generation of observers seems finally to have settled the debate as to its origin. Once again they used the venerable Parkes dish, recently refitted with a multi-beam top end enabling small 'pictures' to be made of the neutral hydrogen distribution in the sky. Mary Putman, Ken Freeman and Brad Gibson were Mount Stromlo members of a large team undertaking an all-sky survey with this new and improved equipment.[24] Their work revealed a faint arm of gas emerging from the Magellanic Clouds on the *opposite* side to the Stream.[25] Features of this arm were also apparent in the original Mathewson, Cleary and Murray survey. Such a 'leading' arm is a clear indication that gravity is at work. Gravitational forces commonly displace matter both towards and away from the attracting body, as evidenced by the ocean tides as they wash around the Earth with a bulge on either side. There is no way in which ram-pressure stripping could cause a similar effect—there is no such thing as a 'forward' wake. So the Galaxy's preferred method of destroying its near neighbours seems now to have been identified—gravity is the culprit.

☆

Mathewson's directorship was also marked by considerable entrepreneurial flair. In fact, he showed his intentions in this regard while still only the Acting Director. On 17 November 1977, just seven weeks after Eggen's departure, he arranged a half-day seminar at the Observatory for 40 senior officers and scientists of the ANU. The afternoon commenced with three presentations highlighting the research interests of the Department, followed by six talks detailing its instrumental aspirations.[26] Light refreshments in the Yale–Columbia Dome brought the proceedings to an agreeable conclusion. Mathewson's equipment wish-list was a comprehensive one embracing optical and infrared array detectors, computers, a micro-densitometer and—most expensive of all—renewal of the bid to the Universities' Commission for a 1.5-metre reflector. The scope of the Acting Director's ambitions for the Observatory could hardly have been lost on the Vice-Chancellor, Tony Low, nor on his deputy, Ian Ross, who carried the Chancelry's brief on matters of astronomical policy.

The seminar also included a presentation on TEST, a project that Mathewson had recently initiated with his erstwhile home institution, the Radiophysics Division.[27] TEST was a dual-purpose acronym, initially denoting a Two-Element Synthesis Telescope, but later, hopefully, a Three-Element one. Mathewson developed his idea in the context of a Radiophysics proposal seeking about $5 million in government funding to build an Earth-rotational synthesis telescope at Parkes. The existing 64-metre dish was to be augmented by up to a dozen smaller antennae, the whole array being called the Australian Synthesis Telescope (AST). The proposal was still at a preliminary stage, there being considerable debate within Radiophysics as to the optimum antenna configuration. Furthermore, it faced stiff funding competition from several rival bids.

Mathewson's TEST proposition was that, while waiting for the AST bid to be finalised, the two existing Parkes dishes (the 64-metre and the 18-metre) should be converted into a simple two-element synthesis system. He pointed out that such an approach would yield valuable high-resolution observations of southern galaxies almost immediately, while also providing hands-on experience with the aperture-synthesis data-analysis techniques that would be required to operate the AST. Mathewson's proposal was accepted by the CSIRO. Given the somewhat frigid relationship between Mount Stromlo and Radiophysics at the time, the University officers attending the Stromlo seminar would doubtless have been interested to note this early indication of a thaw.

There was a second prong to Mathewson's proposition: TEST II, which involved the early construction of a 15-metre prototype AST antenna, which could then be linked to the existing two to increase the power of the simple array. Mathewson flagged an expenditure of $400,000 over the following three years, to be met jointly by CSIRO and the ANU. The 64-metre and 18-metre dishes were indeed successfully linked to create TEST I. Aperture-synthesis observations began in 1980, and two years later were taking up a sixth of the time on the 64-metre dish. Detailed kinematic studies were made of the neutral hydrogen gas in two external galaxies, NGC 1313 and 4945, as well as of an interesting region of the Milky Way.[28] NGC 1313 appears to have undergone a collision with a smaller satellite galaxy.

In the event, the AST itself was never built. On becoming Chief of Radiophysics in 1981, Bob Frater decided that the proposal had been severely under-budgeted and it lapsed, together with any implementation of TEST II.[29] It was soon superseded by a much more ambitious scheme which would lead to the development of the ground-breaking Australia Telescope at Culgoora, commissioned in 1988.[30]

The initiatives flagged by Mathewson in November 1977 were much in evidence during the following twelve months. By the end of 1978, while still acting in the post, he was able to report that many of his initial goals, and some additional ones, had already been attained:[31]

- The TEST I project was proceeding apace, the ANU and CSIRO each having contributed three full-time positions to create a project team. The rail-track of the Parkes 18-metre dish had been doubled in length to 820 metres for the new mode of operation.
- Working relationships between the ANU and the AAO were much improved. A multi-institutional site committee had been established to deal with interface problems at Siding Spring, and the annual AAO scientific symposium had been held at Mount Stromlo in March. More to the point, Stromlo astronomers were making great use of the AAT itself, having been awarded 43 per cent of the Australian share of observing time for the year.
- The Observatory Services Unit put in place by Olin Eggen had been dismantled, and all of the Observatory's activities recombined under a single administrative structure.[32] Special budgetary arrangements continued, however, and a new Astronomy Policy Advisory Committee was set up, chaired by the Deputy Vice-Chancellor, Ian Ross.

- A Digital Equipment Corporation VAX 11/780 had arrived in August, the first of this new type of computer installed anywhere in Australia, or, indeed, in any observatory in the world. A Perkin Elmer PDS micro-densitometer arrived in September.
- A bid made to the ANU's Large Equipment Fund for a high-resolution echelle spectrograph for the 40-inch telescope at Siding Spring had been successful, and an order placed with Boller & Chivens.[33] (The instrument would be commissioned in 1980.)
- The Observatory's public outreach activities, all but discontinued by Eggen, had been resumed with some vigour. The 9-inch 'Oddie' refractor on Mount Stromlo was made available to senior students at Canberra's secondary colleges for use on class projects. Portable telescopes played a similar role in the Coonabarabran Shire, and a successful Open Day had been held at Siding Spring. Vince Ford was making regular radio broadcasts with the blessing of the Observatory.

It had indeed been a busy year.

There remained one pressing problem, however, in that the Universities Commission funding for a 1.5-metre telescope to replace the moribund 50-inch was still proving elusive. Mathewson adopted two strategies. The first was simply to refurbish the 50-inch. Work on this commenced in 1978, but its urgency diminished with the success of his second, more ambitious, plan, and all renovation activity ceased a few years later.[34]

Mathewson launched his second scheme on 20 January 1978, when he convened a working party of Australian astronomers to discuss the building of an instrument 'so novel in design that it belongs to the next generation of telescopes'.[35] It was to be called 'The Universities' Telescope', since his hope was to involve a substantial number of institutions, thus spreading the funding burden across many budgets. The telescope's novel design was an important financial ingredient. It was to incorporate 'three major advances', each of which would serve to reduce its cost:

> The Universities' Telescope . . . will have a 3-metre thin mirror (diameter to thickness ratio ~ 20:1), an alt-azimuth mounting, and a building which rotates in azimuth with the telescope. These innovations lead to a much lighter than usual telescope, of simple mechanical design, which could be made in Australia.

Mathewson estimated that the cost of the telescope and building would be \$3 million, compared with \$12 million for an instrument of the same diameter built in the conventional way. Another meeting followed, on 3 May 1978, at which representatives of the other universities, the CSIRO and the AAO were brought together to explore funding possibilities. The multi-institutional approach soon languished, however, when it became apparent that the other universities simply had insufficient disposable funds. It was then that Mathewson, in a characteristically audacious move, decided that the ANU should proceed alone. He simply changed the project's title to the 'Advanced Technology Telescope' (ATT).

His initiative was especially bold in that it came at a time of increasing financial stringency within the ANU itself. The expansionary days of the 1960s were well past. In 1974, the Whitlam government had assumed financial responsibility for the whole of the Australian higher education sector. This necessarily diluted Commonwealth interest in, and support of, the University which had, until then, been the only one with a specifically national mandate.

The Vice-Chancellor, Tony Low, was well aware that times had changed. Addressing a graduation ceremony in September 1975, shortly after assuming office, he warned that the University's years of expansion were over; no longer would it be able to rely on the allocation of new resources to achieve its hopes and dreams.[36] His foreboding proved well justified. The University's budget declined by over 10 per cent, in real terms, between 1975 and 1981.[37] The new austerity was felt particularly keenly in the Institute of Advanced Studies, the collection of research schools which gave the University its unique reputation. Fundamental research, by its very nature, depends on exploring ideas and, should they prove promising, opening up whole new areas of study. While the ANU had been able to operate in this way during the glorious 1960s, nothing new could now be funded without a painful contraction of some existing activity.

It hardly seemed an opportune time for an incoming Mount Stromlo Director to propose that the University acquire an expensive new telescope but Mathewson was undeterred. One small ray of hope shone through the gloom. In Low, the University had a Vice-Chancellor who was well aware that freezing all new endeavours would sound the death-knell of the University's research ethos. In 1980, to the consternation of some research school directors, he initiated an annual 1 per cent tax on each school's budget to create a pool for funding some new projects. It was by taking skilful advantage of opportunities such as these that Mathewson advanced his 2.3-metre initiative, and he proved adroit at winning the confidence of successive Chancelry administrations. The 1977 seminar was not the only time he invited the University's senior officers to Mount Stromlo to report progress, and to ensure that his aspirations remained clearly in focus.[38]

Mathewson was prepared to scale down his telescope's diameter somewhat from the original 3 metres. He became aware that Owens-Illinois, of Ohio, still had in stock a 95-inch Cer-Vit mirror blank. Cer-Vit is an extremely low-expansion glass-ceramic material developed by Owens-Illinois. The firm's blanks were used in about a dozen large telescopes built during the 1970s; several, including the AAT, were 4-metre-class instruments.[39] The blank still in stock was the last sizeable one remaining from the production run, and Mathewson bought it on 24 May 1979. Implementing his 'thin-mirror' concept, he promptly had it sliced in two through mid-plane. He sold the other half the following year for a price not dissimilar to that which he had paid for the entire blank. It proved a most profitable transaction.

A site for the 2.3-metre was chosen after careful site testing of several locations along the Siding Spring ridge during 1979–80. The mountain is rather prone to strong winds—a factor that had not become apparent in the testing of the 1960s—and wind measurements were added to the usual seeing tests in making the final selection.[40]

The contract to figure the telescope's mirrors was won by Norman Cole, of Tucson, in 1979. The finished primary is 2.3 metres in diameter and has a focal ratio of f/2.05. It is 22 centimetres thick at the edge and 16 centimetres in the centre. By the conventions of the time, this was astonishingly slim. Telescope designers traditionally relied upon a mirror's thickness to minimise the effects of flexure on the precise optical surface as the telescope is moved about to point to objects in different parts of the sky. With such a thin mirror, the optical figure must be maintained by a sophisticated mechanical mirror-support system.

An altitude-azimuth mount (as in a gun turret) was the second novel feature of the 2.3-metre design. Traditionally, an optical telescope has an 'equatorial' mount, one of its two axes being aligned parallel to that of the Earth so that a star may be tracked by driving this 'polar' axis at one revolution per day. It was the advent of digital computers that made the change to 'alt-az' mounts inevitable. The computer's over-riding role in telescope control had, in fact, been recognised some years before—in the design of the AAT for instance, even though it was still built with a conventional mount. Since a computer has no difficulty simultaneously controlling motion about two axes, no compelling reason remains for retaining a polar axis. The advantages of an 'alt-az' arrangement are considerable: the whole mechanical design is much more straightforward, and flexure can be better quantified and compensated. Most important of all, when operating on a limited budget, the mechanical structure can be made smaller and lighter, as can the whole building, since the telescope moves in a more confined space when tracking stars. The alt-az arrangement was not entirely new at the time the 2.3-metre was built; it had already been used for giant radio dishes like those at Jodrell Bank and Parkes, as well as for the Multi Mirror Telescope on Mount Hopkins.[41]

The 2.3-metre telescope's third unusual feature is its rotating building. A traditional telescope dome has a hemispherical roof that rotates to line up the open slit with the top end of the telescope tube. In the case of the 2.3-metre, however, financial imperatives dictated that the building enclose the telescope as snugly as possible. This was achieved by having the whole structure rotate in step with the instrument. The building travels on four two-wheel bogies which together support a total mass of 200 tonnes.

While each of these three innovative concepts had been implemented previously, the ANU's 2.3-metre telescope was the first to combine them in the design of a single instrument. It was also the first to employ control computers in such a comprehensive way. Nearly every function of the telescope's operation, its instrumentation, the data collection and the observing environment (including monitoring the weather) became part of an integrated system that placed at the astronomer's fingertips an observational tool of unprecedented sophistication.

The entire management of the 2.3-metre project and most of the design work took place in-house. Indeed, a great deal of the mechanical fabrication was also done in the workshops of the Observatory and the ANU's Research School of Physical Sciences. John Hart was the engineer principally responsible for the telescope design, Herman Wehner for that of the rotating building and Gary Hovey for the control systems—although it is somewhat invidious to cite just three names in a project which involved virtually all of the Observatory's engineering and technical staff. Ted Stapinski, the Observatory's Chief Electronics Engineer, had a great deal of involvement in the early design work, and Ed Simmonds, formerly of the ANU's Buildings and Grounds Division, served as Site Engineer at Siding Spring. Many contracts were let—two dozen for specialised optical and engineering tasks, and over 100 for the supply of goods or services—three-quarters of them going to Australian firms.[42] The only major mechanical elements fabricated overseas were the high-precision drive (made by the Swiss company Maag) and the telescope bearings (by Rothe Erde of Germany).

The project's successful outcome is a lasting tribute to Mathewson's vision, and to the skill and dedication of the Mount Stromlo technicians, engineers and scientists who brought that vision to reality. It is a tribute, too, to the capability of Australian industry. The acquisition cost was $6.6 million.[43] This figure should be read in context, however, since the project also consumed the lion's share of the Observatory's in-house engineering and technical effort over a five-year period. Nevertheless, the telescope was built for only a fraction of what a 2.3-metre instrument of conventional design would have cost.

In parallel with the telescope's development, the design and fabrication of its auxiliary instruments were also taking place at Mount Stromlo. There are three focal positions (two at Nasmyth and one at Cassegrain), and each has its own distinctive device. Their completion lagged that of the 2.3-metre by some years—not surprisingly, given the enormous demands placed upon the technical staff by the major project itself.

First to be completed was a double-beam spectrograph, designed by Alex Rodgers in 1982, and built by Peter Conroy, Gabe Bloxham and their team.[44] It is an extremely efficient device, which separates the star's light into two spectral domains so that observations at longer and shorter wavelengths may be obtained simultaneously. The instrument occupies one of the Nasmyth foci. It came into operation in September 1987, and has always been much in demand by observers.[45] The second Nasmyth focus is also occupied by a versatile device developed by Rodgers and his colleagues, a combined imager and low-resolution spectrograph commissioned in October 1991.[46]

The 2.3-metre Cassegrain focal position was reserved for infrared equipment, and its secondary mirror permits rapid 'chopping' between the program star and a neighbouring patch of clear sky. This switching technique is used to help overcome a major problem associated with observing at infrared frequencies—the fact that a high background of radiation is given off by the telescope itself, and everything else at room temperature in the surroundings. It is for this reason that infrared instruments also need to be cryogenically cooled to very low temperatures. An existing infrared spectrometer designed by Harry Hyland and Terry Jones was deployed at the Cassegrain focus initially.[47] With the advent of infrared array detectors, however, a much more sophisticated device was developed by Peter McGregor and his team.[48] McGregor obtained his PhD at Mount Stromlo in 1981, and returned to join the staff in 1984; he is now one of the world's leading instrumental astronomers. His 2.3-metre instrument was named *CASPIR*—Cryogenic Array Spectrograph-Imager—and came into service in 1993. It now rivals the double-beam spectrograph in its popularity with observers.[49] *CASPIR* promptly achieved worldwide fame by producing one of the most spectacular images of the Comet Shoemaker-Levy 9 collision with Jupiter on 19 July 1994.[50]

The control room of the 2.3-metre Advanced Technology Telescope
Photo: MSO

The Prime Minister of Australia, Bob Hawke, inaugurated the 2.3-metre ATT on 16 May 1984. It was a notable occasion, drawing to Siding Spring a large audience of national, shire and university dignitaries, representatives of the contracting firms, and astronomers. The Prime Minister was effusive in his praise of the project, describing in some detail the innovative features of the telescope, and highlighting the contribution made by Australian industry and, of course, by the University itself.[51]

After the formal opening ceremony, Mathewson escorted the Prime Minister on a tour of the new telescope, and of the Observatory at large. Included was a display of the double-beam spectrograph which featured a descriptive model since the spectrograph itself was still some way from completion. It fell to Alex Rodgers to explain its key features, including the ability to record light from the two spectral regions simultaneously. Rodgers and Hawke were acquainted from student days at University House and, as the Prime Minister was listening to the description of the spectrograph's two arms, he suddenly noticed that, on the display, these had been neatly labelled 'LEFT' and 'CENTRE-LEFT'.[52] He seemed not entirely pleased to have the enjoyment of his day out interrupted by this reminder of what awaited him back in the corridors of power! Nevertheless, enjoy himself he undoubtedly did. Refreshments at the Siding Spring Lodge brought his visit to an agreeable conclusion, and the staff there revelled in the opportunity of entertaining a Prime Minister of such charisma and charm.

Mount Stromlo's 2.3-metre telescope received an Engineering Excellence Award (1985) from the Canberra Division of the Institution of Engineers, Australia. It also attracted worldwide attention, influencing the design of many larger telescopes built elsewhere.[53] The 1990s, in fact, would witness an astonishing revolution in optical telescope technology. The 4-metre reflectors which had been at the cutting

The 2.3-metre Advanced Technology Telescope at Siding Spring Observatory
Photo: Bob Cooper, ANU

edge in the 1970s and 1980s were suddenly overshadowed by a spate of new telescopes of at least twice that diameter, incorporating a plethora of radical design features. For all its modest size, the ANU's 2.3-metre can be seen as having offered, in the early 1980s, a visionary foretaste of things to come.

☆

Mathewson accompanied the building of the 2.3-metre—which alone would have made his directorship noteworthy—with an equally ambitious international venture in space astronomy. The late 1970s saw great advances in that field, with new-generation satellite telescopes such as the International Ultraviolet Explorer (IUE) and the *Einstein* X-ray Observatory (both launched in 1978) revolutionising the study of high-energy astrophysics.[54] Even more exciting was the imminent prospect of an orbiting 2.4-metre reflector—the Hubble Space Telescope (HST), as it was finally named. The HST's launch by Space Shuttle was expected as early as 1983.[55]

At the time, Australia had fallen well behind in terms of space endeavour, although it had been an active participant in the decades following the Second World War by virtue of a joint British–Australian launching program at Woomera.[56] Indeed, in 1967 Australia had become only the third nation, after the Soviet Union and the United States, to launch a satellite from its own territory. The Woomera program had been wound down during the 1970s, however, and in 1980 the range was closed. Australian scientists were reduced to seeking 'guest observer' collaborations on the new international facilities, rather than having access 'by right' as did researchers from nations actively involved in developing and launching satellites.

Nevertheless, Mike Dopita and Ian Tuohy had managed to take advantage of the IUE and *Einstein*, participating in observing programs which shed new light on the physics of supernova remnants.[57] Long interested in these objects, Mathewson joined Dopita and Tuohy in making a comprehensive survey of the remnants in the Magellanic Clouds, including their X-ray properties as measured by instruments on *Einstein*.

More importantly, however, Mathewson decided, early in his directorship, that Australia should re-capture a major participatory role in space, and he threw himself into organising such a campaign with almost evangelical zeal:

> I believe that Australia is snoozing in blissful oblivion to the Space Age, and that we are soon in for a rude shock when Space Age science and technology developed by other countries puts us into the scientific third world.
>
> The commencement of the Space Shuttle flights by NASA later this year offers the greatest single leap in our ability to observe the distant universe since the days of Galileo . . . the Shuttle will be the focus of the world's most exciting astronomical research for the remainder of this century.
>
> Unfortunately Australian astronomers will find it very difficult to get observing time on Space Shuttle facilities, no matter how good their scientific programs are. Competition for

> telescope time is fierce and, unlike Europe, Australia has no Principal Investigator status with NASA or financial input into the Shuttle program. It is very ironic, at overseas conferences on the uses of space telescopes, to see US and European astronomers putting forward their top priority projects, some of which have been initiated by ground-based work by Australian astronomers![58]

The vehicle Mathewson chose for his crusade was a NASA facility called STARLAB, which was conceived as an adjunct to the HST. Although STARLAB's telescope was to be much smaller in diameter (only 1 metre), it would have a field of view 100 times larger than the HST—about the area of the full Moon. It was designed to fulfil a major survey role—identifying the objects interesting enough to warrant more detailed study—and would complement the HST in much the same way as the 1.2-metre Schmidt cameras complement the 5-metre and 4-metre reflectors at Palomar and Siding Spring.

NASA planned to use the Space Shuttle to deploy STARLAB on to a Space Platform—another NASA concept at the time—orbiting 500 kilometres above the Earth's surface. The telescope would observe for six to twelve months at a time, and then be returned to Earth for refurbishment and updating between missions. STARLAB had a planned working life of perhaps ten such missions spread over twenty years.

Just as Mathewson became Director, the STARLAB project had stalled—the victim of financial cutbacks to the American space program. Mathewson saw this as a golden opportunity. A financial contribution large enough to enable NASA to reinstate the project, but still manageable for a country of Australia's size, would provide a perfect *entrée* to the big league of space research. He travelled to NASA's Goddard Space Flight Centre for a STARLAB meeting to present Australia's case for entry into the program, and received a warm response. Thus encouraged, he promptly organised a two-day Space Astronomy Workshop, held at Mount Stromlo on 25–26 February 1980, with Bob O'Connell, the STARLAB Project Team Leader, as a keynote speaker. Although other Australian astronomers with space interests made contributions at the meeting, its focus was unambiguous. It became the launch pad for an Australian STARLAB campaign.

On the eve of the Workshop, Canada signalled an interest in joining the discussions and hastily sent a delegate. A tripartite proposal soon emerged, in which the United States would be responsible for developing the Space Platform and for meeting all costs associated with the first two launches, Canada would build the telescope, and Australia the instrument package—planned to be a versatile assembly having both imaging and spectroscopic capabilities. A Design Study Group, chaired by Mathewson, was set up to refine and cost the proposal.[59] In December 1980 it reported an estimated capital cost of A$49 million for the telescope and instrument package, to be shared equally by Australia and Canada.[60] Each country's contribution to the annual operating costs was estimated at A$1.4 million. Observing time on the telescope would be shared equally by all three participating nations.

In parallel with this planning activity, Mount Stromlo was busily enhancing its space credentials. More overseas observing collaborations were initiated, and in March 1980 a Programmable Command Station for the British Ariel-6 satellite was set up on Stromlo. This was operated by Ian Tuohy in return for participation in the satellite's X-ray astronomy program. The Observatory was thus positioning itself to host the anticipated Australian STARLAB Project Office, although it was always envisaged that the bulk of the instrument package fabrication would be contracted out to Australian industry.

The proposed tripartite division of responsibilities for STARLAB suited Mount Stromlo admirably. The most technically demanding feature of the instrument package was to be a large, two-dimensional, solid-state detecting device called a Photon Counting Array (PCA). The Observatory already had an active involvement in PCA research and development, led by Alex Rodgers and Ted Stapinski.[61] The STARLAB package presented exciting new challenges, however, since the telescope's half-degree field of view required extremely large PCAs—comprising some 60 million picture elements. Furthermore, 'space-qualification' places rigorous constraints on the design and fabrication of all equipment to be flown; it must be robust enough to survive the launch and to operate in the temperature and pressure extremes of the space environment.

Mathewson's campaign for government funding for the project proceeded remarkably speedily, although an initial submission to Cabinet in 1981 was unsuccessful. Faced with this rebuff, he organised a 'STARLAB Industrial Symposium' in December of that year. This elicited the enthusiastic support of more than a dozen high-tech companies operating in Australia. Notable among these was the French space firm MATRA, with which Mount Stromlo signed a joint project agreement. The CSIRO and the Defence Research Centre at Salisbury in South Australia also became involved. Such a level of industrial support was hard to ignore, and the 1982 federal Budget allocated an initial $3.3 million for instrument package design studies.[62]

In 1983 a formal 'Letter of Intent' was signed by the three participating nations as the project moved into top gear.[63] An Australian Project Office was established at Mount Stromlo—in demountable accommodation—with Ted Stapinski as Manager, and a new company, AUSPACE Pty Ltd, was registered to act as the lead contractor.[64] An 'Initial Design Review' was held in June, at which Australia's progress report on the design of the instrument package received the unqualified endorsement of its two international partners.

Given the level of planning and organisation which Mathewson devoted to it, the STARLAB project deserved a better fate. In early 1984, Canada withdrew its funding following a government decision to focus on more commercially-oriented space applications such as communications and remote sensing. Approaches to France and Japan failed to fill the gap, and the tripartite STARLAB agreement was abandoned. Less than two months later, however, Mathewson himself received an approach—from NASA and the European Space Agency (ESA). They were planning yet another orbiting facility at the time—the Far Ultraviolet Spectroscopic Explorer (FUSE), also known as *Lyman*—and they wished to explore the possibility of incorporating the Mount Stromlo PCA detectors.[65] This proposal was quickly taken up, and a letter of intent signed in December. Thus the Australian element of the abandoned STARLAB project proceeded almost unabated, as did its funding by the Department of Science and Technology—later renamed the Department of Industry, Technology and Commerce (DITAC). In total, the Commonwealth government awarded over $8 million to the Stromlo enterprise during 1982–85. The Observatory was perhaps fortunate in having Barry Jones as Minister for Science and Technology at the crucial time of Canada's withdrawal. His enthusiastic support for STARLAB had been evident as early as 1981 while his party was still in opposition.[66] John Button, Minister for Industry and Commerce, who took over the formal supervision of Australia's involvement in the project, was equally enthusiastic.

By the mid-1980s, Mathewson's crusade for a renewed Australian space enterprise had taken on a life of its own, with Mount Stromlo's endeavours generating considerable related activity in industrial, government and research circles. In March 1984, a 'National Space Symposium' was held, co-sponsored by the Department of Science and Technology and AUSPACE. Later in the same year, at the instigation of another space enthusiast, Ken McCracken, the CSIRO set up an Office of Space Science and Applications.[67] Barry Jones continued to take a keen interest in all these developments and, at his suggestion, the Australian Academy of Technological Sciences and Engineering (who had just elected Mathewson a Fellow) commissioned a special report, *A Space Policy for Australia*, which appeared in 1985.[68] This led to the formal establishment of an Australian Space Program the following year, under the direction of an Australian Space

Board. The Australian Space Office (ASO) was set up in 1987, with a modest inaugural budget of $5.25 million and a mandate to support projects 'of national significance'.[69]

With a national policy now in place, Mount Stromlo felt it was able to transfer some of the hands-on tasks to other entities. In August 1986, the Observatory staff working in the FUSE–*Lyman* Project Office were transferred to AUSPACE Pty Ltd, as was the DITAC funding contract.[70] Thus the focus of technical development moved away from the mountain, although Stromlo astronomers continued their involvement in overall project direction and in providing scientific advice.

Paradoxically, the success of Mount Stromlo's endeavours in heightening the national awareness of space issues militated against the achievement of its own astronomical objectives. By 1988 the Australian Space Program, reflecting a shift in government policy, had narrowed its focus in much the same way as the Canadian program had a few years earlier.[71] Purely scientific endeavours fell from favour, and support for Australia's involvement with FUSE–*Lyman* was all but terminated. When ESA also withdrew its funding in November of that year, the tripartite project agreement suffered the same fate as its predecessor.[72]

In view of the effort that had been put into an Australian detector system, however, a 'space-qualification' mission was arranged, to demonstrate the credentials of the fledgling national space industry if nothing else. A package called *Endeavour*—consisting of a 10-centimetre binocular reflector fitted with ultraviolet-sensitive PCAs—was carried aboard its namesake Space Shuttle in March 1995. The mission obtained images of the nearby galaxies M31 and M101 and of the star formation region 30 Doradus in the Large Magellanic Cloud. Mike Dopita did the scientific interpretation of these observations.[73] This work was, however, but a pale shadow of Mathewson's original STARLAB vision. It is interesting to speculate on what the outcome might have been had the project proceeded. The detectors required were much larger than any yet attempted, and their production would have been extremely challenging. Perhaps this was another case where defeat at the political level may have been something of a blessing in disguise.

Be that as it may, Mathewson's vigorous advocacy had a major impact on Australian science policy at the time. The structures established in the mid-1980s to support the Australian Space Program have changed considerably over the years, but the high-tech capability of Australian industry so clearly established in the STARLAB project persists.[74] For example, AUSPACE played a key role in the development of FedSat—launched from Japan in December 2002—

the first Australian-built satellite to be deployed in over 30 years.[75] As for access to astronomical space facilities, in spite of Mathewson's forebodings Stromlo researchers have proved more than adept at winning observing time by the sheer merit of their scientific proposals—as will become apparent in the next two chapters.

☆

Although the 2.3-metre and the STARLAB ventures placed great demands on staff time and energy, they were far from being the only Observatory accomplishments during the Mathewson years. Even in the instrumental domain, some additional smaller projects were accommodated. One was the relocation, in 1981, of the Uppsala Schmidt Camera from Mount Stromlo to Siding Spring so as to escape the ever-increasing light pollution of the growing Canberra.[76] Moving some of the larger telescopes had also been mooted in earlier years, but it was eventually decided that the Uppsala was the only instrument for which the benefits of relocation outweighed the costs.

Most importantly, however, Mount Stromlo's research productivity went from strength to strength. One of Mathewson's proudest boasts was yet another doubling in the annual output of scientific papers during his term of office.[77] The breadth of research endeavour also increased markedly with the passing years—so much so that a comprehensive coverage is impossible in a general history of this nature. The following account is thus highly selective, attempting merely to provide something of the flavour of the research interests of the times.

Stellar astronomy remained a high priority, both in observational and theoretical work. The focus was firmly on evolutionary development within our Galaxy as a whole, with studies of individual stars contributing clues to that intriguing story. For instance, stellar clustering is a common phenomenon, and it seems likely that most stars—if not all—are born in clusters—it seems that it is only when a protostellar gas cloud is sufficiently dense and massive that a star-forming collapse can occur. Any newborn cluster will gradually lose stars into the general surroundings, and the 'looser' the cluster, the more rapidly this happens. Our Sun no longer appears to be part of any particular cluster, and one might ask whether any other stars in our neighbourhood are fellow escapees from the star-formation event that made the Sun. One telltale piece of evidence would be an extremely similar chemical composition of the stellar surface layers. In 1981 Mike Bessell found just such evidence for the two companions in α Centauri, the star nearest to our Solar System—just 4.4 light years distant.[78] Combining this result with what is known about the stellar motions, Bessell concluded that it is indeed likely that this double star and our Sun were born in the same event.[79]

Catching clusters of stars in the act of being born from their protostellar clouds of gas and dust is not an easy task. For one thing, the gravitational collapse process is a fairly brief one compared with the star's subsequent nuclear fuel-consuming lifetime. Furthermore, protostars commence their collapse as very cool objects, emitting little light in the visible part of the electromagnetic spectrum. A third difficulty is that the clouds out of which the stars are born contain large quantities of dust particles, and these screen much of the action from our view. The second and third of these difficulties are best addressed by observing, not in visible light, but in the infrared; cool objects shine more brightly at longer wavelengths, and infrared light is less affected by the dust absorption. Thus, by working at these longer wavelengths, astronomers are gradually identifying various types of object as playing a role in the birth process.

The 'Bok globules', for instance, have long been suggested as stellar nurseries. These small, round, dark nebulae are seen in silhouette at many points along the Milky Way. They were first commented on by Bart Bok and Edith Riley in 1947, and were much studied by Bok towards the end of his career.[80] One fine example is found in the Southern Coalsack.[81] Called Globule 2, it is a tiny object just one light-year across (a quarter of the Sun's distance from the next nearest star).[82] In the early 1980s, Terry Jones, Harry Hyland and their colleagues made an exhaustive infrared study of Globule 2, including photometry, spectroscopy and polarimetry of the stars lying behind it.[83] They found its gas and dust to be extremely cold—just 10° above absolute zero—and, rather surprisingly, it seems to have a magnetic field. The globule is very opaque at its centre; only one part in 20,000 of light entering the object emerges from the other side. Its total mass is about ten times that of the Sun. It remains uncertain whether Globule 2 is currently collapsing to form one or more protostars, but it is evident that no real stars have formed yet, since this infrared study would certainly have detected any that had.

Peter Wood made several illuminating theoretical studies of Mira variables, a class of stars whose periodic light modulation is caused by an intrinsic stellar pulsation. The prototype of the class—**o** Ceti, or Mira ('Wonderful')—is believed to have been the first variable star ever noticed (by the German theologian Fabricius in 1596). The Miras are distinguished by having very long pulsation periods—eleven months for Mira itself—which indicates that they are physically extremely distended, with radii perhaps 500 times larger than our Sun.

Lowmass stars like the Sun go through the Mira phase shortly before they die, and exhibit many intriguing properties as they do. A good deal of mass loss occurs at that time—sometimes resulting in the formation of a spectacularly beautiful planetary nebula—and the remaining stellar core shrinks a hundred-fold in size to become a white dwarf. Wood's model calculations provided explanations for many of the intricate physical processes involved.[84] Mike Dopita and his colleagues, who observed nearly 50 planetary nebulae in the Magellanic Clouds, threw further light on the dynamics of the mass loss process.[85] They were able to measure the velocity of the gas expansion in these objects, and related their results to the properties of the dying star, particularly its mass.

Although astronomical records are generally less newsworthy than those on the sporting field, they sometimes provide important new insights into previously unexplored domains of knowledge. In 1984, Mike Bessell and John Norris found the red giant star CD –38°245 to have a lower 'heavy element' abundance than any previously observed—it was deficient by a factor of 30,000 with respect to our Sun.[86] This implies that CD –38°245 was born soon after our Galaxy itself was formed, so the abundances of the elements in its atmosphere can tell us much about the nature of the primordial gas in the Universe, and about the kind of stars formed in the very first 'generation' of star formation. There is some evidence that these were quite massive objects, perhaps 500 times heavier than the Sun. This record stood for nearly twenty years, but Mike Bessell and his colleagues have recently found a star, HE 0107–5240, with an even lower 'metal' abundance, down by more than a factor of 100,000 relative to the Sun.[87]

A record of a totally different kind was established in 1982, when Bruce Peterson was part of a team using the AAT to study quasars. They found one (PKS 2000–330) to be the most distant and most luminous object yet discovered in the Universe.[88] The new redshift record, 3.78, was especially interesting since it was thought at the time that quasars might simply not occur at redshifts much greater than 3.5.[89] This work led to the realisation that the suspected 3.5 limit was an artificial one—a so-called 'selection effect'—caused by the techniques that had previously been used to hunt for distant objects. Discovering the latest 'most-distant-object-in-the-Universe' always brings a certain sense of accomplishment, even if one knows that one's record is likely to be short-lived—there have been several holders of the title since PKS 2000–330. At the time of writing, the honour belongs to a galaxy called HCM 6A which has a redshift of 6.56.[90] Such a large value indicates that the galaxy's light we are now seeing started on its journey quite early in the Big Bang expansion of the Universe—its size then was only 13 per cent of what it is today.

Such research on external galaxies and quasars assumed an increasingly high profile at Mount Stromlo during the Mathewson years. As in the stellar research, observational and theoretical studies went hand in hand. Peter Quinn, a graduate student, used computer modelling to investigate the dynamical effects of collisions between galaxies.[91] His calculations showed that such events can sometimes cause the two interacting bodies to merge into a combined system resembling, in appearance, the class of galaxies known as 'ellipticals'. The edges of the combined galaxy, however, betray the effects of the collision in the form of faint 'shells', rather reminiscent of the traces left by successive waves on a sandy beach. Quinn's theoretical work enjoyed immediate observational confirmation since, at just that time, two AAO astronomers—David Malin and David Carter—were working on elliptical galaxies, obtaining photographs of them which went much 'deeper' than any previously recorded.[92] To everyone's surprise, and Quinn's delight, the normal-looking ellipticals, which had previously been thought to have perfectly smooth outlines, showed just the type of faint outer shells that the modelling explained. This provided some of the best evidence yet obtained that galaxy mergers really do occur in nature.

When quasars were first recognised as being extragalactic objects, it soon became apparent that they emit enormous amounts of energy from extremely small volumes of space. For instance, the first quasar ever discovered, 3C 273, was found to be 100 times as luminous as our entire Milky Way galaxy, yet all of this energy is coming from a region scarcely 100 times the size of our Solar System.[93] It is now widely accepted that a quasar must be powered by a massive black hole residing in its core, perhaps a million times, or even some hundreds of million times, the mass of the Sun. Although the emission mechanism is complex, and there is still considerable debate as to the details, it undoubtedly involves the release of gravitational energy as surrounding matter falls into the black hole. Entire stars may be swallowed by the monster, but they are unlikely to be digested whole; rather, they will be tidally disrupted by the enormous gravitational fields they encounter as they fall inexorably to their doom. Two theoreticians recruited to the Mount Stromlo staff in the early 1980s, Geoff Bicknell and Bob Gingold, calculated three-dimensional numerical models to investigate the likely consequences of a star like the Sun passing close by such a massive black hole.[94] They found that the stellar material was squeezed to the point where shocks and nuclear reactions developed, adding to the complexity of the physical processes involved.

The most fundamental goal to which any astronomer can aspire is to understand the origin, development and fate of our Universe as a whole. During the 1980s, Visvanathan made notable progress in scaling the expansion of the cosmos. As we saw in the previous chapter, Edwin Hubble in the 1920s demonstrated that the further away from us a galaxy lies, the faster it seems to be receding from us.[95] The constant of proportionality between distance and speed—the Hubble Constant—is named in his honour. Its reciprocal has the units of time, and it may be thought of as giving the 'expansion age' for the Universe. Determining the value of Hubble's Constant is a real challenge since it involves observing objects that are extremely remote. In the early 1980s its value was uncertain by a factor of two, in spite of the best endeavours of some of the world's most famous astronomers. In fact, a strenuous debate was raging between those who had measured the Universe as being about 10,000 million years old, and those who believed it to be twice that age. In a substantial infrared study of 52 galaxies in six clusters, ranging in distance from 55 to 250 million light years, Visvanathan obtained an expansion age almost exactly midway between the two then-fashionable values—earning him few friends in either camp![96] History—and the Hubble Space Telescope—have vindicated his result, as we shall see in Chapter 11.

☆

The Mathewson years were marked by growth in other respects as well. Staff–student numbers reached an all-time high—rising from 105 in 1978 to 140 in 1985.[97] The establishment of the STARLAB Project Office had accounted for an additional sixteen people, while Mathewson's vigorous promotion of the Graduate Program saw student numbers double (from thirteen to 27). Mathewson was a keen advocate of education in astronomy and, in 1981, as President of the Astronomical Society of Australia, initiated a summer school for astronomy students which has become a popular annual event. The relocation of the STARLAB staff to AUSPACE Pty Ltd in 1986 initiated a decline in the Observatory's population which has continued slowly since then. The 2001 *Annual Report* listed 106 people at year's end, 21 of whom were graduate students.[98]

At its peak, Mount Stromlo sat uncomfortably as a department within the ANU's Research School of Physical Sciences (RSPhysS). Its staff–student complement and its budget were not dissimilar to those of the smaller research schools themselves.[99] There was, however, an unwillingness to grant the Observatory full school status, due not least to the financially straitened climate of the time. Mathewson proposed that Mount Stromlo and Siding Spring

Observatories break away from the School and become an independent 'centre' within the Institute of Advanced Studies. With the strong support of Ian Ross, he worked for several months with Jane Flecknoe from the Chancelry to draw up guidelines for the separation of 'The Observatories'. The new entity would be given its own one-line budget, but it would not have the same membership rights on the University's academic boards and funding committees. The separation from RSPhysS was amicable enough in principle, although the financial details were vigorously debated since both the Observatories and the School were overspent.[100] Agreement was eventually reached, and the ANU Council approved the separation in October 1985. The new arrangements took effect from 1 January 1986.[101]

With so much achieved, both scientifically and administratively, there seemed little doubt that Mathewson would continue as Director at the conclusion of his initial seven years in office (April 1986). On 7 November 1985, however, he convened a meeting of the academic staff and, to general surprise, announced that he would not be seeking a second term—his directorship had been a strenuous one, and he wished to have an opportunity to resume his research career full-time.

This necessitated some quick thinking within the Chancelry. The new administrative arrangements were due to be implemented within a matter of weeks, and procedures for making a directorial appointment had scarcely been thought through. Nevertheless, a week to the day after Mathewson's announcement, the Deputy Vice-Chancellor, Ian Ross, convened another staff meeting to address the various issues arising.[102] He made some important points. The foremost was that the Vice-Chancellor had decided to take advantage of the directorial *interregnum* to hold an external review of the Observatories by a small, high-profile committee—two eminent astronomers from overseas, one from elsewhere in Australia and a head (or former head) of one of the ANU's other research schools. No search for a new Mount Stromlo Director was to be undertaken until the review report was received, hopefully by mid-1986. Instead, an Acting Director would be appointed from among the Stromlo staff, to serve for perhaps a year from the conclusion of Mathewson's term. Alex Rodgers was subsequently invited to undertake this task. Ross also signalled that the Chancelry wished to look carefully at the Observatories' financial arrangements before a new Director was appointed, and that in the meantime 'expenditure may be subject to controls not normally used'.

Paying tribute to the entrepreneurial achievements of Don Mathewson's directorship, Ross said:[103]

Consider the four Directors of MSSSO since it became part of the ANU: Woolley, Bok, Eggen and Mathewson. Their styles and their contributions could hardly have been more different. Also the times in which they worked were different. That last is important—in good times all things are possible, and some of the people we now remember as larger than life became that way because their creative period coincided with a period when opportunities were opening up. A lesser number can be identified for whom there was no apparent age of opportunity, but who nevertheless created opportunities not foreseen. Don Mathewson became Acting Director just at the time that the anchors were thrown out on University growth. Nevertheless his eight years in office have already transcended everything that has gone before, with the 2.3-metre telescope, STARLAB, FUSE–*Lyman*, and the updating of the Observatories' instrumentation (to mention just a few matters).

His times have admittedly been robust ones—at times a bit too robust. I am sure that it has all been personally exhausting for him. So exhausting that I can understand why he might want to stand down. Personally I wish he would continue, but I cannot—nor can the Vice-Chancellor—reasonably ask him to do more.

That is why we are meeting today, with common cause of making a new, but the best possible, future for the Observatories.

☆

In early 1986, Comet Halley afforded Mathewson one last entrepreneurial opportunity as Director. Staff at the Observatory were quite astonished by the intense public interest generated by this once-every-76-years interloper from the outer regions of our Solar System. It was known in advance, and well advertised, that this return would offer quite a poor spectacle compared with Halley's previous two appearances, in 1835 and 1910. The dispositions of the Earth and the comet in their respective orbits were such that at no time during Halley's inward passage to perihelion in late 1985, or on its return journey in early 1986, did the two come closer to each other than 63 million kilometres. By contrast, the Earth was separated from the comet by only one-third of this distance in 1910, and by only one-eighth in 1835. Nevertheless, the public imagination was on fire, and Mount Stromlo astronomers were kept busy for weeks responding to invitations to talk about the comet in schools, clubs, groups and on radio broadcasts.[104] A commercial airline even arranged sightseeing flights.

The comet's closest approach occurred in early April 1986, and Mathewson issued numerous invitations to VIP groups to visit the Observatory over a period of about a week. Many parliamentarians, embassy staff and leaders of commerce and industry accepted. The visits had to be carefully handled, since almost everyone came with the expectation of seeing a really fine image of Comet Halley through Mount Stromlo's largest telescope, the 74-inch. This would have been logistically impossible, given the large number of guests, but in any case the small field of view of a large telescope makes it quite inappropriate for observing such an extensive object as a comet. By far the best instrument to use is a good pair of binoculars. The Observatory had many such pairs available (rather more at the start of the week than at its conclusion), but it took a good deal of tact to persuade some VIPs that they were not being fobbed off with a second-rate experience.

The night of Monday 7 April 1986 was noteworthy. The Prime Minister, Bob Hawke, was among the guests. The evening was co-hosted by the Observatory's then major computer supplier, the Digital Equipment Corporation (DEC), and, in addition to the outdoor binocular activities, a special presentation was arranged in the Observatory's lecture theatre. It began with a talk, copiously illustrated with slides describing where comets come from, how they orbit within the Solar System, and the recorded history of Comet Halley in particular. Then came the technological climax of the evening—on-line viewing from the Observatory's 2.3-metre reflector at Siding Spring. In practice, of course, it was a small lens attached to the 2.3-metre that was producing the image, for the reason just explained. The image was digitised at Siding Spring and transmitted via a computer link to a DEC-supplied workstation in the Stromlo lecture theatre. This computer then reconstituted the signal for video display.

Comet Halley duly appeared on the screens, to polite applause. Then came a question from the audience, 'Can you set on anything else?' There was just a hint of scepticism attached. After all, the audience had seen lots of slides that evening; how could they be sure the image before them now was really coming from Siding Spring and not one stored somewhere in the computer's memory? 'What would you like?' asked Mathewson. 'How about Saturn?' 'Sure, we'll just see if it's around at this time of the year.' (Professional astronomers seldom know what planets are currently visible in the night sky, to the disabused astonishment of the general public.) 'It is,' responded the questioner, 'I saw it last night.' Things were getting serious. There followed an urgent conversation across the phone link to Siding Spring. The *Astronomical Ephemeris* was consulted; Saturn's coordinates for that night were keyed in at the 2.3-metre control desk, and a tense few minutes ensued as the telescope moved automatically to the new position and its detector recorded a digital image for transmission down the line. Then, right on cue, Saturn's familiar ringed spectacle appeared on the lecture theatre screens—to the great relief of all of the Stromlo staff involved.

An enthusiastic cheer rang out from the VIP audience. They were convinced. As everyone moved on to the refreshments a feeling of general good humour prevailed. Everyone had had a great night, no one more so than Bob Hawke who was heard to remark what his sentence on the Director would have been had Saturn *not* appeared—'Off with his head!' The Observatory did itself no harm whatever by this intense public relations effort; and when the DEC company officials took their leave somewhat after midnight, they left their workstation behind. That was the deal.

Following the completion of his term of office, Mathewson did indeed return to research work with renewed vigour. Together with Vince Ford and his student Markus Buchhorn, he used the Siding Spring telescopes and the Parkes dish to undertake a photometric, spectroscopic and radio study of 2500 spiral galaxies in the local Universe, in an attempt to identify their bulk motion.[105] The massive database obtained remains much in use by researchers around the world.

After his retirement in 1994, Mathewson continued an active involvement as Vice-President of the International Astronomical Union until his term of office ended in 1997.

chapter ten

Dark matter: the Rodgers directorship, 1986–1992

As Don Mathewson's term ended, the University implemented its foreshadowed review of the Observatory's activities.[1] Alex Rodgers became Acting Director in May 1986, for what would be a thirteen-month *interregnum*. The review itself was carried out by a panel of four—David Craig, a former Dean of the ANU's Research School of Chemistry, Bob Frater, Chief of the CSIRO's Radiophysics Division, Donald Lynden-Bell of Cambridge University, and Allan Sandage of Mount Wilson Observatory. They received copious documentation in advance of a brief, but intense, visit to Mount Stromlo and Siding Spring in June 1986.

Their report was generally complimentary concerning the Observatory's work.[2] It suggested a number of research emphases for the future—observational cosmology, the interstellar medium and radio projects using the Australia Telescope (the aperture synthesis array which the CSIRO was about to commission). It further recommended that Mount Stromlo take the lead in a national Large Optical Telescope (LOT) initiative. On the matter of the directorship, the Committee recommended that 'the new appointment should be from the present tenured staff', citing the availability of strong internal candidates and the tight financial climate at the time. The review report received general endorsement both from within the Observatory and more widely.[3]

Acting on the Review Committee's advice, the ANU Council established an Electoral Committee to select a new Director, which by year's end had issued Mount Stromlo's senior staff with a call for applications.[4] The procedures were not rushed, even though there was considerable support for Rodgers' appointment. Ken Freeman took the unusual step of submitting an application himself, not because he wanted the job (as he made quite explicit in his covering letter), but to ensure he had ample opportunity, during his interview, of urging that the Acting Director be confirmed in the post.[5]

The Electoral Committee met for the last time in May 1987—during a Canberra gathering of the Astronomical Society of Australia—and the ANU Council formalised the appointment on 12 June when Rodgers took office as Mount Stromlo's sixth Director, for an initial term of five years.

Alex Rodgers was born in Newcastle in 1932, and grew up close to the BHP steelworks where his father worked. The only way he could afford to attend university was by bonding himself to the New South Wales Education Department. He studied science at what is now the University of New England, graduating with honours and the H.C. Russell Prize in 1953. His timing was excellent. The ANU's youthful Research School of Physical Sciences was by then enrolling students, and Richard Woolley, as an Honorary Professor, was able to offer ANU Scholarships for study at Mount Stromlo. Rodgers won one of these, which enabled him not only to embark on his chosen career, but also to buy his way out of his teaching bond. He commenced as a graduate student in March 1954, and obtained his doctorate in 1958 with a thesis entitled 'A Photometric Study of the Southern Coalsack'.

Apart from a few periods of study leave, Rodgers would spend his entire career at Mount Stromlo Observatory. After obtaining his PhD, he was appointed, successively, as a Research Fellow, Fellow, Senior Fellow, Professorial Fellow and Professor. No other astronomer has been able to boast such a long and dedicated association with Stromlo.

Some students emerge from their PhD studies with an admiration for their mentors which lasts a lifetime—for others, the reverse is true! Alex Rodgers was one of the former. He held 'Dick' Woolley in the highest regard, and looked on him as a role model throughout his subsequent career—a framed photograph of Woolley had a permanent place on his desk. There were some notable similarities between the two men: both adopted with enthusiasm the 'new' astrophysics which, in the middle decades of the twentieth century, overtook the more descriptive approach that characterised traditional astronomy; both rose to direct observatories; and both were keenly interested in the provision of optical telescope facilities.

It is somewhat puzzling, given Rodgers' upbringing and socio-political leanings, that a 'British establishment' figure such as Woolley should have so won his admiration. He certainly displayed little sympathy for things British in other matters. In the battle for control of the Anglo-Australian Telescope, for instance, he adopted a staunchly Australian stance, and was more inclined than most to view the dispute as one in which 'perfidious Albion' was seeking to impose her will on an erstwhile colony! Neither was Rodgers blind to Woolley's idiosyncrasies—indeed, we are indebted to him for many of the tales which make the Woolley legend at Mount Stromlo such a rich and colourful one. But he was quick to jump to his former Director's defence when others denigrated him, as they sometimes did, and he evinced a personal sense of hurt when people read more than he felt was justified into some of Woolley's more outrageous statements.

Rodgers had a somewhat ambiguous personality; it was hard to predict the stance he would take on any particular issue. He had an affinity with the rebellious elements of society, and a certain larrikin streak—he took great pride in the fact that he had a convict ancestor. By contrast, his invariable habit of wearing a jacket and tie when observing seemed not entirely tongue-in-cheek. Though egalitarian by conviction, he allowed himself to be addressed by many at the Observatory as 'Professor Rodgers', even when such formality had come to seem quite old-fashioned. Though socialist and republican by inclination, he was genuinely delighted to entertain royal and vice-regal visitors to the Observatory.[6] He also enjoyed accepting reciprocal invitations to Government House, although sometimes he would arrive at functions in his rather dilapidated 'second' car, so as to savour the footman's look of disdain as he handed over the keys! While he espoused strong views on many issues, he was wary about becoming too emotionally involved with social causes. Rather, he was inclined to respond to what he saw as society's shortcomings with the quirky sense of humour that made him very entertaining company at the morning tea table.

Humour was quite a feature of observatory life from time to time. It was during Alex Rodgers' term that an *Alternative Annual Report* began to appear, courtesy of the students. Continuing for four editions, it was an irreverent satire of the authentic document which exposed the eccentricities and foibles of staff and students alike. No one was spared. For instance, Ken Freeman's research eminence had, over the years, resulted in his being much in demand as a keynote speaker at international conferences. He enjoyed these events enormously and accepted as many invitations as possible (he still does). Consequently, his overseas absences had become legendary. In a mock listing of staff conference attendance for one year, the *Alternative Report* gave him just a single entry: 'Prof. K.C. Freeman visited Mt Stromlo and Siding Spring Observatory, 23 June, 3.00–6.15pm'.[7]

After completing his PhD, Rodgers took a period of overseas leave, spending a year at the Mount Wilson and Palomar Observatories on a Carnegie Fellowship.[8] There he was immersed in an environment of advanced technology that had propelled American West Coast astronomy into a position of world leadership. He became an enthusiastic admirer of this technical excellence, and formed scientific associations which he developed in later years. One was with Bev Oke of Caltech, who spent a sabbatical leave at Mount Stromlo in the early 1980s, during which he and Rodgers collaborated on the early design work for the 2.3-metre telescope's double-beam spectrograph.[9]

In view of such associations, and of Rodgers' admiration for American expertise, one might have expected him to be an enthusiast for the multinational approach to the acquisition of major new telescope facilities. (Large international collaborations had begun to emerge during the 1970s, and were commonplace by the 1990s.) But this was not the case. In spite of his empathy for the American way of *doing* astronomical research, his stance in matters of astro-politics was one of fierce Australian independence. To the end of his life, he entertained the hope that Australia would go it alone in building a LOT on home soil and, as we shall see, this quest consumed much of his time and energy as Director. Paradoxically, however, this nationalistic stance on the large telescope issue stood side by side with a rather disparaging attitude towards some other landmarks on the Australian astronomical scene. He had little time for the Astronomical Society of Australia, for instance, and published nothing in its *Proceedings*, apart from a half-column article in the first issue. By contrast, he enjoyed being a life member of the Astronomical Society of India and made sure that this distinction was highlighted in his *curriculum vitae*.

During his long career at Mount Stromlo, Alex Rodgers acquired an intimate, first-hand knowledge of every facet of the Observatory's operations. He was at home with everyone who worked there—especially the technical staff—and enjoyed nothing more than escaping from his office for an afternoon and spending it in the workshops, getting his hands dirty with the nuts and bolts of instrument design and development, or in the optical shop, exploring the delights of the 'flight simulator' that had somehow found its way on to Gabe Bloxham's computer.

Rodgers also excelled in student education. He was the Convener of the Graduate Program for many years, and he personally supervised several excellent PhD theses.

Engineering was one of his great loves—both at work, and tinkering with his rather idiosyncratic car at home. Against the advice of John Hart, Mount

Stromlo's Chief Mechanical Engineer, he had bought a second-hand, rotary-engine RO-80 sedan made by the German NSU company—the start of an interesting man–machine relationship that would last for the rest of his life. Among other things, he became quite adept at replacing the car's engine, an operation required on two or three occasions—one of them following a motoring holiday to South Australia from which he, his wife Ruth and the RO-80 all returned by rail! The floor of the Yale–Columbia dome at Stromlo proved a perfect venue for this operation, and for several happy bonding sessions with Hart, another rotary engine enthusiast.

Rodgers held responsibility for the Observatory's instrumentation for many decades, and argued strongly that Mount Stromlo's engineering capability was a vital part of its identity. His first development project was a spectrophotometer, built in the mid-1960s for the 50-inch telescope.[10] Several notable PhD theses were accomplished with this instrument.[11]

The 74-inch telescope was Rodgers' favourite. Installed by Woolley in his final year at Stromlo, it was Australia's major spectroscopic research tool until the advent of the 3.9-metre AAT in the 1970s. Rodgers used it as a test-bed for many of his technical projects. He led Mount Stromlo's migration from the tedious photographic spectroscopy of the 1960s into the semi-electronic age, commissioning a fine new Boller & Chivens spectrograph with a Carnegie image tube in 1969.[12] This revolutionised the Observatory's spectroscopic capabilities, making the 1970s an extremely productive period for the 74-inch. He then took the Observatories into the new era of fully electronic detectors. There were some choices to be made between competing technologies, each with advantages and disadvantages. Rodgers chose the photon-counting route and built systems (PCAs) which were in constant use until about 1990.[13] By that time, charge-coupled devices (CCDs) were seen as offering clear technical advantages, and the Observatory switched to using them during his directorship.[14]

Another developing technology Rodgers embraced was the use of 'adaptive optics'. It is a great irony that the light from distant galaxies, having travelled unimpeded for several thousand million years in reaching us, has to spend the last millisecond of its journey being buffeted this way and that by the Earth's unsteady, gaseous atmosphere. Thus the wavefronts of the light, which arrive outside the atmosphere as perfectly flat planes, look more like crumpled sheets of paper by the time they reach our ground-based telescopes. The turbulence of the atmosphere at any particular time dictates how sharp or blurred the astronomical image will be. This is known among astronomers as 'the seeing'. One obvious way around the problem, of course, is to place the telescope high above the atmosphere on an orbiting satellite. The Hubble Space Telescope has proved just how well this solution works—but it is not a cheap one!

Another approach is to try to *repair* the damage caused by the atmosphere. This can be done by incorporating a deformable element in the telescope optics, which can be adjusted instant-by-instant to make the wavefronts flat again. Fortunately, the turbulent blobs in our atmosphere have a size and stability which make this approach feasible, if technically demanding. The technique is known as 'adaptive optics' (AO).

With the help of a strategic developments grant from the University, Rodgers launched an AO project at Mount Stromlo during his directorship.[15] After a good deal of laboratory development work, a functional eight-element system was placed in service on the 74-inch telescope in 1994.[16] AO technology was in its infancy at the time, and the early systems developed around the world did little more than demonstrate the feasibility of the technique. Great advances have occurred over the last decade, however, and today the world's largest telescopes all specify an AO capability. At the time of writing, Stromlo has just won a multi-million dollar contract to design and construct an imager to work on the Gemini AO facility in Chile.

Some of the instruments that Rodgers produced worked better than others—his best ones came in his latter years. As noted in the previous chapter, he built two highly successful devices for the ANU's 2.3-metre telescope—a double-beam spectrograph and an imager. These were completed while he was Director, coming into service in 1987 and 1991 respectively. Rodgers' last instrumental triumph was to complete the transformation of the decrepit Great Melbourne Telescope (built in 1868) into an outstanding, computer-controlled, wide-field imaging telescope which would be used for the MACHO project described later in this chapter.

☆

As already noted, Rodgers campaigned actively for an Australian LOT while he was Director. The world's two largest instruments at the time were the 5-metre Hale Telescope, built at Palomar in the 1940s, and a Russian 6-metre reflector commissioned at Mount Pastukhov in 1975.[17] Considerable difficulty had been experienced in producing the primary mirrors for these two giants, and since technical complexity and cost both increase dramatically with aperture, telescopes with a diameter of about 4 metres had come to be regarded as providing the best value for money. Half a dozen such instruments, including the

Anglo-Australian Telescope, were built around the world during the 1970s. However, by the mid-1980s a dramatic change of outlook was developing. A spate of radical new design concepts emerged, each promising to make apertures of 10 metres or more quite practicable.

One approach was to abandon the convention of making a telescope's primary mirror out of a *single* piece of 'glass'. In 1984, a Californian consortium launched a project to build a 10-metre reflector consisting of 36 hexagonal sub-mirrors, each figured in such a way that they would operate together as a single (segmented) primary when intricately positioned on a common mount. It worked. The first Keck Telescope began scientific observations on Mauna Kea, Hawaii, in 1993, and a twin was commissioned there in 1996.[18]

Another approach was to use a 'spin-casting' technique to produce single-surface 'meniscus' primaries that were much larger than had previously been possible, though extremely thin. The blank is cast in a rotating mould inside a giant oven, and the combination of centrifugal force and gravity gives the molten mass of glass a surface of the right shape for a telescope mirror. The resulting thin mirrors are too flexible to retain their shape unaided, and must be supported at dozens of points across their area. 'Actuators' placed at these locations continuously fine-tune the mirror surface to the precise shape required.

The European Southern Observatory (ESO) consortium adopted this technology for an ambitious project—the rather prosaically named Very Large Telescope (VLT)—approved in 1987.[19] Located on Cerro Paranal in Chile, it consists of four reflectors which may be operated individually, in combination or (eventually) as an interferometer. Each 8.2-metre diameter Zerodur primary is only 175 millimetres thick, and is controlled by 150 actuators. Together, the four VLT reflectors have an equivalent diameter of 16 metres. The first of the four began operating in April 1998, and the last in September 2000.

Another major international facility, the Gemini Observatory, was conceived at much the same time, and adopted a similar meniscus-mirror approach.[20] It was initiated by the United States, the United Kingdom and Canada, but the consortium now includes seven nations. They operate two 8.1-metre reflectors, one in each hemisphere. 'Gemini North' was inaugurated in June 1999, on Mauna Kea, and 'Gemini South' in January 2002, on Cerro Pachon in Chile.

By the late 1980s, these impending developments presented Australian astronomy with a major challenge. There was general agreement that acquiring access to a 10-metre class LOT was a high priority if the nation's astronomers were to remain competitive on the world scene. Addressing the major facility requirements for the coming decade, *The Future of Australian Astronomy*, a 1989 review of the discipline, highlighted a need for 'urgent site testing to assess the viability of a very large optical/infrared telescope in Australia'.[21] Balancing this requirement, however, the review acknowledged that siting decisions were ones which Australia might have to share with other nations. Commenting on the increasing scale and cost of leading-edge astronomical facilities, it recommended that 'in any new major astronomy project, the prospects for international collaboration be actively pursued. In the case of the very large optical/infrared telescope, 50 per cent of funding should be sought from overseas'.[22]

On becoming Director, Rodgers mounted a strong LOT campaign. He had definite views about the most appropriate strategy. As already noted, he adopted an individualistic approach. His argument was a considered one: Australia had an enviable international reputation for excellence in astronomical research—one out of all proportion to its size. While accidents of history during World War II played a part in this, by far the major contributing factor was a string of successful telescope acquisition campaigns during the postwar decades—the 74-inch at Mount Stromlo in the 1950s, the Parkes dish in the 1960s, the AAT in the 1970s and the Australia Telescope in the 1980s. These cutting-edge facilities had alternated, decade by decade, between the optical and the radio domains, enabling Australian researchers to excel on a broad front. Rodgers advocated a continuance of this successful strategy. His ambition was to build, during the 1990s, a 10-metre class LOT on Australian soil, to be operated as a national facility. He commissioned a scale model of an ambitious 14-metre instrument, designed along Keck lines, which graced the Observatory foyer at Stromlo for the duration of his directorship.[23]

There was a strong counter-argument to the Rodgers strategy. Increasingly, astronomers around the world were demanding that expensive new facilities be located at the best astronomical sites available, irrespective of national boundaries. The geography of the Australian continent—in particular, the low elevation and disposition of its mountains—made for atmospheric conditions which were neither as cloud-free nor as scintillation-free as those at sites like Mauna Kea, in Hawaii, or the desert mountains of northern Chile. Many Australian astronomers were already convinced that they should follow the lead of other nations, particularly those in Europe, and locate any new LOT offshore.

Rodgers contended, however, that Australia's remoteness from the world's major research centres was a significant disadvantage, and that locating its major telescopes at home had, in fact, been the nation's scientific lifeline. Any move offshore, he claimed, would result in the slow demise of the nation's relevance on the world astronomical scene. He further argued that the ready availability of engineering services, and even Australia's political stability, were factors that needed to be considered. It had frequently been argued, for instance, that such factors had contributed much to the relative success of the AAT operation, even though observing conditions at Siding Spring were less than ideal.

Accordingly, Rodgers initiated a project to establish whether the Australian continent offered any acceptable location for a LOT. Many places could already be ruled out by the testing done in the late 1950s, but he felt that a fresh approach was now desirable. Of the two chief climatic factors defining a good site—cloud-free conditions and good 'seeing'—he regarded the former as the more fundamental. He entertained the hope that adaptive-optics techniques would eventually make it possible to remedy the deleterious effects of seeing and, as we have noted, he initiated an AO development at Mount Stromlo. His first concern, therefore, was to locate a site enjoying a high proportion of cloud-free nights, since this is a *sine qua non* for observing acceptability. He established a collaboration with the Australian Bureau of Meteorology to analyse its archival satellite data for cloud cover over the continent, and by 1992 the desert region around Lake Eyre had been identified as the most promising site.[24] The nearest sizeable mountains are in the North Flinders Ranges, and Rodgers chose Freeling Heights with an altitude of 951 metres for more detailed on-site investigations.[25]

Meanwhile, in 1992 the astronomical community's debate on how best to meet Australia's LOT needs acquired a sharper focus when financial difficulties encountered by one of the three original Gemini participants—Canada—created a 10 per cent gap in the project's funding.[26] Australia was invited to step into the breach, and many were inclined to accept. Rodgers (and others at Mount Stromlo) felt that any such decision would be premature—and the Gemini proponents' case was not helped by a suggestion that, although an Australian 10 per cent involvement would be most welcome, this might not be large enough to justify a seat on the Gemini Board, Australia's interests instead being handled by the British Board member. In the event, this early Gemini opportunity was not taken up, and Chile, Argentina and Brazil joined at a combined 10 per cent level.[27]

Alex Rodgers' term as Director concluded with Freeling Heights still a live option in the LOT debate, but the matter was finally settled during the eighteen-month *interregnum* before the arrival of the next Director. Gary Hovey and Peter Wood outfitted an expedition, and Mark Carson and Alister Graham conducted on-site testing there during June–July and November–December 1993.[28] They were equipped to collect meteorological data, and also to make objective measures of the seeing with a 'differential image motion monitor' developed by

Wood. The operation was far from trouble-free—illness caused a site-tester to be hospitalised in Adelaide at one stage—but enough data were collected to demonstrate that observing conditions at Freeling Heights were far from comparable with those at the world-class sites in northern Chile.[29]

When the expedition's results were digested, there was almost universal acceptance of the fact that the Australian continent affords *no* satisfactory site for a LOT, thus clearing the way for the incoming Director to pursue an Australian participation in one or other of the major international consortium projects then coming on-stream (see Chapter 11).

The Rodgers directorship was distinguished by Mount Stromlo's participation in a major international experiment—the MACHO Project—which addressed one of the most baffling astronomical puzzles of the day.[30] Over the previous 50 years, there had been a growing realisation that the things we can 'see' in the Universe—the planets, the stars, the gas, and so on—account for less than 10 per cent of all the material that must be there.[31] Fritz Zwicki, a rather irascible Swiss astronomer who spent his career working at Caltech, first raised this problem in the 1930s. He pointed out that the mass associated with the luminous material in the giant Coma Cluster of galaxies was far too small for its gravitational pull to hold the galaxies of the cluster together against the disruption of their energetic motions.[32] The implication was startling. Not only must there be unseen matter in the cluster—lying either within or between its member galaxies—but this material must be by far the largest component, contributing more than 90 per cent of the total mass. At the time, Zwicki's conclusions did not attract the attention they deserved.

By the 1960s, however, galaxy mass investigations were much in vogue, with astronomers making strenuous efforts to determine not only their *total* masses, but also the mass *distribution* within them. Of particular interest were spiral galaxies. Their stars and emission nebulae exhibit a largely circular rotation about the galactic centre, and by measuring how the rotational velocity changes with radial distance—the so-called 'rotation curve'—one is able to build up a picture of the galaxy's mass distribution. The physics involved is the same as that governing the motion of the planets in our Solar System—indeed, observing the planetary motions was the way in which the mass of our Sun was first determined. Ken Freeman was the first to suggest that the rotation curves of some spiral galaxies imply the presence of invisible material.[33]

These investigations acquired added impetus with the development of aperture-synthesis radio telescopes capable of making high-resolution 21-centimetre observations. Neutral hydrogen gas is the most ubiquitous constituent of spiral galaxies, and it typically extends two to three times as far from the galaxy centre as do the stars and nebulae. Thus 21-centimetre rotation curves can be observed to quite large radii—over 100,000 light years in some cases. Virtually all of a galaxy's luminosity comes from well within such radii, and it had always been assumed that the same would apply for the galaxy's mass. Thus the observed rotational velocity in these outer regions should decrease with radius according to a well-known relationship, first discovered by Kepler, which describes the orbital velocities of the planets in our Solar System.

In 1974, Mort Roberts of the National Radio Astronomy Observatory in the United States presented a conference paper detailing the rotation curves of fourteen galaxies.[34] Startlingly, many of them showed *no* decrease at large radius, suggesting that the invisible mass associated with galaxies resides in vast surrounding halos. The interpretation of the observations was far from straightforward, however, and the existence of these massive coronae was still being debated a decade later. In 1983, Agris Kalnajs was able to suggest an alternative explanation for *some* of the observed rotation curves—one which so stunned a conference audience that the proceedings volume records their reaction almost entirely as a string of exclamation marks![35]

Nevertheless, as the years went by, evidence continued to accumulate for the existence of some mysterious invisible constituent of the Universe, which began to be referred to as 'dark matter' or 'missing mass'.[36] In the case of our own Galaxy, John Norris collaborated with Mike Hawkins (of the Royal Observatory, Edinburgh) in studying the motions of individual stars located at vast distances out into the Galaxy's halo—up to 200,000 light years.[37] They showed that these objects too were feeling the gravitational pull of unseen material extending out to at least half this distance. Other studies of small satellite galaxies in the vicinity of our own yielded similar results.[38]

Additional evidence came from satellite-borne X-ray telescopes, which revealed the presence of a great deal of extremely hot gas (up to 100 million degrees) surrounding some galaxies, and pervading clusters of galaxies.[39] While this material is not sufficiently abundant in itself to account for the missing mass, it can be used as yet another diagnostic probe in determining its location. The X-ray properties once again showed that galaxies, particularly large ones, have massive dark halos.[40] Paul Nulsen, a Mount Stromlo Research Fellow in the late 1980s, was involved in many investigations of the behaviour of the hot gas in galaxy clusters.[41]

On an entirely different front, cosmologists at the time had become much attracted to a theory for the origin of the Universe—'inflation'—which required its current observed density to be almost exactly that needed to 'close' it.[42] That is to say, the expansion of the Universe will neither continue forever nor go into reverse leading to an eventual collapse; rather, the velocity of the expansion will decrease unerringly to zero (but at infinite time). The 'critical density' needed to achieve this delicate balance is, once again, at least a factor of ten greater than can be accounted for by all the visible material in the Universe—providing another (albeit somewhat esoteric) incentive for believing in the existence of dark matter.

By 1985, the missing mass conundrum was attracting such widespread interest that an entire conference was organised to discuss all the observational evidence and its implications.[43] With his long-standing interest in the dynamics of galaxies, Ken Freeman had become intimately involved. For instance, he had joined with his student Claude Carignan in a careful study of four key galaxies, finding convincing evidence that they have dark halos, comparable in mass to their luminous material.[44] Freeman was one of the conference organisers, and he presented a paper reviewing the dark halo properties inferred from the observed rotation curves of spirals.[45]

The greatest challenge, of course, was to identify this missing mass. Some argued that the most likely explanation was that galaxies are surrounded by vast numbers of discrete, but dim, astronomical objects—white dwarf stars, or small black holes, or even planet-sized bodies. A halo of 1,000 million million Jupiters surrounding our own Milky Way would explain its dark matter properties very nicely.

Others suggested that the answer was more likely to be found in the realm of particle physics, with the unseen material consisting of myriads of tiny entities belonging to one or more previously unknown particle species left over from the Big Bang creation of the Universe itself. This was not a far-fetched suggestion, since the Universe was already known to be pervaded by a number of such Big Bang remnants. The most famous is the cosmic microwave background (CMB), but since it is composed of *massless* particles—photons—this particular remnant is of no help in the present context.[46] The unknown 'dark matter' particles would certainly need to have a mass, and there are reasons for believing that it might be a considerable one, by particle standards. They would also need to partake in only the weakest known forms of physical interaction, or they would surely reveal their presence in some visible way. Thus they came to be known as WIMPs, an acronym denoting 'weakly interacting massive particles'.[47]

Not to be outdone, those supporting an explanation in terms of discrete, astronomically sized entities also invented an acronym. Noting that these bodies seemed to surround galaxies in halos, they called them MACHOs—'massive astronomical compact halo objects'.[48] So began an intense but friendly rivalry between the MACHO and the WIMP proponents, each hoping to claim ownership of the unidentified missing mass in the Universe.

The Observatory's involvement in the now-famous Stromlo MACHO experiment can be traced back to yet another of Ken Freeman's overseas visiting appointments. In 1984 he was at Princeton when a Polish astronomer working there, Bohdan Paczynski, proposed an imaginative experiment to put *one* of these two rival dark matter theories to the test.[49] His idea was that a vast cloud of MACHOs inhabiting a halo around our Milky Way, although too dim to be seen themselves, might still become observable by magnifying the light of more distant stars. This 'lensing' effect is a consequence of Einstein's theory of relativity. The mere presence of a massive body warps the geometry of the space around it, and this warping results in the bending of any light path passing close to the mass. It was demonstrated during a total eclipse in 1919 that our Sun warps the space around it as predicted by Einstein, and this provided one of the early confirmations of his theory.[50] Thus, when there is an almost perfect alignment between a distant star, a dark matter object and an observer, the MACHO will act as a lens, increasing the brightness of the distant star severalfold.

The Paczynski experiment required an adequate backdrop of stars located beyond, but not too far beyond, the MACHO halo of our Galaxy; the Magellanic Clouds are perfect for the purpose. The chance of a Cloud star being aligned with a MACHO at any instant is extremely small—less than one in a million. Furthermore, both the MACHO and the distant star will be in motion, so any alignment will be temporary, with the lensing event lasting between a couple of hours and a couple of years, depending on the mass of the MACHO. Thus the requirements of the experiment became clear. Several million stars in the Magellanic Clouds needed to be observed on a regular basis, at least every night and preferably several times a night, looking for instances where a star's brightness rises and then falls again, reproducing the quite specific 'light curve' calculated by Paczynski. The expectation was that, over the course of a decade, perhaps a few dozen such lensing events might be observed, providing some insight into the mass and spatial distribution of the MACHOs. Such a data-intensive experiment was unprecedented but, as Freeman realised, quite feasible in principle using modern observing and computing techniques.

The momentum for conducting Paczynski's experiment began to grow. A group led by Charles Alcock at the Lawrence Livermore National Laboratory (LLNL) had been involved in looking for comets in the outer Solar System—a project with technically similar observing requirements—and they became interested. So too did a group led by Chris Stubbs at the University of California's Center for Particle Astrophysics (CFPA) at Berkeley.

In mid-1989, Freeman received informal approaches about the possibility of a Mount Stromlo collaboration.[51] Since the Magellanic Clouds lie in the far southern sky, the experiment needed to be done at a Southern Hemisphere observatory, and Freeman's early involvement with Paczynski made him an obvious starting point for inquiry. He discussed the matter with two other Stromlo colleagues who were interested in dark matter, Bruce Peterson and Peter Quinn, and both were enthusiastic. He then approached Alex Rodgers, whose interest was also aroused.

The proposed experiment would be very demanding of telescope time, requiring several years' dedicated use of a sizeable instrument. Fortuitously, a potentially suitable one was lying idle at Mount Stromlo—the 50-inch former Great Melbourne Telescope (GMT) on which refurbishment work had been suspended in the early 1980s. Following discussions at Stromlo with the LLNL and CFPA researchers, Rodgers offered to complete the refurbishment of the GMT and to dedicate it to the MACHO experiment for an initial period of four years (later extended by another four), as Stromlo's contribution to a three-way collaboration. The CFPA took responsibility for developing two detector mosaics, each comprising four 2048 × 2048 CCDs. These were the world's largest electronic cameras at that time, imaging a field half a degree across—about the size of the full Moon. The LLNL partners provided the computer resources, and developed the sophisticated software codes needed to handle the prodigious amount of data reduction generated by the experiment.

The three parties signed a formal Memorandum of Understanding in November 1990.[52] The MACHO Project team initially comprised a dozen researchers from the three institutions, with Freeman, Peterson, Quinn and Rodgers representing Mount Stromlo. Somewhat later (1994) one of the original LLNL team members, Tim Axelrod, who was much involved with the development of the data-handling software, moved to Canberra and joined the Stromlo staff.

The refurbishment of the GMT was reactivated in 1990, and the MACHO collaborators commenced observations in July 1992.[53] Not surprisingly, they were not alone. Two other groups also attempted the same crucial experiment. A French team used photographs taken at the ESO Schmidt telescope in Chile for

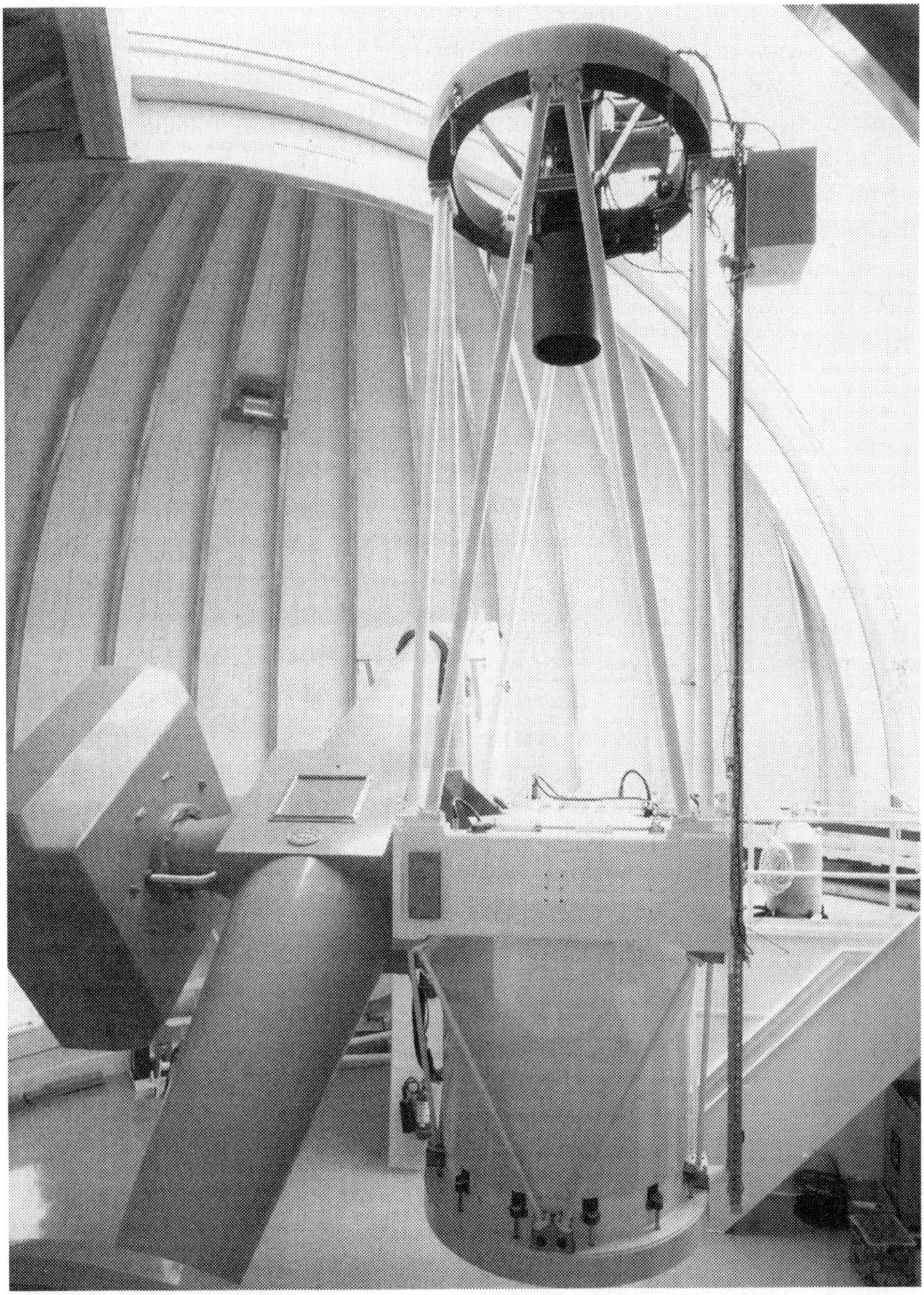

The Great Melbourne Telescope rebuilt for the MACHO Project in the 1990s
Photo: Keith Smith, MSO

their search (the EROS Project), and a Polish–American consortium conducted a smaller-scale CCD project (OGLE), also in Chile.[54]

There was great excitement when the analysis of the first year's data yielded possible microlensing events, including one in which the Magellanic Cloud star had brightened and then faded by an extremely convincing factor of seven over a period of about a month. The announcement of this became the cover story in the prestigious science weekly, *Nature*.[55] The brightening characteristics indicated that the unseen 'gravitational lens' has a mass about one-tenth that of the Sun—making it an excellent MACHO candidate. The other two groups also began reporting events at about this time.[56]

The MACHO experiment's observations were concluded in December 1999, by which time over 200,000 million individual photometric measurements had been made. About fifteen microlensing events were observed for Large Magellanic Cloud stars, two for the Small Cloud and several hundred towards the centre of the Galaxy.[57] The full implications of the data are still being assessed. For one thing, uncertainty has arisen about the exact location of some of the dark objects responsible for the lensing. In the case of one of the Small Magellanic Cloud events, the culprit appears *not* to lie in the halo of our Galaxy at all, but rather in the Small Cloud itself.[58]

The lensing events occurred less frequently than would be expected were MACHOs to comprise *all* of the missing mass in the Galaxy's halo—and that hypothesis can now be rejected with a high level of statistical confidence. On the other hand, quite enough were seen to show that MACHOs are responsible for *some* of the dark matter (about

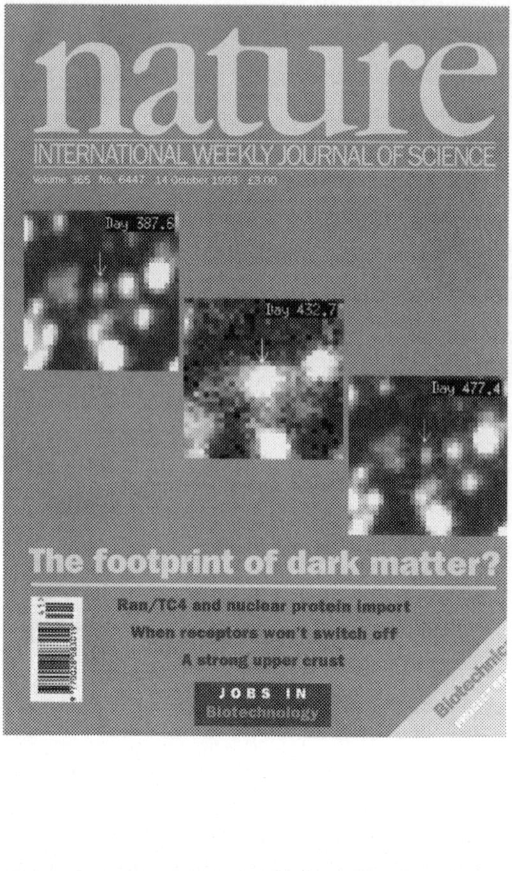

The first MACHO event, seen in 1993, became the cover story in Nature *magazine, a coveted honour*
Image: *Nature*

20 per cent).[59] The individual lensing objects seem to have a typical mass about half that of the Sun. There is a strong possibility that they are, in fact, one of the commonest of all astronomical entities—white dwarf stars. This might seem like a mundane anticlimax, but it would still come as something of a surprise to find so many of these burnt-out stellar remnants occupying so large a halo around our Galaxy. Where could they have come from? Have they been thrown out from our Galaxy's disc; are they the leftovers of past satellite–galaxy mergers; or have they been formed *in situ* by some unknown process? And what of the 80 per cent of the missing mass that is still unaccounted for? As is frequently the case, the MACHO experiment has posed as many new questions as it has answered old ones.

A useful first step in addressing these new riddles would be to obtain confirmation, by independent means, that the objects seen in the MACHO experiment are indeed white dwarfs. This is now being attempted using the Hubble Space Telescope. A team led by Jeremy Mould, including Ken Freeman, Tim Axelrod and Gary Da Costa (a Mount Stromlo graduate of the 1970s who returned as a Senior Fellow in 1995), has been awarded a substantial amount of observing time to look for faint moving objects in a wide area of the sky.[60] In 2001 they announced the identification of five stars as strong candidates for halo white dwarf status.[61] Their origin, however, still remains something of a mystery. There is obviously still a long way to go before we can say we understand fully the nature of the dark matter in our Universe.

The MACHO Project had scarcely begun when the observing conditions on Mount Stromlo came under a significant threat from a proposed suburban development. Although the Observatory had, for some decades, been enduring gradually rising levels of light pollution, most of Canberra's expansion was by then taking place on the city outskirts well to the north and south of Stromlo. In the early 1990s, however, the Australian Capital Territory government adopted a policy of 'urban infill', releasing pockets of unused inner-city land for additional housing.[62] To Mount Stromlo's dismay, a 1,000-dwelling development lying literally at the foot of the mountain was proposed, which would have resulted in a significant increase in night-sky brightness.[63]

The year 1993 saw lengthy, and at times tense, discussions that embroiled the highest echelons of the University and the government. Expert lighting consultants were engaged to examine the issue, and the whole debate became a very public one.[64] One result was that the MACHO Project, which afforded a perfect example of the type of science that would be put at risk, received a great deal of early publicity.

The matter was finally resolved satisfactorily—owing mostly, one suspects, to the timely arrival of an economic downturn that produced a glut on the Canberra home site market! The formal outcome was that the government offered a ten-year moratorium on the development in question to allow the MACHO Project to run its course—an offer the University gratefully accepted.[65]

On 24 February 1987, a celestial event was seen that set telephones ringing around the world—the explosion of a supernova (designated SN 1987A) in the Large Magellanic Cloud (LMC).[66] Supernovae are quite rare phenomena. In our own Milky Way galaxy, such an explosion occurs perhaps every hundred years on average, but many of the outbursts are hidden from our view by the obscuring clouds of interstellar dust that pervade the Galaxy's disc. The last Milky Way supernova seen was Kelper's in 1604—before the invention of the telescope—although the radio source Cassiopeia A is believed to be the remnant of another unrecorded outburst that occurred later in the seventeenth century.[67] Thus astronomers have been waiting expectantly for nearly 400 years to *see* the next supernova in our own system, and this one in the neighbouring LMC, being the next best thing, was greeted with great excitement! The supernova attained a magnitude of 2.9, the brightest event since that of 1604.

SN 1987A proved to be an example of a Type II supernova—the type caused by core collapse in a massive star at the end of its short, luminous life. Convincing evidence for this emerged when neutrinos from the event (albeit only nineteen of them) were detected at laboratories located in Japan and North America.[68] This was the first stellar catastrophe occurring near enough to be observable by these 'neutrino observatories', and the 1987 detection confirmed a great deal of the numerical modelling that had been done by supernova theorists. To a large extent, neutrinos control what happens within the collapsing core of a massive star when it ends its life. They carry away 10,000 times more energy than emerges as visible light, and they can escape from the dense core much more easily than the photons. Thus they provide information on the physical processes at work which is obtainable in no other way.

The outburst of SN 1987A caused a frenzy of activity in the astronomical world—especially, of course, at observatories in the Southern Hemisphere. Regular observing schedules were suspended as all available telescopes were diverted to recording every aspect of this once-in-a-lifetime event.[69] Mount Stromlo astronomers and telescopes played their part. A dozen Australian researchers joined with colleagues in South America and South Africa to monitor the supernova's brightness and colour changes as the days, weeks and months went by.[70] Peter McGregor obtained infrared observations;[71] Peter Wood calculated evolutionary models for the precursor star;[72] and Wood, Paul Nulsen, Mike Bessell and Mike Dopita were part of a team that used the AAT to determine the angular size of the expanding supernova cloud during the first nineteen months of its life.[73] Since the size is so small—only 20–40 thousandths of a second of arc, which is much smaller than the 'seeing' disc due to atmospheric turbulence—they needed to employ a sophisticated observing technique called 'speckle interferometry'. An expansion of the supernova image was clearly evident, the corresponding outward gas velocity being over 2,800 kilometres per second.

Many research conferences on SN 1987A were held around the world, and in June 1988—about 500 days after the outburst—one took place in Canberra.[74] Mike Dopita was invited to present the introductory overview, outlining what had been learned concerning the physics of supernova explosions from the more than 500 scientific papers that had even then been published in international journals.[75]

Because of the proximity of the LMC, the precursor star of SN 1987A could be identified on pre-event photographs. It had been of 12th magnitude, which means that the explosion caused it to brighten by a factor of about 4,000. Its blue colour indicated that it was also a hot star, which is unusual since most supernova outbursts are believed to occur when the precursor is in a cool, red giant phase of its evolution. The progenitor was found to have had a considerable mass—about 15–20 times that of the Sun. The supernova light curve proved to be an unusual one as well, rising to maximum much more slowly (over a few months) than is usual for Type II supernovae. Nevertheless, the detailed observations of the event confirmed a good many pre-existing theories. In particular, there was good support for the widely held belief that it is the supernova deaths of such massive stars that are responsible for ejecting the first 'heavy elements' into the primordial hydrogen–helium gas out of which collapsing galaxies form.[76]

Stars of lower mass—such as our Sun—end their lives in a much less spectacular fashion. Material leaves their surface in a type of gentle wind while they are in the final red giant phase of their evolution. Eventually a substantial outburst occurs which sweeps up some of this surrounding material into a gaseous 'shell'. This is sometimes quite spherical in form, but dumbbell-shaped or irregular shells can also occur. The mass loss exposes the hot inner core of the dying star, and the intense ultraviolet light that it radiates excites the gas in the shell, transforming it into a luminous cloud called a planetary nebula (PN). The remnant core of the star then contracts a hundredfold in radius over a period of about 10,000 years, becoming a white dwarf star. The PN shell expands and dissipates over much the same interval. Although many beautiful examples of planetary nebulae are known in our own Galaxy, it is extremely difficult to estimate their distances. This frustrates any attempt to determine the exact physics of the ejection process using these Milky Way objects.

The Rodgers years saw a major research thrust that overcame this problem by observing PNe not in our Milky Way, but in the Magellanic Clouds. Those involved were Mike Dopita, Peter Wood, Mike Bessell, their students Steve Meatheringham and Emanuel Vassiliadis, and colleagues from other institutions. They observed about 150 Cloud PNe, obtaining quantitative data for the shell properties for a good sample in each of the LMC and the SMC.[77] Some sophisticated observing techniques were required, since the advantage of having known distances for these objects is offset by the fact that the distances are large—160,000 and 200,000 light years to the Large and the Small Clouds respectively. Thus the PNe appear much smaller and fainter than do those in our Milky Way. Once again, speckle interferometry was pressed into service to determine the radii of some of the smaller shells.

The observations were complemented by detailed theoretical modelling—of the gaseous shell by Dopita and Meatheringham, and of the dying star by Vassiliadis and Wood.[78] Thus a comprehensive picture emerged of the whole PN phenomenon. The remnant stars were found to be typically about two-thirds the mass of the Sun, but up to twice that value in some cases. The amount of mass observed in the PN shell showed even greater variation, from less than a tenth of the solar mass to more than half.

With the launch of the Hubble Space Telescope in 1990, the team members were successful in winning time for an observing program to study these Magellanic Cloud PNe, taking advantage of the Hubble's much finer resolution.[79] They began a series of detailed investigations of individual nebulae, commencing with two catalogued as SMP 83 and SMP 85. The former proved to be a real heavyweight, having a shell about half the mass of the Sun and a remnant star of just over a solar mass, which was exceedingly hot—170,000 degrees. SMP 85, by contrast, had a stellar remnant of only two-thirds the Sun's mass, and a temperature of 45,000 degrees. The nebula shell proved to be quite young—formed less than 1,000 years ago—and of quite small mass. The team continued its HST observations for several years, eventually investigating nearly 20 Cloud planetaries in this way.

Australian astronomers, and Mount Stromlo researchers in particular, have proved quite adept at winning observing time on the HST. In one of the early application rounds, Australia was awarded 10 per cent of the total time and 6 per cent of Principal Investigator time, three-quarters of which went to Mount Stromlo.[80] Such time grants have enabled many exciting projects to be carried out, especially following the 1993 mission to repair the HST's flawed optics. For example:

- Gary Da Costa has been able to exploit the telescope to extend his work on dwarf spheroidal galaxies. These small systems are very common in the Universe, although their faintness makes them difficult to detect. The nine or more that surround our own Galaxy show a great diversity of evolutionary histories, making them important in investigating galaxy formation. Da Costa and his team have used the HST to investigate dwarf spheroidals surrounding the Andromeda Galaxy (M31) as well, demonstrating that they too exhibit a similar diversity.[81]
- Ralph Sutherland (a Mount Stromlo graduate of 1993 who returned as a Research Fellow in 1997) and his colleagues have examined the Orion Nebula, a celestial nursery inhabited by newly born stars and the vast interstellar gas cloud out of which they have recently been formed. They have studied over 40 such objects, obtaining high-resolution HST images in which many appear still half-shrouded by their dusty placentae.[82]
- Peter McGregor participated in a study of the curious star η Carinae.[83] This object has been varying irregularly in brightness for centuries; during 1837–60 it became, for a time, the second brightest star in the sky. It is thought by some to be perilously close to suffering a supernova death. The HST team observed an unexpected doubling in brightness during 1998, the largest change for several decades.
- Mount Stromlo's HST research has embraced objects as close as the planets in our Solar System, and as remote as the distant quasars. One investigation obtained ultraviolet images of the cloud features on Jupiter, including the Great Red Spot, while another acquired multi-wavelength spectra of PG 0946+301, an object so distant that the light we now see from it started its journey when the Universe was only 45 per cent of its current size.[84]

☆

Another research topic which flourished at Mount Stromlo during the 1980s, and in which interest continues, is the study of 'active' galaxies. This field of endeavour grew out of the early discovery by radio astronomers that some galaxies emit a great deal of radio 'light', whereas others are relatively 'radio-quiet'. As their observations improved, it became apparent that the explanation for much of this radio emission needed to be sought in the very core of the objects concerned. While such galaxies may display large, diametrically opposed radio lobes, these are fed by highly collimated jets of energetic gas, expelled by a mysterious central 'engine' at speeds which sometimes approach that of light.

Such galaxies shot into prominence, of course, when it was discovered that they can sometimes appear as 'quasi-stellar radio sources' (quasars), which are so luminous in the radio domain that we can see them close to the outer limits of the observable Universe.[85] By contrast, other active galaxies reside in close proximity to our own. The nearest is Centaurus A, lying just 11 million light years distant—one of the radio sources first identified using the observations of the Radiophysics sea interferometer at Dover Heights.[86]

If a galaxy has such an energetic core, it is described as being 'active', and the core itself is called an 'active galactic nucleus' (AGN).[87] Many questions spring to mind. Are such galaxies *always* active, or do they go through *intervals* of activity at various stages of their lives? Will *all* galaxies become active sooner or later? Was activity in galaxies more or less common when the Universe was young? What is the mysterious physical process that energises the non-stellar emission, the jets and the radio lobes? Today there is almost universal agreement that the answer to this last—and most fundamental—question is that the culprit is a massive black hole lying in the AGN. Its mass might be 100 million suns, and it has a voracious appetite for the material in its surroundings. Whole stars may be disrupted and swallowed. It is the release of gravitational potential energy as this material falls faster and faster into the abyss that fuels the spectacular displays that surround such galaxies.

Over the decades, researchers have tried to construct a self-consistent picture of an AGN which will account for all their observed properties—not only at radio wavelengths, but in the optical, infrared, X-ray and γ-ray domains as well—a so-called 'unified model'.[88] This has proved to be far from straightforward. Instead of all such galaxies falling into a common pattern, hosts of different characteristics have been observed. There is now a veritable zoo of active galaxy types, some having fantastic-sounding names—Seyfert galaxies (types 1 and 2), LINERS, blazars, ULIRGs, quasars, broad- and narrow-line radio galaxies, starburst galaxies, and so on.

In spite of all this observational complexity, theorists still believe that there is only one basic physical process capable of releasing energy at the required rate—accretion on to a massive black hole. How can a *single* mechanism produce such a variety of observational consequences? Part of the answer lies in the way that the black hole swallows the surrounding material. Conservation of angular momentum requires that, as each chunk of mass falls closer and closer to the hole, it orbits around it faster and faster. Thus, instead of a cloud of debris falling directly to its doom, it forms a sort of rotating disc about the black hole, and gradually spirals its way down. This 'accretion disc', as it is called, frequently obscures the innermost regions of the AGN from our view. The bewildering variety of active galaxy types simply reflects the range of *angles* at which our line of sight intersects this disc. The great hope of theorists working on AGNs is that, taking account of our angle of view, they will be able to devise a unified model which will reproduce the properties of not just *one*, but *all* of the various observed classes.

Observing active galaxies and making such theoretical models has been a preoccupation of many Mount Stromlo researchers: Mike Dopita, Geoff Bicknell, Ralph Sutherland, Charlene Heisler, Carole Jackson, Peter McGregor, David Carter, Ian Tuohy, Marthijn de Kool, Paul Francis and Curtis Saxton, together with their students Rodney Smith, Colin Coleman, Neil Killeen, Ian Evans, Robert Cameron, Saul Caganoff, Roger Brissenden, Anton Koekemoer, Steven Tingay, Mark Allen and Lisa Kewley.

In the early 1990s, the Observatory's AGN enthusiasts decided to organise a major international meeting of the kind Mount Stromlo had hosted occasionally in the past.[89] They proposed that their conference should launch a new series of symposia to be held every few years on topics of immediate interest to the Observatory.[90] Accordingly, the event was styled 'The First Stromlo Symposium: The Physics of Active Galaxies'.[91] It took place at the Australian Academy of Science's Becker House in June–July 1993. The meeting was an outstanding success, attracting 120 participants, two-thirds of them from overseas. Unified schemes were much on the agenda, as was the discovery and interpretation of high-energy γ-rays from some of the most violently active nuclei. Further Stromlo Symposia have been held as promised. Subsequent topics have been: 'The Nature of Elliptical Galaxies'; 'The Galactic Halo: Bright Stars and Dark Matter'; and 'Planetary Nebulae: Their Role in the Universe'.[92] On a smaller scale, Stromlo hosts an annual AGN workshop each December which, since her sad but courageous death in 1999, has been dedicated to the memory of Charlene Heisler.[93]

★

Several months before the conclusion of Rodgers' five-year term (June 1992), the University appointed a broad-based Electoral Committee to consider the directorship. Such is normal ANU practice, and the Committee has wide discretion. It may recommend the reappointment of the incumbent Director, or that the position be filled 'by invitation' to someone else, or it may opt to advertise the post. The Committee is expected to consult widely before deciding on its course of action, and a Liaison Committee is also appointed to garner the views of staff in the area concerned as part of this process.

The Electoral Committee met for the first time on 15 October 1991. It considered several reports, including the Review of the Observatory conducted five years earlier, and it interviewed the Liaison Committee and the incumbent Director. Its decision was to proceed to open advertisement.

This was something of an about-face from the 1986 verdict that the Observatory's interests would be served best by making an *internal* appointment, not least because of the financial constraints at the time. The 1991 Committee obviously felt differently, and one can see several factors which might have shaped its decision. The most obvious is that there had now been two internally appointed Directors in succession. It had been fourteen years since Don Mathewson had first become acting head, and there was a growing perception—particularly in the wider astronomical community in Australia—that the time was ripe for Mount Stromlo to have an infusion of new blood.

Another factor may well have been some tensions that had arisen within the Observatory itself. Paradoxically, the MACHO Project was a major cause. While this was an outstandingly successful enterprise scientifically, the associated refurbishment of the 50-inch telescope had come at a cost, and staff members with other instrumental aspirations felt somewhat aggrieved at the large slice of the Observatory's financial and technical resources being devoted to a single project. Nor was this situation helped by the fact that the contractual arrangements with the overseas collaborators specified that only four nominated Stromlo astronomers were to become members of the MACHO research team. Furthermore, all the 50-inch observing time was to be dedicated to MACHO for the duration of the project. For many decades it had been an article of faith that there should be an open competition for telescope time, with the allocation being determined by a peer-review committee. Not since Woolley's day had a Mount Stromlo Director adopted such a hands-on approach to the distribution of observing resources. All in all, these arrangements and their implementation had engendered a certain divisiveness among the Stromlo staff, and it would be surprising if some hint of this had not surfaced when the Electoral Committee met with its Liaison Committee.

Be that as it may, the Electoral Committee adopted a strictly even-handed approach. It initiated an open advertisement for the post, while requesting the Vice-Chancellor to encourage the incumbent Director to apply, which Rodgers did. The advertisement was placed in November 1991. It detailed the University's expectations, one being that the appointee's research interests should dovetail with the existing observatory facilities and resources—so as to forestall any expectations of massive new spending programs.

There was a wide response. The short-list alone ran to six: three Mount Stromlo astronomers and three from North America. Electoral Committee interviews took place in May 1992, and each of the six short-listed applicants was invited to make a scientific presentation at a mini-symposium held at Mount Stromlo—the ANU's appointment procedures have become much more open and transparent with the passing years. In the event, the selection process proved rather tortuous, for reasons having little to do with either astronomy or the Observatory, and there was an eighteen-month *interregnum* between the end of Alex Rodgers' term and Jeremy Mould's arrival as Mount Stromlo's seventh Director on 13 December 1993. Don Faulkner served as Acting Director during this period. Mould's appointment was for an initial five years.

Rodgers was keenly disappointed at this outcome. A five-year renewal as Director would have enabled him to pursue his aspirations for the Observatory for the remainder of his career.[94] Nevertheless, he accepted the verdict, and resumed his research and other departmental involvements as energetically as ever.

He had long been interested in variable stars, on which he was an expert. In this respect, the MACHO experiment yielded a rich by-product—in searching for lensing events among the observations of some nine million Magellanic Cloud stars, the computer incidentally identified about 40,000 intrinsic variables as well, the vast majority of them new discoveries. They are of many different types, and the analysis of such a magnificent data set enabled many unresolved problems to be addressed. The team produced a dozen research papers on pulsating and eclipsing variables, and Rodgers contributed to this work until the last weeks of his life. One important conclusion was that the oldest variable stars in the Large Cloud seem to have been born a full 4,000 million years later than some of their counterparts in the Galaxy.[95] This has interesting implications for the formation history of the Cloud itself. Peter Wood joined the team in exploring the implications of the MACHO observations of red giant variables.[96] Many questions were settled, but other new problems emerged. For instance, a surprisingly high proportion of these giants appear to owe their variability to the unexpected presence of binary companions.

Rodgers always extended his passion for the Observatory to the University at large. Indeed, he took great pride in the role that his father-in-law, John Dedman, the Labor Minister for Post-war Reconstruction, had played in its founding, and devoted considerable energy and enthusiasm to carrying on this family tradition.[97] He served several terms on the ANU Council, which gave him great satisfaction. He was again elected to the Council in 1996 and served until his death. He was forthright in debate, and contributed many ideas for the advancement of the University. One of his last exchanges of correspondence was with the ANU Chancellor, Peter Baume, in which he proposed that the University establish a major new Research School of Inland Australia.[98] Ecological, demographic, economic and national developments in the intervening five years have done nothing to diminish the value of this suggestion.

Alex Rodgers attended his last meeting of Council in September 1997.[99] He died on 10 October, just weeks short of his statutory retirement date. Working at the Observatory spanned his whole adult life. A gift from his widow, Ruth, has endowed a student travelling scholarship—a fitting memorial to someone who had a career-long involvement with the Graduate Program at Mount Stromlo.[100]

chapter eleven

Big science: the return of Jeremy Mould, 1993–2002

LIKE HIS PREDECESSOR, JEREMY MOULD WAS a product of the Mount Stromlo Graduate Program. He obtained his PhD in 1977 with a thesis entitled 'The Lower Main Sequence and the Atmospheres of M Dwarfs'. Unlike Rodgers, however, he then moved abroad, working for fifteen years at some of the world's most famous institutions—the Royal Greenwich Observatory, Mount Wilson and Las Campanas Observatories, the Kitt Peak National Observatory, and the California Institute of Technology. He held a professorship at Caltech before returning to Stromlo.

As Director of Mount Stromlo, Mould displayed the same dedication to research, to his observatory, to the affairs of the astronomical community and to astronomy in general for which he had become renowned in his earlier career. He reserved no time for personal interests or pursuits. His response to any request or invitation—be it for a meeting, a public lecture, a committee task or an overseas conference—was immediate and invariable. He would consult his personal assistant, Fiona Aplin, and if there was a space in his diary at the relevant time he assented to the request without hesitation. He seemed to have mastered the art of sleeping on international flights. After an early morning landing in Sydney he would fly on to Canberra and immediately report to his office for a full day's work, looking every bit as fresh as if he had simply walked across from the Director's Residence.

Mould was one of the most even-handed and even-tempered people one could ever meet. He seemed entirely at ease with all personalities, and impervious to all provocations. He wasted no time fretting over adverse political outcomes—and there were some during his time as Director—but immediately adjusted his thinking and planning to the new circumstances, salvaging what could be salvaged and devising alternative strategies to achieve as much as possible of his thwarted agenda. His personality was not always conducive to his forming close personal relationships, however; his very equanimity could leave one wondering about his real views on, and assessments of, people and proposals.

With the appointment of the new Director, long-standing animosities between Mount Stromlo and some other sections of the Australian astronomical community seemed to melt away. His eminence, and his even-handedness, won him ready acceptance on to many boards and committees. The month after his arrival, he became a member of the Anglo-Australian Telescope Board—the first Stromlo astronomer to receive such an invitation since Eggen resigned from that body in August 1973. He was also invited to join the Academy's National Committee for Astronomy, and a year later became its Chair, remaining so for several crucial years of decision-making and lobbying. His committee load became more and more formidable with every passing year—the National Committee for Space Science, steering or advisory committees for several Australian and international astronomical centres, an editorial board, organising committees for scientific conferences, as well as more than a full load of board work within the ANU itself. He was elected a Fellow of the Australian Academy of Science in 1998.

Jeremy Mould, Director 1993–2001
Photo: MSO

Mould came to Mount Stromlo having had an intimate involvement with several of the North American and international large optical telescope projects that were conceived in the 1980s. He served on the Science Steering Committee for the Keck 10-metre Telescope, and chaired the Working Group on 'Stars and Star

Clusters in the Galaxy' for the Hubble Space Telescope. Thus he had a ready *entrée* to the 'big science' teams that were rapidly changing the complexion of world astronomy, and his arrival at Mount Stromlo influenced the research ethos there enormously. Although the MACHO Project had provided something of a foretaste of things to come, big science can be said to have been embraced at Stromlo during the Mould directorship. Many of the staff, including several new appointees, participated in *big* international research collaborations, using *big* telescope facilities to work on *big* projects addressing some of the *big* unanswered questions concerning the Universe.

Nowhere is the new ethos more obvious than in the publication lists that appear in the *Annual Reports* of the 1990s. In the year in which Mould arrived, 1993, the average paper had just three or four authors, and no paper had more than seventeen. In the last full year of his directorship, 2000, the average authorship list had trebled to ten, and 17 per cent of papers published had twenty or more authors—the largest list ran to 93. Mould himself wrote or co-authored over 120 scientific papers during his seven years as Director, substantially outdistancing Olin Eggen's record of 99 publications in eleven years. It must be conceded, however, that Eggen was a more solitary researcher, and over 80 per cent of his papers appeared under his own name alone.

Many of the large-team projects that dominated the Mount Stromlo research scene during the 1990s were cosmological in nature, designed to address some of the fundamental gaps in our knowledge of the Universe as a whole. Exactly how big and how old is the Universe? Can we learn anything about its Big Bang origin from the way the galaxies are distributed in space today? Is its expansion slowing down and, if so, is it decelerating enough to trigger an eventual collapse into a 'Big Crunch'? Exploring such mysteries has become one of the most exciting human endeavours of all time, capturing the attention of the public at large as well as consuming the energies of astronomers themselves.

An excellent example of Mount Stromlo's cosmological endeavours during the 1990s was the Director's own chief research involvement, the Hubble Space Telescope (HST) Key Project to obtain an accurate calibration of the extragalactic distance scale.

Although much delayed by the Space Shuttle *Challenger* disaster, the launch of the HST remained as keenly anticipated as ever. In the early science planning, three 'Key Projects' had emerged, one of them being to determine the value of the Hubble Constant with a precision hitherto unattainable.[1] Indeed, the naming of the Space Telescope for Edwin Hubble demonstrates the importance that the astronomical community attached to this particular aspect of its work.

Research groups were encouraged to make proposals for undertaking the Key Project initiatives, and Jeremy Mould became keenly involved.[2] He was one of three co-leaders of an international team of over twenty researchers which, in 1986, was entrusted with undertaking the task of re-establishing the distance scale.[3]

As noted in earlier chapters, Edwin Hubble was responsible for making one of the most fundamental scientific discoveries of all time—that the radial velocities of external galaxies are directly proportional to their distance from us.[4] At first sight this seems to place our Milky Way in a very privileged position; if all other galaxies are receding from us, surely this implies that we are at the centre of some sort of expanding system. That would indeed be the case were the relationship between velocity and distance anything other than one of direct proportionality, but this particular relationship carries no such implication. One can picture the galaxies as being like dots drawn on a sheet of rubber which is being stretched. *Any* dot will see all the other dots moving away from it, and each will see exactly the same proportionality between velocity and distance that Hubble found for the galaxies. In other words, Hubble discovered that the cosmos seems to be undergoing a type of universal expansion, there being no particular centre.

The rubber substrate analogy has another advantage. It is not correct to visualise remote galaxies as receding from us through space—it is space itself that is expanding, carrying the galaxies along with it. The dots are attached to a rubber sheet that is being stretched, not sliding across the surface of a static rubber sheet.

By the same token, it is not really correct to think of the redshift seen in a remote galaxy's spectrum as an example of the Doppler effect. A Doppler shift is a change in wavelength caused by an object's motion; it is used to measure velocity in many practical applications—in police radar for instance. The cosmological redshift is better thought of as being due to the fact that the wavelength of the light concerned has been expanding with the Universe during its long journey. It is larger when it reaches us than it was when it set out simply because the whole Universe has expanded by the same factor. Nevertheless, it is a convenient shorthand to refer to cosmological redshifts as if they did denote actual velocities, and we will usually resort to this common expedient.

Thus the relationship Hubble found can be written down in a simple mathematical way:

$$\text{Velocity of recession} = H_0 \times \text{Distance}$$

The constant of proportionality, H_0, is referred to as Hubble's Constant.[5] The velocity is usually measured in kilometres per second (km/s), and the distance in megaparsecs (Mpc); these units are implied whenever a 'numerical' value for H_0 is quoted in the following.[6] The value of H_0 (in these units) had been found to lie somewhere between 50 and 100, but obtaining it precisely had always been a great challenge, and there were still strenuous disagreements among cosmologists in the lead-up to the launch of the Hubble Space Telescope.[7]

The Hubble Constant is a fundamental ingredient in cosmology. Many aspects of the past and future history of our Universe depend in an intimate way on just how fast it is expanding now. For instance, the reciprocal of H_0 has the 'dimension' of a time, and its value can be thought of as telling us how long it has been since the expansion started—the age of the Universe, in other words. Several possibilities arise. If our Universe is decelerating just sufficiently to come to a standstill at infinite time—a 'flat' Universe—then its current age will be two-thirds the reciprocal of H_0.[8] If it is decelerating more strongly—a 'closed' Universe—it will eventually come to a halt and contract in on itself again, and its present age will be shorter than the two-thirds value. If it is decelerating less strongly—an 'open' Universe—it will continue expanding forever, and its present age will be correspondingly longer. For the special case in which the expansion is neither accelerating nor decelerating, the age of the Universe will equal the reciprocal of H_0 exactly. It is even possible that the expansion of the Universe may be accelerating, in which case its current age will be greater than the reciprocal. Whatever the case, however, the reciprocal of Hubble's Constant is the fundamental quantity that must be scaled appropriately to derive the age.

The value of H_0 can also be used to determine just what average density the Universe would need to have for its geometry to be 'flat'. This is called the 'critical density'—it can be thought of as marking the dividing line between the two possibilities of eventual collapse or expansion forever. The value of H_0 is also intimately related to the production of the chemical elements during the first few minutes after the Big Bang. Most of this primordial matter consisted of the two lightest elements, hydrogen and helium, but very small traces of other light elements such as lithium also emerged.

Given the importance of the Hubble Constant to our understanding of the Universe, and the tireless work of many researchers in trying to determine its value, why was it that, 60 years after Hubble first discovered his law, the value of H_0 was still uncertain by a factor of two? The reason is that the effect being measured—the expansion of the Universe—is an extremely weak one and, close at hand, the motions it causes are much smaller than those due to the gravitational interactions experienced by the various entities involved. For example:

- Our Earth has an orbital velocity of 30 kilometres per second due to the gravitational pull of the Sun, while the velocity due to expansion of the Universe over the Earth–Sun distance is a mere 0.0000000003 kilometres per second—entirely negligible by comparison.
- The Sun is being pulled along at about 200 kilometres per second by the mass of our Galaxy, but the cosmological expansion over Milky Way dimensions amounts to only about 1 kilometre per second.
- The gravitationally induced motions of the 40 or so galaxies forming our Local Group are typically about 100 kilometres per second, comparable with the Hubble expansion over its 5 million light year radius.
- Our Local Group members experience a gravitational pull of 200 kilometres per second towards the large Virgo Cluster of galaxies—*still* a sizeable contamination of the 1,100 kilometres per second cosmological expansion over the 50 million light years separating us from that cluster.

The Hubble motion is certainly real but, on the local scene, it is so tiny as to be swamped by gravitational effects. It is only when one observes galaxies some millions of light years away that the awesome cosmological expansion of the Universe becomes clearly apparent. Distances of 1,000 million light years are required before one can be sure that the contaminating effects of gravity will amount to just a few per cent.

Thus obtaining a precise value for the constant in Hubble's equation involves measuring the recession velocities and the distances for galaxies that are very remote indeed. As we have noted before, the recession velocities of such galaxies are fairly easily obtained from their redshifts, but finding out how far away they are is a daunting observational task.

Distance determination is challenging even for objects in our own Milky Way, and various approaches must be used, depending on how remote the object is. Astronomers have built up a sort of ladder of techniques, starting with distances inside our Solar System and working outwards to the farthest galaxies. Each technique is like an additional rung on the ladder, and it must be firmly tied in with all the rungs below it. Any error in the calibration of one particular technique affects not only the distances obtained by that technique, but those obtained on all the higher ladder rungs as well. It is easy to see, therefore, that the more remote an object is, the greater will be the uncertainty attaching to its distance determination.

Many of the distance techniques are based on 'standard candles'—astronomical phenomena which have a known intrinsic brightness. Since the apparent brightness of an object decreases as the square of distance, measuring the luminosity of such a candle provides a measure of its distance, provided care is taken in correcting for such annoyances as the dimming of the light by interstellar dust clouds on its journey. Since we need large distances to measure H_0, the standard candles need to be exceedingly bright ones. Supernova explosions serve admirably, especially one particular type of supernova (Type Ia), which occurs when a white dwarf in a binary system has material dumped on its surface by mass loss from its companion star. White dwarf stars cannot survive beyond a certain upper mass limit, and when the accreting star reaches this limit it suffers a sudden explosion, resulting in its complete destruction. So much energy is released that such a supernova event can be measured in galaxies well beyond 1,000 million light years away. Thus Type Ia supernovae provide one of the most promising standard candles being used by cosmologists today.

Other, quite distinct, distance measurement techniques also exist. Some depend on observing the dynamical properties of the galaxy concerned. Strictly speaking, these methods tell us about the intrinsic *mass* of the distant galaxy, rather than its intrinsic *luminosity*, but since the more massive the galaxy is, the more luminous it is, generally speaking, such dynamical observations can still yield a distance estimate for the galaxy. Yet another method of measuring the distance of a remote galaxy depends on observing how well resolved, or poorly resolved, are its component stars. The stellar field becomes more and more indistinct for galaxies at greater and greater distances, and this technique takes advantage of this effect.

Although some of these techniques have shown great promise for measuring the distances of remote galaxies, the difficulty has been to tie them into the established bottom rungs of the astronomical distance-scale ladder. Take the Type Ia supernovae, for instance. Such events occur only rarely, typically less than once per century in any one galaxy. Thus the chances of observing an event in the small number of galaxies that lie quite close to us are slim indeed. In order to observe an appreciable number of them, we would need to monitor some of the nearest clusters of galaxies—such as the Virgo Cluster, a grouping of well over 1,000 galaxies lying at a distance of 50 million light years. The challenge, then, is to extend the established bottom rungs of the ladder out to objects like the Virgo Cluster, and so place techniques such as that using the Type Ia supernovae on a firm footing. It was to meet this challenge that the HST Key Project was designed.

The most celebrated rung on the distance-scale ladder is that involving the Cepheid variable stars. These have an intrinsic luminosity which is closely related to their period of pulsation, making them excellent standard candles. Furthermore, they are quite bright stars (though not nearly as bright as supernovae) and can be observed in some of the nearby external galaxies. The story of Ben Gascoigne's observations of Magellanic Cloud Cepheids in the 1950s, and of his calibration of the Cepheid distance scale in general, has already been told in Chapter 6.

The Cepheid technique is so valuable because it spans an enormous range of astronomical distances. The closest Cepheids we can use lie only a few thousand light years from us, in our own Milky Way, while the most distant objects measured (prior to the HST) lie in the galaxy M101, 22 million light years away. It appears almost greedy to want to extend such a large range by another substantial factor, but that was precisely the goal of the HST Key Project. Having the Telescope above the Earth's turbulent atmosphere extends its effective observational reach tenfold, leading to a thousand-fold increase in the volume of space it can explore.

The Key Project team commenced their observations in earnest following the December 1993 HST repair mission. They selected, for Cepheid observations, a list of eighteen galaxies lying at distances ranging from 10 to 80 million light years. In addition, they were able to make use of data for a further eight galaxies observed by other teams using the HST, as well as for five well-studied nearby galaxies. The analysis thus involved a total of 31 objects, an enormous advance on what had previously been possible. Some of the objects on the list were members of the nearest large clusters of galaxies, in Virgo and Fornax.

The Key Project's progress was swift following the HST repair, a fitting reward for all its careful planning and preparation. By the end of the first year, observations had been completed for five of the galaxies on the list.[9] Team members were elated by their success in observing Cepheids in the galaxy M100 in the Virgo Cluster, one of their most distant objects.[10] By the time the project had reached its midway point, there was a growing certainty that the true numerical value of H_0 was approximately 73.[11]

This raised an interesting dilemma. Such a value for H_0 implies an age for the Universe of only about 13,500 million years, even if its expansion has suffered no deceleration since the Big Bang. Such an age is dangerously close to being in conflict with that calculated for the oldest stars. Don VandenBerg, a Mount Stromlo PhD graduate of the 1970s, has spent his subsequent career at the University of Victoria in Canada, making computer models of the interiors of stars. These can be used, among other things, to estimate their age. In 1996 he co-authored a review of the age of the Milky Way globular clusters (some of the oldest stellar systems known), concluding that it is 15,000 million years, with an uncertainty of perhaps 20 per cent.[12] Clearly the Universe itself cannot be younger than its oldest inhabitants. The dilemma was made much worse, however, by the fact that the model of the Universe in vogue at the time was *not* one with zero deceleration. Cosmologists favoured a 'flat' Universe—that is to say, one which is decelerating to such an extent that the age just cited must be reduced by a third. In other words, the Key Project's preliminary value for H_0 corresponds to a Universe just 9,000 million years old—in clear conflict with VandenBerg's globular cluster results. It was already becoming apparent that the Key Project was placing some interesting constraints on cosmological models if they were to accord with the observed properties of the real Universe.

The Project's HST observations drew to a conclusion in September 1998, about five years after they had commenced in earnest. During that time, distance estimates for each of the program galaxies appeared in a series of nearly 30 research papers and, at the program's conclusion, the results were all combined and tied in with the data for five separate long-range distance techniques capable of reaching out into the Universe at large (supernova Type Ia measurements, dynamical observations and the like). The overall conclusions were published in 2001.[13] The outcome was a resoundingly successful attainment of the Project's goal. There was an almost exact concordance between the estimates of H_0 derived from the five independent long-range scaling techniques, indicating that the Project had successfully tied each of them into the well-established Cepheid distance scale. The final result for the numerical value of H_0 was 72, and its calculated precision had indeed been reduced from the previous factor of two to a mere 10 per cent.

The Cepheid distance scale developed in the HST Key Project was based on an adopted distance for the variables in the Large Magellanic Cloud—a strategy almost universally adopted in extragalactic research. The team considered about 80 independent estimates for the LMC distance, made over the previous twenty years, before adopting a value of 50 kiloparsecs (163 thousand light years). This may be compared with the value found by Ben Gascoigne in his pioneering study of LMC Cepheids begun in 1951.[14] His result was 56 kiloparsecs, in remarkably close agreement with the best modern estimates. Subsequent to the completion of the HST Key Project, the fiducial period–luminosity relationship for LMC Cepheids has been refined further by a massive observational study of about

600 variables, undertaken by Kim Sebo and Daya Rawson of Mount Stromlo, in collaboration with Jeremy Mould and several other colleagues.[15]

So what are the cosmological implications of the Key Project's new, precise value for H_0? Even as the project was drawing to its conclusion, a significant downward revision was being made to the age estimates for globular clusters.[16] Observations by the astrometric satellite *Hipparcos* provided many new distance estimates for nearby stars, which caused a flow-on reassessment of those for other objects within our Galaxy, including the globular clusters. The upshot has been that the clusters are now believed to be only 11,500 million years old (still with a 20 per cent uncertainty).[17] Even following this adjustment, however, there is an awkward near-discrepancy between the lowest acceptable age for the clusters—9,500 million years—and the oldest flat-Universe age that the Key Project would allow—9,900 million years. The two are compatible only if the error estimates, in both cases, are stretched to the limit in the appropriate direction—a very uncomfortable requirement. It is almost certain, therefore, that the Key Project has excluded the cosmological model that was in vogue when the Project commenced.

In the event, however, this has hardly mattered. Cosmology does not stand still for long. Even while the HST work was in progress, some other exciting developments were taking place which have had profound consequences for our erstwhile 'standard' view of the Universe. Several Stromlo researchers played a crucial role in this outcome.

☆

In 1995, Brian Schmidt, recently graduated from Harvard, arrived at Mount Stromlo as a Postdoctoral Fellow. He immediately set himself the task of organising an international research collaboration to take advantage of one of the most promising techniques for exploring the Hubble relationship in more distant realms of the Universe—observation of the Type Ia supernovae (SN Ia) just described. We have noted how the HST Key Project was successful in tying this technique into the well-established Cepheid rung of the distance-scale ladder. We should now consider, in a little more detail, what would be involved in exploiting this achievement to advance our understanding of the remote Universe.

There are advantages and disadvantages in using SN Ia events to study cosmology. One advantage is that these explosions are exceedingly bright phenomena. Although they involve only a single star in its death throes, that star becomes as bright as a whole galaxy of perhaps 100 thousand million stars. Thus supernovae can be observed to a considerable distance in the Universe—several thousand million light years. Another advantage is that their intrinsic brightness is predictable—in other words, they make excellent standard candles. That is not to say that all SN Ia events attain exactly the same luminosity. In fact, they do not; however, the intrinsic brightness that any one SN Ia achieves is closely related to the time it takes to fade from its peak into oblivion. Typically, such an exploding star will increase in brightness for a few weeks, and then gradually decline over several months. By carefully measuring this rate of decline, the star's intrinsic luminosity at peak may be calculated, and its apparent brightness at peak may then be used to provide a measure of the distance of its host galaxy. This standard candle distance to any particular event may be estimated with a precision of about 7 per cent.

The most obvious disadvantage of using supernovae, of course, is that (unlike the Cepheids) they are transient phenomena whose occurrence cannot be predicted. Each event must be caught during just those few weeks while it is still increasing in brightness. Another problem is that supernovae are rare events in any particular galaxy, occurring perhaps only once per century, even in large ones.

Given these characteristics, the observing requirements for an effective SN Ia distance-measuring program become clear. Some thousands of galaxies must be observed routinely and regularly in the hope of spotting any supernova outbursts that occur. This requires a survey instrument with a fairly large field of view, and a capacity to record each image in digital form before moving quickly on to the next field. The prodigious amounts of data obtained must be stored and analysed almost in real time if an outburst is to be caught on the rising part of the light curve. The computer asks only one question at this stage: is there, in any of the galaxies on the image, a bright point of light that was not there when that galaxy was last observed? When a supernova has been discovered, its light curve must be monitored for many months, generally by an entirely different telescope, in order to determine its peak brightness and rate of decay. Also, one has to obtain spectra to confirm that the outburst really was of the SN Ia type, and to determine the redshift of the supernova's host galaxy, if this is not already known.

It is easy to see how such a project can involve a large team. Schmidt organised a loose affiliation of over twenty researchers, who called their endeavour the 'High-z Supernova Search'.[18] They work on four continents, making use of about a dozen of the world's finest telescopes, including the HST. As one might expect, other research groups are also exploiting the Type Ia supernovae. One such is the 'Supernova Cosmology Project' being led by Saul Perlmutter of Lawrence Berkeley National Laboratory in the United States.[19]

Schmidt's team had immediate success, discovering over 30 SN Ia events in the first two years, including the then most distant outburst ever seen, SN 1995K.[20] This object has a redshift of 0.478. That is to say, the features in its spectrum have been shifted to 1.478 times their original wavelength, which places the event over 4,000 million light years away.[21] Thus the supernova technique is capable of reaching out to vast distances in the Universe. The current record is held by the outburst SN1997ff, which occurred in an elliptical galaxy with a redshift of about 1.7. The light we see from it started its journey when the Universe was only 37 per cent of its current size, or a twentieth of its current volume, providing us with a glimpse of the cosmos in its extreme youth.

In examining the part of the Universe that is comparatively nearby, the HST Key Project established its rate of expansion *at the current epoch*. The High-z Supernova Search team can ask more penetrating questions. By peering deep into the past as well, they can also examine how the rate of expansion has been changing. The expectation is that it will have been slowing down, since some of the expansion energy of the Universe is constantly being used up in overcoming gravity as its mass content is dispersed into a larger and larger volume. We have already noted that the ultimate fate of the Universe and its space–time geometry—whether it is closed, flat or open—both depend on how rapidly its expansion is being decelerated. This in turn depends on just how much mass the Universe contains. The greater the current average density, the greater the deceleration, and the more likely it is that our Universe will ultimately come to a halt and re-collapse.

We have mentioned earlier that, in the 1980s, cosmologists had a strong preference for a flat model of the Universe—one in which the energy of its expansion and the gravitational pull of its mass are exactly in balance. A flat Universe will eventually come to a standstill—but only at infinite time. There were good reasons for this preference, since a flat geometry is predicted by a successful model of the early development of the Universe, called 'inflation'. This was invented in the early 1980s to overcome some difficulties with the then current cosmological models. One difficulty has to do with the fact that the average properties of the Universe seem everywhere to be the same. Suppose we were to choose two directions diametrically opposite each other on the sky, look as far away as we can in each and compare how the Universe appears in these two remote locations. The two could never have been in touch with each other—by light signal or any other way—since there has only just been enough time, since the Big Bang, for their light to reach us at the midway point between them. So how did these two isolated regions know that they had to resemble each other?

The theory of inflation overcomes this difficulty by supposing that there was an era of extraordinarily rapid expansion in the early Universe when, for a tiny fraction of a second, galaxies were torn apart at an alarming rate.[22] This helps with the problem, since we can now suppose that our two remote, unconnected domains actually *were* in contact before this inflationary event. It was only because of the rapid expansion that they became separated by such a vast distance that it has taken all of the subsequent history of the Universe for the light from each of them to be meeting now at the halfway point.

This may sound as if we've simply replaced one perplexing problem—the need for unconnected parts of the Universe to 'know' about each other—with an equally perplexing one—the need for there to have been an early period of unbelievably rapid expansion. In fact, however, there are other cogent reasons—in the domain of particle physics—for postulating inflation.[23] Cosmologists and particle physicists have built up an impressive case for the theory, neatly tying together some quite fundamental concepts in both fields. One remaining problem, however, is that the mass needed to achieve the flat geometry predicted by inflation is much greater than is seen in the galaxies of stars and nebulae that surround us. It was partly because of this cosmological missing-mass problem that the MACHO Project's hunt for dark matter attracted such interest.

The Type Ia supernova observations are thus extremely significant. They hold the promise of determining just how much mass there is in the Universe, not by looking for the mass itself, but rather by measuring the deceleration it has been causing in the Hubble expansion. The working diagram for deciding the issue is a plot with the 'luminosity' distances obtained for these remote supernovae along one axis, and the redshifts obtained for their host galaxies along the other. For galaxies close at hand, this plot will simply be a straight line, reflecting the present-day Hubble expansion. As supernovae are observed at greater and greater distances, however, their points on the graph will begin to deviate from the line in a way that reflects the fact that the past expansion was more rapid than it is today. The shape of the deviation required by a flat Universe is well known, and one might have hoped that the SN Ia observations would match this particular curve, thus confirming the predictions of the inflation theory.

The world's cosmologists were in for a surprise! Far from confirming the flat-Universe value of deceleration, the most distant supernovae indicated that the Universe is not decelerating at all, but speeding up. This was quite unexpected. For many decades, cosmologists had been working on the assumption that there is *no* process at work in the Universe that can cause such an acceleration. The Universe contains mass—mass will always be self-gravitating—dissipating the mass will always consume energy—providing this energy will always decelerate the expansion. The most extreme situation envisaged was that the mass of the Universe might be so tiny in comparison with its expansion energy that the resultant deceleration could be almost zero. But it would still be a deceleration, however small.

Instead, the High-*z* Supernova Search consortium found an acceleration. In late-1998 they published an analysis of their observations for sixteen remote SN Ia explosions (and 34 nearby events used for comparison purposes).[24] The large redshifts of the distant outbursts show that they occurred at considerable 'look-back' times into the past. The nearest of the sixteen exploded when the Universe was 15 per cent smaller than it is at present; the farthest occurred so long ago that the 'now-to-then' size ratio is almost a factor of two—in terms of volume, the Universe was nearly eight times smaller. These observations demonstrated, with a statistical confidence of over 99 per cent, that the expansion of the Universe is accelerating. At almost the same time, the rival Supernova Cosmology Project team published an account of their work on an even larger sample of remote outbursts.[25] Reassuringly, they reached an identical conclusion.

Such a surprising discovery immediately caught the attention of the astronomical world. The prestigious weekly *Science* proclaimed the two teams' research as the 1998 'Breakthrough of the Year'.[26] Brian Schmidt himself has received many accolades, including the inaugural Malcolm McIntosh Award for Achievement in the Physical Sciences in 2000, the Pawsey Medal of the Australian Academy of Science in 2001, and several named lectureships.

What explanation can there be for this totally unexpected outcome? Fortunately, there is a ready-made one to hand—provided by someone with impeccable credentials: Albert Einstein himself. Einstein developed his theory of General Relativity (GR) a decade or so before Hubble discovered that the Universe is expanding. At the time, everyone assumed that the Universe was essentially static. That's not too surprising—it certainly looks static in the neighbourhood of our Galaxy, just as the surface of the Earth looks flat in the neighbourhood of our house. The fallibility of such local perceptions is revealed only when we can view things on a much vaster scale. Thus, in applying his GR theory to the Universe at large, Einstein was at pains to achieve a mathematical solution which would produce a static model. He did so by introducing a constant term into his equations (called the Cosmological Constant, and denoted by Λ), which had the effect of exactly balancing gravity so as to yield an entirely unchanging Universe.[27] When, a decade or so later, Hubble announced that the Universe is far from unchanging, Einstein is said to have pronounced the inclusion of Λ in his equations as the greatest blunder of his life; he immediately set it equal to zero. There is no theoretical reason why he should have done so. It's just that he considered that the inclusion of Λ detracted from the elegance and simplicity of his theory—so he crossed it out. Most cosmologists since that time have crossed it out as well. But the Cosmological Constant has always been lurking in the wings, and there have been times when some awkward observational results have persuaded cosmologists to resurrect Λ. Until now, however, its reintroduction has always proved temporary.

Thus it is again to Einstein's Λ that cosmologists have resorted in trying to account for the SN Ia search teams' acceleration result. In physical terms, a non-zero Λ can be thought of as introducing a new kind of energy into the Universe—vacuum-energy. This is entirely unrelated to the matter-energy with which we are so familiar, and it causes a repulsive gravitational force even when no material is present.[28] In mathematical terms, a non-zero Λ can be thought of as introducing an additional 'degree of freedom' into our choice of cosmological model. We now need two parameters to describe any particular model, one related to the matter-energy density of the Universe, and the other related to its vacuum-energy density.[29] It is as if we now require a whole sheet of graph paper to describe all the model choices available to us, whereas before a single line drawn on the sheet with a ruler would have sufficed.

One immediate consequence of introducing a non-zero–Λ is to break the nexus between the geometry of the Universe and its ultimate fate. All the statements we have previously made linking 'open' or 'closed' models with universes that expand forever or eventually re-contract must now be modified. All four combinations of these two pairs of characteristics now become possible. A much richer range of model universes lies before us, including some that are not Big Bang universes at all. Einstein's original static model is one such, of course, since time has no beginning and no end in that particular case.

The model which best accounts for all the current observational evidence, including the new SN Ia results, is one in which the total energy density of the Universe still lies at the critical value—that is, its geometry is still flat (as required by inflation)—but only about one-quarter of this total is supplied by matter-energy, the remaining three-quarters coming from the vacuum-energy associated with the reintroduction of Einstein's Λ term. The gravitational effects of the latter outweigh the former at the current epoch, so that the Universe feels an overall repulsive force, which explains its newly discovered acceleration.

This was not always the case, however; the new 'standard' model would still have experienced a deceleration in its very early stages. One can think of the attractive and repulsive forces of matter-energy and vacuum-energy as having been in a contest with each other to decide the ultimate fate of the Universe, with the latter having now attained the ascendancy. The most distant outburst ever seen, SN 1997ff (mentioned above), provides evidence for such a contest. Measurements for this event were obtained by the HST, somewhat serendipitously, rather than in either of the SN Ia search programs. The object is so far away that it samples that early phase in the development of the Universe during which, on the new model, deceleration should still have been occurring. Its extreme faintness leads to some observational uncertainties but, after a careful analysis of all the evidence, the High-z Supernova Search team has concluded that this event does indeed support the early deceleration required in the new standard model.[30] The two teams are seeking to detect several more equally remote supernovae in the hope of placing the issue beyond doubt. It may well be that our long search for the best possible cosmological model to represent the real Universe is at last reaching some finality.

One particularly satisfying attribute of the new standard model is that it restores the age of the Universe to a respectable 14,000 million years (even after the HST Key Project's revision of H_0), thus removing the worrying conflict with the ages of globular cluster stars.

☆

A third major observational thrust in the field of cosmology involves making large-scale maps of the Universe to study the 'clumpiness' of its various constituents. Although cosmological models assume a system that is perfectly smooth and homogenous, this description hardly fits the Universe we see about us. It is full of galaxies, and clusters of galaxies, and even superclusters, giving it a patchy appearance even on quite vast scales.

Once again we must look to the very early Universe for an explanation, since quantum fluctuations occurring during the inflationary era are thought to have caused the patterns we see about us today.[31] Such an explanation would require that a blotchy pattern also be evident in the cosmic microwave background (CMB) radiation—the present-day remnant of the cosmic flash as it was just 100,000 years after the Big Bang.[32] There was enormous excitement in 1992 when sensitive observations with the COBE satellite detected just such a mottled appearance in the CMB.[33]

This glimpse of the clumpiness in the early Universe can be compared with its current galaxy distribution, yielding additional insights into the development of the cosmos since the Big Bang. Thus, in recent years, astronomers around the world have initiated many programs to map the large-scale distributions of the Universe's various constituents. One particularly ambitious project was launched in 1997 and is being carried out by yet another large international team of about 30 researchers.[34] Matthew Colless, who joined the Mount Stromlo staff in 1993, is one of the co-leaders, while Bruce Peterson, Carole Jackson, Simon Driver and Roberto De Propris are team members.

The project became feasible with the commissioning of a magnificent new instrument at the Anglo-Australian Telescope. A two-degree-field (2dF), multi-fibre facility, it can sample an area of sky about sixteen times that of the full Moon, and is able to record spectra of 400 objects simultaneously.[35] The project team used this instrument during 1997–2001 to obtain the redshifts of over 220,000 galaxies. Their '2dF Galaxy Redshift Survey' involves ten times as many objects as in the previous largest study, and the volume of the Universe sampled is ten times greater. The data have been used to map the distribution of galaxies in two fan-shaped slices of the Universe, extending out to about 2,500 million

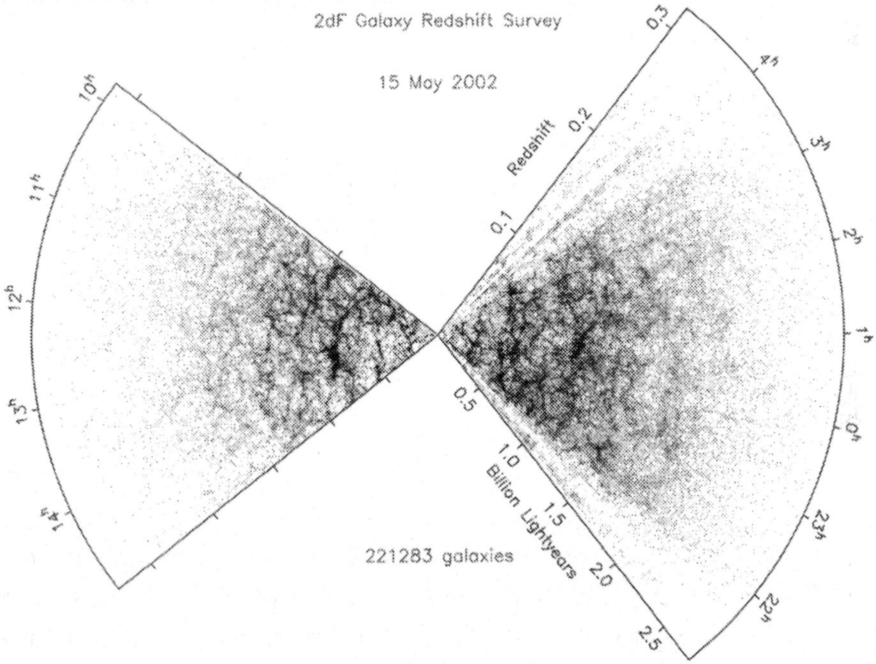

Two fan-shaped wedges of the Universe containing over 220,000 galaxies
Image: The 2dF Galaxy Redshift Survey Team

light years.[36] In both cases, the point of the fan lies at our own Galaxy, and the Hubble relationship is used to estimate the radial distance to any particular galaxy from its redshift. The maps obtained show a fascinating pattern of clumps, voids and filaments. All these structural features are considerably smaller than the total volume explored, however, so the team can claim to have sampled a statistically representative volume of the Universe. On the vast scale of their work, the cosmos at last appears smooth and uniform, in keeping with the models.

Combining these galaxy observations with the CMB data, the survey team have been able to place another set of constraints on the cosmological parameters of the real Universe.[37] Their results are in remarkable agreement with the those inferred from the SN Ia observations. That is, we seem to live in a Universe having a flat geometry, for which matter-energy is about one-quarter responsible, and vacuum-energy about three-quarters. Yuzuru Yoshii (a Fellow at Mount Stromlo in 1992–93) and Bruce Peterson have been able to provide additional support for this broad picture from an analysis of galaxy number counts with increasing distance.[38] Altogether, this is a particularly reassuring outcome, since a common 'standard' model has now emerged from a number of superb scientific investigations based on independent methods.

Can we assume, therefore, that Einstein's Cosmological Constant is at last here to stay? This would be a brave prediction, given its previous chequered history. Its inclusion has certainly circumvented some current problems, but many questions still need to be addressed. For instance, the SN Ia research relies on the physics of supernova explosions being the same today as in the distant past. Is this a safe assumption? Just what are the other physical implications, if any, of our introducing this mysterious vacuum-energy? Can we be sure that it is the *only* type of 'massless' energy at work in the real Universe—might there not be *many* different types waiting to be discovered? And so on.

Nevertheless, the new 'standard', inflationary, non-zero–Λ model of the Universe is looking very promising indeed, and new sensitive experiments in the pipeline will ensure continuing rapid progress. As the Galaxy Redshift Survey team has remarked, 'it is clear that the era of quantitative cosmology has arrived'.[39] We live in exciting times, cosmologically speaking, and Mount Stromlo Observatory can be well satisfied with the part it has been able to play in this great human endeavour.

The high-profile cosmological projects that distinguished the Mould directorship were accompanied by a host of other research endeavours in Mount Stromlo's long-standing fields of interest in stellar, interstellar, galactic and extra-galactic astronomy. We cannot do justice to all that was accomplished, but the following selection illustrates something of the breadth of the last decade's work.

- Peter McGregor, Mark Allen and a Cornell colleague, Phil Nicholson, used the CASPIR instrument on the 2.3-metre at Siding Spring to obtain some widely acclaimed infrared observations of the spectacular collision of Comet Shoemaker-Levy 9 with Jupiter (in July 1994).[40]
- In a re-analysis of the spectrum of Comet Halley, Bruce Peterson and colleagues found that its gaseous coma has much the same carbon isotope ratio as does the Sun, contrary to some earlier results.[41] Thus comets seem to have formed in the same event, and from the same gaseous material, as did other objects in the Solar System.
- Peter Wood has become an acknowledged world expert on the behaviour of stars in the late stages of their lives, including the various kinds of pulsation that such stars experience, and he receives many invitations to present conference reviews in this field.[42]
- Jon Bell and Ben Stappers chose the field of pulsars for their PhD research.[43] These rapidly spinning objects are the collapsed remnants of intermediate-mass stars which have ended their lives as supernovae. The remnant shrinks until it becomes a 'neutron star', a body only about 10 kilometres in size, composed almost entirely of neutron particles. Many pulsars emit a lighthouse-like radio beam (hence the name), which can be used to time their spin with extreme precision. If the object is a member of a binary system, it can be 'spun-up' until it is typically rotating at a thousand times a second. Because of their extreme properties, binary pulsars make excellent 'laboratories' for testing the predictions of physical theories such as Einstein's Theory of General Relativity.
- John Norris has played a leading role in an extensive program studying stars with extremely low 'metal' abundances, indicating that they are among the earliest formed in our Milky Way.[44] Thus they provide important clues to the physical processes at work in galaxy formation. Mike Bessell is also keenly interested in such objects; he is part of a team which, in 2002, discovered a star having the lowest iron abundance on record—less than 0.000005 that of the Sun.[45]
- John Norris, Gary Da Costa, Ken Freeman and Ken Mighell brought the

Mount Stromlo 74-inch and the AAT to bear on the continuing problem of what has caused the variations in the chemical make-up of stars in the globular cluster ω Centauri.[46] They observed additional chemical elements and many more stars in an effort to unravel the complexities of this giant cluster's formation and subsequent evolution.[47]

- Freeman also participated in a study of the motions of nearly 10,000 stars in the same cluster, using photographic plates taken with the Yale–Columbia 26-inch refractor over an interval of more than 50 years.[48]
- Infrared observations have provided strong evidence that the Milky Way has a bar feature in its central regions—as many spiral galaxies do—with the long axis of the bar pointing towards us. Sylvie Beaulieu, Ken Freeman, Agris Kalnajs, Prasenjit Saha, Roger Fux and Maartje Sevenster have all been involved in investigations of this. Their approaches have been quite varied—using planetary nebulae as tracers of the dynamics near the Galaxy's centre; interpreting the MACHO Project's results for the microlensing events seen in that direction; and constructing numerical models.[49]
- Helmut Jerjen, Agris Kalnajs and a Swiss colleague, Bruno Binggeli, discovered that an apparently featureless dwarf elliptical galaxy in the Virgo Cluster actually exhibits a very faint spiral pattern.[50] This totally unexpected result—which emerged in an analysis of a VLT image—implies the presence of a stellar disc in the system. There is no other known example of such behaviour in elliptical galaxies.
- Ken Freeman and Magda Arnaboldi have been using planetary nebulae as tracers in exploring the dynamics of the outer regions of elliptical galaxies.[51] The technique proved so successful that Freeman initiated and led an international project to build a special instrument for this purpose. It was designed and fabricated, in large part, in the Mount Stromlo workshops, and commenced service at La Palma Observatory in 2001.[52]
- John Norris and his colleagues have been studying lithium in some of the oldest Milky Way stars. This light element was produced, in small amounts, in the original Big Bang.[53] The team is seeking to quantify just how much lithium was made in this way, since this provides yet another constraint on the cosmological model that must be adopted to describe the real Universe.

☆

Given the increasingly important role that the world's largest astronomical telescopes were playing in Mount Stromlo's research endeavours in the 1990s, it is not surprising that Australia's lack of any formal link with a major international facility was being viewed with increasing concern. On becoming Director, Jeremy Mould lost no time in assuming a leadership role in this regard.

A vehicle for addressing the matter was to hand. For some time, the Academy's National Committee for Astronomy (NCA) had been seeking to formulate a strategy for the future. In the month of Mould's arrival (December 1993), it decided to undertake a thorough, year-long 'discipline review' under the auspices of the Australian Research Council (ARC).[54] The review was chaired by Harry Hyland, the then President of the Astronomical Society of Australia and Chair of the NCA. No less than eight sub-committees were established, and Mould convened the one on 'Future Facilities'.[55] The lengthy, well-documented review report contained many recommendations. There was no doubt, however, as to the astronomers' top priority:[56]

> The first priority of the astronomical community remains the same as it was in 1989, *viz.* to obtain significant access to a large optical/infrared telescope. To achieve this end the Review Committee believes that Australia should immediately accept the European Southern Observatory's (ESO) invitation to join ESO and participate in the world's premier astronomy project, the Very Large Telescope (VLT).[57] Through ESO, the Australian share of a $500m project is affordable, feasible and timely. Membership of ESO (~$5m per annum) will maintain Australia's place among its peers for the foreseeable future, promoting Australian leadership of the organisation in some technologies.

That Australia had received such an invitation from the ESO, a hitherto entirely European consortium, was a significant compliment and, now that almost everyone accepted that the Australian continent afforded no world-class LOT site, there was strong enthusiasm for accepting the invitation and thus acquiring access to the VLT.

The review's timing was excellent, coinciding with a government call for proposals in a new Major National Research Facilities program.[58] A total of $60 million was to be made available over eight years, and more than 60 applications were received. Astronomers were elated when their $28 million ESO bid received top ranking from two expert committees set up by the government—but their joy was to be short-lived. In a pre-election announcement in December 1995, Prime Minister Paul Keating overturned the committees' advice, awarding the funds to a larger number of more moderately priced projects—one based in each of the seven Australian capital cities![59] Public reaction to this statement was so adverse that the Science Minister, Senator Peter Cook, felt compelled to mount a salvage operation.[60] Jeremy Mould and the government's Chief Scientist, Michael Pitman, were sent to ESO's Garching (Germany) headquarters in February 1996, returning with a draft agreement. Cook exchanged letters with the ESO and, in the context of the election campaign itself, announced that ESO funding would be made a priority in the August 1996 budget—should the government be returned in the poll on 2 March.[61] The Keating government suffered a resounding defeat.

The new Liberal–National Coalition government led by John Howard loudly decried the budgetary situation it had inherited, in common with most incoming governments. It announced a comprehensive two-year program of cuts totalling $8 billion, and the new Science Minister, Senator Peter McGauran, warned that there could now be no certainty about funding for ESO membership.[62] Mould led the nation's astronomers in a vigorous rearguard campaign. In a meeting with McGauran on 11 April, he highlighted the significant scientific *and commercial* benefits that membership would bring. Being eligible to bid for the construction of VLT components was expected to earn Australian industry tens of millions of dollars worth of contracts. Mould came away feeling encouraged, but in May McGauran wrote to the ESO's Director General sounding the proposal's death knell.[63] The letter held out the possibility of another ESO bid 'at some time in the future', but this was little consolation given that the opportunity for winning initial VLT contracts would by then be a thing of the past. Mould reacted characteristically. While not concealing his disappointment, he announced that 'Australian astronomers are now regrouping for a further campaign'.[64]

The NCA and its Large Telescope Working Party remained in active negotiation with the ESO, and in June 1997 the ESO Council approved a protocol whereby cooperative arrangements could be forged with non-member states, enabling them to provide enhancements to ESO facilities in return for guaranteed access and rights to use these facilities, as non-members.[65]

Meanwhile, in May 1997 a new large telescope option emerged when another major consortium, the International Gemini Project, faced a funding crisis precipitated by the impending withdrawal of Chile, a partner nation at the 5 per cent level.[66] Gemini called for expressions of interest in taking up this share. Mould persuaded the ARC to make such an offer, which was received with interest. In the event, Chile managed to pay its contribution at the eleventh hour (August 1997), and the emergency was averted. Nevertheless, the brief prospect of Australia's joining the collaboration had generated some enthusiasm, and the Gemini consortium decided to reaffirm its invitation, even though this would now involve the project's total participation rising to 105 per cent! On

18 February 1998, David Kemp, the Education Minister (with responsibility for the ARC), announced the government's support, and a revised Gemini agreement was signed in May, earning Australia 4.76 per cent of observing time on the twin 8.1-metre telescopes.[67] An Australian Gemini Office was established at Mount Stromlo, with Gary Da Costa serving as Project Scientist.[68] An Australian Gemini Steering Committee was also set up, with Lawrence Cram, of the University of Sydney, becoming the first Australian member of the Gemini Board.

This was not the astronomical community's first preference—a 4.76 per cent share of *two* 8.1-metre reflectors is hardly comparable with a similar share of the *four* 8.2-metre instruments forming the VLT array—but it did at least provide some access to one of the world's leading-edge optical facilities. Most disappointing of all, perhaps, was the fact that the final decisions at government level seemed to have taken little account of the scientific, or indeed of the commercial, merit of the case, but were shaped rather by political timings having more to do with Australia's three-year electoral cycle.

Nevertheless, Australian—and Stromlo—astronomers have become extremely proactive in instrument development projects for both these major facilities, earning considerable respect for the technical competence of the nation's observatories.

- Mount Stromlo, in collaboration with the Institute for Astronomy of the University of Hawaii, has contracted to design and build a Near-infrared Integral Field Spectrograph (NIFS) for the Gemini North Telescope during 1999–2003. This $4.5 million project is being led by Peter McGregor as Project Scientist and Jan van Harmelen as Project Manager.[69] With this instrument, the Gemini Telescope will attain the same spatial resolution, at infrared wavelengths, as the Hubble Space Telescope achieves in the optical.
- Matthew Colless is a co-Principal Investigator in a tri-institutional consortium (Stromlo, AAO, University of New South Wales) which has been developing instrumentation for the ESO's VLT.[70] They first undertook a conceptual design study, completed in 1996, for a multi-object spectrograph called AUSTRALIS. More recently, they were awarded a design and fabrication contract for a fibre positioner called OzPoz, due to come into service in April 2003. These contracts have a combined value of $3.7 million.
- At the time of writing (November 2002), a Mount Stromlo project team led by McGregor, van Harmelen and John Hart has just been awarded

> a $6.3 million contract for an Adaptive Optics Imager on the 8.1-metre Gemini South Telescope in Chile.[71] This sophisticated camera will be part of an instrumental arrangement that removes the 'blur' caused by the Earth's turbulent atmosphere.

Clearly, Mould's representations to successive governments concerning the commercial value of the astronomical community's proposals were in no way exaggerated. Contracts worth tens of millions of dollars will indeed be achieved, in spite of the governments' timidity in supporting the astronomers' aspirations.

Work on these overseas projects has been undertaken in parallel with many other development tasks for the telescopes at Siding Spring. One major device, built to operate on both the ANU's 40-inch reflector and the 3.9-metre AAT, is a wide field imager (WFI)—one of the largest electronic cameras in the world.[72] Gary Da Costa was Project Scientist, and the WFI came into service in September 2000. Another telescope improvement at Siding Spring was the building of a 'tip-tilt' device which allows the 2.3-metre telescope to make first-order adaptive-optic corrections when using the CASPIR infrared camera.[73]

The Mould directorship was a time of great development in many other respects as well. One achievement resulted in yet another change of name for the Observatory. In August 1998, Mount Stromlo attained full research school status within the ANU's Institute of Advanced Studies, becoming the Research School of Astronomy and Astrophysics. Although the stature and fame of the Observatory's work over the decades had been rewarded by an increasing measure of independence within the ANU, this ultimate prize was still not awarded lightly. Stromlo's bid was scrutinised minutely for three years before being approved by Council.[74]

One of Mould's first moves as Director (in March 1994) was to establish formal links with the ANU's School of Mathematical Sciences, by creating an Astrophysical Theory Centre (ATC).[75] Theoretical pursuits in Australian astronomy have always been somewhat overshadowed by the nation's world-class observational and instrumental exploits, and this was one of a number of measures taken within universities to redress this imbalance. The ATC had the added benefit of forging a closer relationship between the ANU's graduate and the undergraduate teaching programs, to their mutual benefit.

The ATC was initiated chiefly through the affiliation of *existing* staff in the two areas, but eventually it became possible to make a number of new, fixed-term appointments as well.[76] An active visitor and conference program was also developed. In the event, the enthusiasm for the Centre's creation soured somewhat as

The Sun Room at the Mount Stromlo Observatory Visitor Centre Photo: MSO

time went by, owing partially to some personality conflicts, and the formal ATC arrangement went into abeyance in 1999. By then, however, Mount Stromlo had established practical links with the undergraduate departments concerned—not least by making a number of joint appointments—so the original aims of the ATC have been at least partially fulfilled. Furthermore, a new national affiliation of astrophysical theorists has been formed—the Australian National Institute for Theoretical Astrophysics (ANITA). Mike Dopita chairs the ANITA steering committee, and Geoff Bicknell also serves on it.

Mould was also instrumental in establishing a fine new Visitor Centre at Mount Stromlo, to showcase the research endeavours of the Observatory (and indeed of the ANU at large), and to help satisfy the public's fascination with the mysteries of the Universe.[77] Visitor facilities at Siding Spring had long surpassed those at Mount Stromlo itself, so the new Centre was overdue. It has a large exhibition hall, and a restaurant which—having one of the most spectacular views in Canberra—boasts a large and growing clientele. One of the most popular displays is a Sun Room where the solar disc is imaged by the lens of Duffield's original Sun Telescope. The pioneer Director's daughter, Miss Joan Duffield, who has been a most generous benefactor of the University on several occasions, donated the cost of mounting this display. Another of her gifts endowed a named chair.[78] At a ceremony attended by Joan on 20 June 2000, Ken Freeman was installed as the inaugural Duffield Professor of Astronomy, a worthy recipient of this high honour.

During the 1990s, Mount Stromlo gave strong support to a cross-institutional activity to assess the potential of the Antarctic continent for astronomy. Although the popular image of Antarctica is not immediately suggestive of a 'good observing site', the continent does have several advantages—long intervals of total darkness; extremely cold conditions, which are helpful in making infrared observations; a large region at 3,000–4,000 metre altitudes; and, on this high plateau, much calmer and drier conditions than prevail on the blizzard-plagued coast.[79] Of course, formidable logistical challenges present the other side of the picture.

JACARA's Automated Astrophysical Site Testing Observatory at the South Pole
Photo: MSO

In 1994, Mount Stromlo joined with the University of New South Wales in establishing a Joint Australian Centre for Astrophysical Research in Antarctica (JACARA).[80] A good deal of testing equipment was developed, and this has been deployed at the South Pole base, as a first step towards its eventual relocation to higher altitude.[81] Several staff have participated in short-term polar expeditions, and a Lockheed Automated Geophysical Observatory was acquired to enable testing to continue for the whole dark season. Stromlo staff members closely involved in this venture include Mike Dopita, Ralph Sutherland, Gary Hovey, Peter Conroy, Peter Wood, Brett McLinden and Mark Jarnyk.

By the turn of the decade, however, the Observatory's involvement in the Antarctic project was being overshadowed by its growing list of new national and international commitments, and in 2001 Mount Stromlo withdrew from active participation in the site testing, handing over its interest in the test equipment to the University of New South Wales.[82]

Two new 'foreign' developments have been initiated at the ANU's observatory sites during the last decade. A precision laser satellite-ranging telescope was installed at Mount Stromlo, near the Oddie Dome, in 1998. It was built by Electro Optic Systems Pty Ltd of Queanbeyan, and is being operated by them for Geoscience Australia.[83] At Siding Spring, a British educational foundation, the Dill Faulkes Educational Trust Limited, is currently building a 2.0-metre reflector, which will serve as a remote-observing facility for school students around the world.[84] This Siding Spring installation will be the twin of another on Maui in the Hawaiian Islands.

At the conclusion of Mould's initial five-year term, the University, to general acclaim, renewed his appointment. Two years later, however, he announced his intention to move on.[85] He left office on 31 January 2001. In his first year as Director, he had developed a comprehensive ten-year Strategic Plan for the Observatory. Now, seven years later, he was able to reflect on progress in its implementation. While some matters had not quite followed the script (notably winning membership in Gemini rather than ESO), he could justifiably claim that his directorship had achieved as much as anyone could have hoped—and now fresh challenges beckoned. The post to which he moved was indeed a challenge: the directorship of the United States' National Optical Astronomy Observatory carries a mammoth responsibility for several major facilities in North and South America.

John Norris became Interim Director for what turned out to be yet another eighteen-month *interregnum* and after an international search, the Electoral Committee recommended the appointment of Penny Sackett as Mount

Stromlo's eighth Director.[86] Sackett's previous post was at the Kapteyn Astronomical Institute in Groningen, the Netherlands, but her career has been wide-ranging, embracing appointments in physics as well as in astronomy, and in government organisations as well as in academia. She has broad research interests in both galactic and extragalactic astronomy; her immediate enthusiasm is the search for extra-solar system planets. Sackett took up her appointment on 22 July 2002.

The period following Mould's departure has been one of considerable fluidity, especially in respect of financial matters. Government decisions in recent years have created a climate of increasing competitiveness and unpredictability in research funding. In particular, the University has, after a good deal of hesitation, agreed to a proposal for its Institute of Advanced Studies to relinquish a fraction of its block-grant funding, in exchange for eligibility to bid in ARC grant competitions.[87]

Sometimes, of course, such times of change create opportunities. One such arose when the government announced the creation of a prestigious new Federation Fellowship program. Recipients are funded entirely by the scheme, and are required to relinquish all other formal ties and responsibilities in order to devote themselves entirely to research. Mike Dopita was awarded a Fellowship in the inaugural round, bringing great distinction to the Observatory.[88]

To date, Mount Stromlo has fared extremely well in the new competitive funding regime, a tribute to Norris's energetic stewardship during the *interregnum*, and to Sackett's infectious enthusiasm during her first few months in office.[89] The Observatory's successes have included (alone or in various consortia): a $23.5 million grant in the Major National Facilities Program, which will, *inter alia*, fund an augmentation of Australia's share of Gemini; a $5.6 million Systemic Infrastructure Initiative grant for a wide-ranging upgrade of telescope facilities at Siding Spring; additional ARC grants worth some millions of dollars; and the $6.3 million Gemini AO Imager contract already mentioned. Such an outstanding start augurs well for the future of the Observatory.

☆

The story is told of an academic who was applying for promotion in a university where it had become apparent that the committee used a rule-of-thumb of twenty published papers for an application to succeed.[90] Many of the academic's contributions were important and substantial pieces of work but, sadly, when he counted them, he found he had only nineteen. Nevertheless, he bundled them up and went ahead with his application. Rather nervously, he attached a note to the pile: 'Please weigh; don't count'.

How does one weigh the 'value' of a piece of fundamental scientific research? Providing such a measure has developed into something of an industry over recent decades. One can tell how much impact a paper has had, so the argument goes, by noting how frequently other people reference it. An Institute for Scientific Information (ISI) now exists, and they (or their computers) laboriously examine the references at the end of every paper in over 3000 scientific journals. They then compile lists of them all, indexed against each and every original paper. The resulting compilation—the Science Citation Index—is consulted eagerly by appointment and promotion committees the world over.

From time to time, the ISI examines its own database in order to recognise individual researchers whose papers consistently appear at the top of the citation lists—'Citation Laureates' as it calls them. During 2001 they held a symposium at the Australian Academy of Science to honour 33 Australian scientists whose work over the years 1981–98 had earned them this distinction.[91]

Nine of the Citation Laureates were astronomers, and six of them were researchers at Mount Stromlo Observatory—Mike Bessell, Matthew Colless, Mike Dopita, Ken Freeman, Jeremy Mould and Bruce Peterson. The other three astronomers so honoured were Warrick Couch of the University of New South Wales, Dick Hunstead of the University of Sydney and Dick Manchester of CSIRO's Australian Telescope National Facility. It is nothing short of astonishing that Mount Stromlo, employing just two dozen of the 47,000 researchers currently working in Australia, and spending only 0.2 per cent of the national scientific research budget, should nevertheless claim 18 per cent of the Citation Laureate awards for work done, in all fields, during the last two decades.[92]

Other measures of Mount Stromlo's scientific acclaim may also be cited. Five staff members have been awarded the Pawsey Medal of the Australian Academy of Science (for research in physics) during the last 30 years—Ken Freeman, John Norris, Peter Wood, Mike Dopita and Brian Schmidt. Six students of the Graduate Program (or supervised externally by Mount Stromlo staff) have won Hubble Fellowships since the inauguration of this prestigious award in 1990—Xiaohui Hui, Heather Morrison, Sean Ryan, Yong-Ik Byun, Mary Putman and Martin Bureau. Cambridge and Leiden are the only other non-American schools that can boast a similar record.[93]

Measures such as these provide graphic illustrations of the scientific success that has attended the labours described in this book. The hopes and expectations

Mount Stromlo's six Citation Laureates. L to R: Jeremy Mould, Mike Dopita, Ken Freeman, Mike Bessell, Bruce Peterson and Matthew Colless Photo: MSO

of Coombs in the founding of the Australian National University have nowhere been better fulfilled than at Mount Stromlo. The vision of Duffield in establishing the Observatory has been fully vindicated.

What has been the secret of Mount Stromlo's success? Many answers to that question have been attempted, including one by the Director with whom this chapter has been chiefly concerned. On the occasion of a significant anniversary towards the end of his term, Jeremy Mould wrote:[94]

> In Mount Stromlo Observatory's seventy-fifth year we can and should celebrate its uniqueness as an astronomical institution. Stromlo is the Research School's astronomy campus. Like all campuses it is a meeting place. Interacting in the one location, one finds a research institute, a telescope sampling the dark matter in the Galaxy, an instrumentation engineering laboratory, a graduate program, a visitor centre, and on-campus residences. From California to Cambridge to Chile, there is nowhere which can boast more than two-thirds of these facilities. No wonder visitors consider Stromlo a little piece of paradise.
>
> Each of these facilities, and especially the superb remote observatory at Siding Spring, plays its role in supporting the primary element of the School's mission within the ANU,

to add to our knowledge about and understanding of the Universe. Each of the facilities is resourced leanly by the block grant from the Institute of Advanced Studies, with the exception of the visitor centre and the residential areas, which are self-supporting. The 1995 review of the IAS found that in its fifty-year history and in all its serious fields of endeavour, it had achieved international prominence. At the end of 75 years Mount Stromlo Observatory is a model for the IAS in sustaining international competitiveness.

What is the institution's secret? And what values, therefore, are the most important to retain, when everything is changing more rapidly than ever before? In common with other top academic institutions, challengingly high expectations of all members are a key part of the IAS ethos. Where else in Australian academia do you find students treated like postdocs, postdocs treated like faculty, faculty treated as research group leaders, and directors empowered to set the direction of the University? Maximising the responsibility we put in the researchers' hands is a winning strategy, but it is a course which sails very close to the strong winds of accountability that we feel in the present climate. It must be demonstrated excellent performance that sustains a School which continues to plot its course. [This anniversary] year has been a strong one from this perspective.

Aerial view of Mount Stromlo Observatory Photo: MSO

Epilogue
January 2003

HOW SHOULD A BOOK COVERING ALMOST 80 years in the continuing life of a famous scientific institution conclude? Should it attempt a summary of major achievements? Should it try to predict the institution's prospects for the future? Should there be some reflection on our vastly increased understanding of the Universe and what this means for the human race?

The last thing that the authors of the present history would have wished was that their tale should come to a *dramatic* conclusion—but nature has left them no choice. On the afternoon of Saturday 18 January 2003, a bushfire of unprecedented speed and ferocity swept towards Canberra from the mountains to the west. On its way, it overwhelmed Mount Stromlo, consuming all the major domes and the oldest of the heritage buildings. This 'Epilogue', written within a month of 'Black Saturday', describes the events of the most traumatic day in the eight-decade history of the Observatory—and provides a brief glimpse of what is, even now, rising from the ashes.

☆

During the summer of 2002–03, Australia experienced a prolonged drought, particularly severe in the southeast of the continent. In many places it was the worst in living memory. The month of January, usually a high-risk one for bushfires, was awaited with apprehension. This fear was not misplaced; fires broke out in every Australian state.

Canberra—as the 'bush capital'—is especially prone to wildfire attack. The Namadgi National Park, a mountainous reserve of natural bushland comprising 45 per cent of the area of the Australian Capital Territory, lies to the southwest of the city, and comes within 10 kilometres of suburbia at its nearest point. Across the border there is more of the same, with the much larger Kosciusko National Park stretching southwards into the alpine region of New South Wales. There are also three extensive pine plantations to the west of the city, the nearest of which surrounds Mount Stromlo Observatory and abuts the nearest suburbs in the Weston Creek area. Between the wilderness and the city lies the Murrumbidgee River valley, always thought of as a natural containment line for any fires approaching from the west, although, since it is quite a small stream at this point, and its valley has been retained in its natural bushland state, the protection it offers is perhaps more psychological than real. The Stromlo Pine Forest lies on the city side of the Murrumbidgee Corridor.

Most bushfires arise from natural causes, such as lightning strikes, and in early January outbreaks began to occur to the west of the city. By Monday 13 January, fire crews from across Canberra and elsewhere in Australia were battling three in particular. Some of Canberra's water reservoirs lie in the Namadgi Park and water-dump helicopters were in action. The battle took a dramatic turn that day, when one of the helicopters accidentally ditched into the Bendora Dam, and began to sink with its Queensland pilot trapped in his seat. As it happened, the Chief Minister of the Australian Capital Territory, Jon Stanhope, and the Emergency Services Director, Peter Lucas-Smith, were assessing the fire situation at the Dam at the time and assisted in the pilot's rescue.[1]

Thereafter, the fires made daily headlines in the *Canberra Times*. By Wednesday, two of the three Namadgi outbreaks had combined into a single blaze, and the third was still raging. An even larger fire was burning in New South Wales just across the Territory's northwestern border. Throughout the week, interstate volunteer fire fighters arrived to boost the defensive effort; by late Friday, 450 were deployed. The weather predictions towards the end of the week were dire—a four-day period of searing 40°C temperatures and 35 kilometres per hour northwesterly winds. Saturday's headlines brought grim news: 'Bushfires break through'.[2] Wind gusts of up to 65 kilometres per hour had caused 'spot fires' well ahead of the main fronts, and within 12 kilometres of the outskirts of the city.

The threat to Canberra was just one element of a much larger picture; equally dangerous fires spread over a wide area, particularly to the south in the alpine district. The emergency services were stretched to the limit, and were forced to redirect their efforts from trying to control the firefronts to protecting property. The authorities issued increasingly urgent advice on what action residents should take in the event of their homes being threatened.

In the end, it all happened very quickly. The firestorm generated its own fierce wind and attacked with devastating force. The Murrumbidgee Corridor was breached at about midday, and the storm swept up the mixed pastoral–forest land towards Canberra reaching the suburbs by mid-afternoon. The Weston Creek area suffered horrific devastation. Four people died; hundreds were injured, many seriously; thousands had to be evacuated; and about 500 homes were destroyed, including nearly 250 in Duffy, the suburb lying at the foot of Mount Stromlo. The four fatalities also occurred in or near Duffy.

The emergency services were completely overwhelmed. Indeed, had they had ten or even a hundred times the number of appliances to deploy, it is doubtful whether they could have prevented the catastrophe. Seasoned fire chiefs later commented that, in all their experience, they had never witnessed such a phenomenon. The storm was quite capricious. One man and his son were on their roof hosing down their residence when a fireball came and knocked them to the ground, only to leap-frog their home, leaving it unscathed, and race on to destroy the house behind. Stories of despair, heroism, miraculous escape and individual and community compassion would fill the pages of the newspapers for weeks.

Preliminary damage estimates ran to $250 million, making this the second most costly fire attack in Australia's history. Thirty farms in the region were destroyed, with the loss of thousands of animals. Virtually all of the Namadgi National Park was burnt.

The Stromlo Pine Forest, and with it the Observatory, was overwhelmed as the fire raced eastwards. The order to evacuate was given at 2.30 p.m. By the end of the afternoon, most of the heritage areas had been destroyed. No fire-fighting effort could have saved them. Many Observatory staff had chosen to live in Weston Creek because of its proximity to the Observatory. Four staff couples lost their homes, and several others only narrowly managed to save theirs as they fought spot fires around their properties in quite terrifying circumstances.

It was the next day before the emergency authorities permitted any access to the Mount Stromlo site. When the Director and Associate Director, Penny Sackett and John Norris, and the Observatory's Executive Officer, Vince O'Connor, were able to inspect the damage, a horrific sight awaited them. The domes of the 74-inch reflector, the 50-inch Great Melbourne Telescope, the 'Reynolds' 30-inch reflector, the Yale–Columbia 26-inch refractor, and the 'Oddie' 9-inch refractor were all gutted, and the telescopes themselves irreparably damaged. The lenses or mirrors of the 50-inch, the 26-inch and the 9-inch were found lying on their dome floors. The 1920s Commonwealth Solar Observatory buildings (which now housed Administration, the Mechanical Design Office and the Library) were also completely destroyed, as were the workshops. Perhaps the most heart-breaking single loss was that of the NIFS instrument, four years in the building and undergoing final testing prior to its delivery to Gemini North in Hawaii in just six months' time.[3] Seven of the twelve residences on the mountain were destroyed, leaving their occupants—many of them graduate students—bereft of all their possessions.

It is noteworthy that the three most recently constructed major Observatory buildings survived: the Duffield Building (completed in 1964), the Woolley (1995) and the Visitor Centre (1997). The Duffield and the Woolley buildings house the astronomers' offices. Damage was slight: broken windows and a little water from the fire sprinklers. The Computer Section staff and the network servers are also located in the Woolley Building and back-ups are routine. Thus the Observatory's electronic environment was quickly restored, with minimal loss of information, following the return of staff to the Woolley and Duffield buildings on 10 February 2003.

The 74-inch Telescope after the January 2003 fire Photo: Kim Rawlings

The firestorm seemed to have swept over the entire site, and it is surprising that anything at all survived. Why was it, for instance, that the Visitor Centre was left intact although the adjacent 74-inch was gutted, and even a wooden crate stored underneath the Centre burnt? Why did Agris and Leslie Kalnajs's house fail to catch alight although its garden was completely destroyed, as was the residence next door? Although the heritage buildings of the old Commonwealth Solar Observatory were completely destroyed, the wooden park benches in their enclosed courtyard survived entirely unscathed, including the memorial bench to Su Khouri, the Director's Secretary in the 1970s. The dynamics of the firestorm must have been absolutely bizarre.

The ANU's support was immediate and unstinting. The Vice-Chancellor, Professor Ian Chubb, convened a meeting of all Mount Stromlo staff in the Chancelry at 9.00 a.m. on Monday, 20 January 2003. He announced that everyone at the University had but a single agendum: the re-establishment of the Research School of Astronomy and Astrophysics at its Mount Stromlo site with the minimum possible delay. Staff members were found immediate desk space on campus, and some of the Observatory's computer servers were brought down from the mountain to establish a temporary network. Since the disaster, Mount Stromlo staff meetings have been held daily. Five working groups have also been established to plan key aspects of the reconstruction.

The Observatory staff have been absolutely amazed by the expressions of sympathy and concern that have been flooding in, from all over the city, the country and the world. Even more overwhelming have been the offers of assistance and practical help in the post-fire recovery. One of the most touching gifts was five dollars from the piggy-bank of a nine-year-old girl who had received advice on her school project a few days before—now it was her turn to help the Observatory, she wrote. At the other end of the scale, the ANU's Research School of Physical Sciences and Engineering and the Australian Defence Force Academy have made workshop facilities available, allowing work on Mount Stromlo's instrumental contracts to proceed with a minimum of disruption. Offers of replacement items for the library have been numerous and substantial.

While the fire losses are massive, it is worth noting that much remains. The Research School's Siding Spring Observatory is entirely intact. It is worth reflecting, too, that the Observatory has sited all major new telescopes at Siding Spring since it joined the University in 1957. The spirit of the Mount Stromlo staff remains high. The new Mount Stromlo will not be an exact replica of the old—nor would anyone want it to be—but quite enough of its heritage landmarks will be restored to leave no doubt in anyone's mind that it is still the same observatory.

A survivor of the fire surveys the devastation from inside the burnt out Administration Building
Photo: Helen Frame

Support has also been forthcoming at the highest possible political level. The Prime Minister, John Howard, inspected the damaged Observatory on Tuesday, 21 January, and the Minister for Science, Peter McGauran, was another early visitor. On 4 February, McGauran issued a media release detailing the Prime Minister's reaction:

> Tragically, the famous Mount Stromlo Observatory was decimated. One of my saddest experiences of going around in the aftermath of the fires was looking at the mangled remains of those wonderful old telescopes—in particular, the Victoria telescope.
>
> The scientific community has suffered an enormous loss. In an appropriate way, Mount Stromlo must and will be rebuilt, because it is a very important element of our national identity and of our excellence in science and astronomy and because it is a very important symbol of our national capital.[4]

Penny Sackett's leadership in the trauma of such a major loss, so early in her directorship, has been truly inspirational. Almost her first words to the assembled staff on 20 January 2003 were: 'Fortunately, the Research School's three most valuable assets remain entirely intact—its people, its reputation, and its spirit'. Indeed they do, and everyone is absolutely determined that Mount Stromlo Observatory will rise from the ashes stronger and better than ever.

appendix 1

Observatory Directors

W. Geoffrey Duffield
Director, Commonwealth Solar Observatory
1 January 1924–1 August 1929

William B. Rimmer
Officer in Charge, Commonwealth Solar Observatory
1 August 1929–4 December 1939

Richard v.d.R. Woolley
Commonwealth Astronomer
Director, Commonwealth Observatory
4 December 1939–7 December 1955

Arthur R. Hogg
Acting Commonwealth Astronomer
8 December 1955–7 March 1957

Bart J. Bok
Director, Mount Stromlo Observatory
7 March 1957–19 March 1966

Arthur R. Hogg
Acting Director, Mount Stromlo Observatory
20 March 1966–31 March 1966

S.C.B. 'Ben' Gascoigne
Acting Director, Mount Stromlo Observatory
1 April 1966–30 June 1966

Olin J. Eggen
Director, Mount Stromlo and Siding Spring Observatories
1 July 1966–30 September 1977

Donald S. Mathewson
Acting Director, Mount Stromlo and Siding Spring Observatories
1 October 1977–4 April 1979

Donald S. Mathewson
Director, Mount Stromlo and Siding Spring Observatories
5 April 1979–4 May 1986

Alexander W. Rodgers
Acting Director, Mount Stromlo and Siding Spring Observatories
5 May 1986–11 June 1987

Alexander W. Rodgers
Director, Mount Stromlo and Siding Spring Observatories
12 June 1987–11 June 1992

Donald J. Faulkner
Acting Director, Mount Stromlo and Siding Spring Observatories
12 June 1992–12 December 1993

Jeremy R. Mould
Director, Mount Stromlo and Siding Spring Observatories
13 December 1993–30 January 2001

John E. Norris
Interim Director, Research School of Astronomy and Astrophysics
1 February 2001–21 July 2002

Penny D. Sackett
Director, Research School of Astronomy and Astrophysics
22 July 2002–

appendix 2

Staff astronomers at Mount Stromlo

THIS APPENDIX WAS COMPILED IN DECEMBER 2002. Astronomers are listed in order of arrival at Mount Stromlo. Public service grades were in use in the Commonwealth (Solar) Observatory until the transfer of the Observatory to the ANU in January 1957. Institute of Advanced Studies grades have been in use thereafter. The tabulation lists appointees to *scientific* posts before the change and to *academic* posts thereafter, which may introduce some inconsistencies. The most senior position attained is listed; this will not usually have been held for the entire duration of the appointment span shown. The Observatory's *Annual Reports* have been used to obtain the appointment spans. This may introduce some minor errors since some Reports provide staff listings *at a specific date in the year in question*, and others *for the whole of the year*. Seconded and adjunct appointments (including those of Uppsala and Yale–Columbia Observers) have been listed; visiting, emeritus and honorary appointments have not. The earliest *Annual Reports* provide somewhat sketchy staffing details, and other references have also been consulted, such as C.W. Allen, 'The beginnings of the Commonwealth Solar Observatory', *Rec. Aust. Acad. Sci.*, vol. 4, 1978, pp. 27–49. There may still be some inadvertent early omissions. Some early researchers at the Commonwealth Solar Observatory would be better described as atmospheric or radio physicists rather than astronomers; they have nevertheless been included.

STAFF APPOINTED BY THE COMMONWEALTH DEPARTMENT OF THE INTERIOR

W.G. Duffield	1924–29	Foundation Director
W.B. Rimmer	1925–45	Officer-in-Charge, Chief Assistant
A.L. Kennedy	1925–27	Assistant
W.H. Love	1925	Research Fellow
C.W. Allen	1926–51	Principal Research Assistant
A.J. Higgs	1926–41	Assistant
A.R. Hogg	1929–66	Professorial Fellow, Deputy Director
L.G.H. Huxley	1930–31	(Radio Research Board) Assistant
F.J.P. Wood	1935–37	(temporary) Assistant
S.E. Williams	1936–37	Research Fellow
R.G. Giovanelli	1937–39	Research Fellow
R.v.d.R. Woolley	1939–55	Commonwealth Astronomer, second Director
D.W.N. Stibbs	1940–42, 1945–51	Senior Scientific Officer
S.C.B. Gascoigne	1941–80	Professor, Assistant Director
H.J.McK. Abraham	1944–78	Senior Fellow, Head of Time Service
J.M. Baldwin	1944	(temporary) Assistant
G.J. Odgers	1944–46	Assistant
G.H. Woodhouse	1944	(temporary) Assistant
D.F. Martyn	1949–56	on secondment from CSIR
K. Gottlieb	1945–66	Fellow (Research Engineer)
C. Halliday	1945–46	Assistant
B. Dawson	1945–46	Research Fellow
G.V. Simonow	1946–53	Scientific Officer
D.G. Thomas	1946–55	Scientific Officer
C.A. Maddern	1947–48	Scientific Officer
M.L. Woods	1947–50	Research Fellow
P.W. Bowe	1948	Research Fellow
G. Hagemann	1949–54	Scientific Officer
R. Brockman	1949–50	Scientific Officer
R.A. Henry	1949–55	Scientific Officer

J. Burr	1949–53	Scientific Officer
C.S. Gum	1949–50	Research Fellow
B.F. Hall	1950	Scientific Officer
H.R.F. Gollnow	1950–76	Senior Fellow
A. Przybylski	1950–78	Fellow
G.H. de Vaucouleurs	1951–57	Research Fellow, Yale–Columbia Observer
W. Buscombe	1952–67	Senior Fellow
C. Jackson	1952–63	Yale–Columbia Observer
D.W.C. Baker	1955	Scientific Officer

Staff appointed by the Australian National University

B.J. Bok	1957–66	Professor, third Director
T. Dunham Jr	1957–65	Professor
A.W. Rodgers	1957–97	Professor, sixth Director
B.E. Westerlund	1957–67	Uppsala Observer, Professorial Fellow
G. Lyngå	1961–63, 1965–69	Uppsala Observer
T. Walraven	1963–64	Senior Fellow
L. Searle	1963–68	Senior Fellow, Assistant Director
C. Roslund	1963–65, 1970	Uppsala Observer
D.J. Faulkner	1965–98	Senior Fellow, Associate Director
C. Perry	1965–66	Research Fellow
D.S. Mathewson	1966–94	Professor, fifth Director
O.J. Eggen	1966–77	Professor, fourth Director
K.C. Freeman	1967–present	Duffield Professor
M.E. Dixon	1967–73	Research Fellow
K. Serkowski	1967–70	Senior Research Fellow
D.H.P. Jones	1969–73	Senior Research Fellow
M.S. Bessell	1969–present	Professor, Associate Director
R.D. Cannon	1969–74	Research Fellow
B.A. Peterson	1969–present	Senior Fellow
A.V. Peterson	1969–72	Postdoctoral Fellow
R.S. Stobie	1970–74	Research Fellow
A.R. Hyland	1970–90	Senior Fellow

M.J. Disney	1970–75	Research Fellow
B. Karlsson	1970–73	Uppsala Observer
N.V. Vidal	1972–76	Research Fellow
K.A. Bern	1972–76	Uppsala Observer
A.J. Kalnajs	1973–present	Senior Fellow
G.E. Kron	1974–77	Senior Research Fellow
J.E. Norris	1974–present	Professor, Interim Director
B.C. Cogan	1974–80	Research Fellow
N. Visvanathan	1975–97	Senior Fellow
E.B. Newell	1975–84	Senior Research Fellow
M.A. Dopita	1975–present	Professor, Federation Fellow
P.R. Wood	1976–present	Senior Fellow
S.M. Simkin	1977–80	Senior Research Fellow
A. Bosma	1978–81	Research Fellow
T.J. Jones	1978–82	Research Fellow
I.R. Tuohy	1979–90	Senior Research Fellow
P.D. Nicholson	1979–82	Postdoctoral Fellow
P. Seitzer	1979–81	Yale–Columbia Observer
G.V. Bicknell	1980–present	Senior Fellow
D. Griersmith	1980–81	Postdoctoral Fellow
J.R. Rohde	1980–83	Yale–Columbia Observer
R.A. Gingold	1981–88	Research Fellow
C.R. Jenkins	1981–83	Postdoctoral Fellow
N.A. Sharp	1981–84	Postdoctoral Fellow
P.A. Ianna	1981–2002	Yale–Columbia Observer
W.L.M. Peters	1982–85	Research Fellow
D. Carter	1982–87	Postdoctoral Fellow
E.M. Green	1982–85	Postdoctoral Fellow
K. Kawara	1983–84	Research Fellow
D. Saunders	1983–84	Yale–Columbia Observer
P.J. McGregor	1984–present	Senior Fellow
R.J. Buta	1984–86	Postdoctoral Fellow
P.E.J. Nulsen	1985–89	Research Fellow
J. Sommer-Larson	1986–89	Postdoctoral Fellow
J.P. Reynolds	1986–89	Research Fellow
M.T. McGinn	1988–89	Postdoctoral Fellow
R.J. Lavery	1988–91	Postdoctoral Fellow
G.H. Smith	1988–89	Postdoctoral Fellow

P.J. Quinn	1989–95	Fellow
M.D. Gregg	1989–92	Postdoctoral Fellow
T.E. Harrison	1989–92	Postdoctoral Fellow
J.N. Loveday	1989–92	Postdoctoral Fellow
K.J. Mighell	1989–92	Postdoctoral Fellow
R.D. Ekers	1990–present	Adjunct Professor
P. te Lintel Hekkert	1990–93	Postdoctoral Fellow
G.S. Stringfellow	1990–93	Postdoctoral Fellow
Y. Yoshii	1992–93	Fellow
J.R. Mould	1993–2001	Professor, seventh Director
M.M. Colless	1993–present	Senior Fellow
M. Arnaboldi	1993–97	Research Fellow
P. Saha	1993–96	Research Fellow
T.S. Axelrod	1994–2002	Senior Fellow
E. Sadler	1994–present	Adjunct Fellow
C. Stubbs	1994–present	Adjunct Fellow
L. Ferrario	1994–96	Research Fellow
A. Lançon	1994–95	Postdoctoral Fellow
G.S. Da Costa	1995–present	Senior Fellow
B.K. Gibson	1995–98	Postdoctoral Fellow
H. Jerjen	1995–present	Research Fellow
H.M. Schmid	1995–96	Postdoctoral Fellow
B.P. Schmidt	1995–present	Fellow
J. Li	1996–2000	Research Fellow
D. Bersier	1996–99	Postdoctoral Fellow
H. Sung	1996–97	Postdoctoral Fellow
C. Heisler	1997–99	Research Fellow
R.S. Sutherland	1997–present	Research Fellow
P.J. Francis	1997–present	Fellow
M. de Kool	1997–2002	Research Fellow
Z. Kuncic	1997	Postdoctoral Fellow
S.J. Wagner	1997–98	ARC International Fellow
J. Bland-Hawthorn	1998–present	Adjunct Fellow
M. Sevenster	1998–2001	Postdoctoral Fellow
R. Smith	1998–2000	Postdoctoral Fellow
G. Van de Steene	1998–2000	Postdoctoral Fellow
R. Fux	1999–present	Postdoctoral Fellow
C.A. Jackson	1999–present	Research Fellow

K. Sebo	1999	Postdoctoral Fellow
E. Corbett	1999	ARC International Fellow
C. Saxton	2000–present	Postdoctoral Fellow
S.P. Driver	2001–present	Fellow
R.P. Saglia	2001–02	IREX Fellow
R. de Propris	2001–present	Postdoctoral Fellow
E.C. Sung	2001–02	IREX Fellow
P.D. Sackett	2002–present	Professor, eighth Director
F.H. Briggs	2002–present	Adjunct Professor
N. Christlieb	2002–present	Postdoctoral Fellow
J. Fischera	2002–present	Postdoctoral Fellow
D.H. Jones	2002–present	Postdoctoral Fellow

appendix 3

Stromlo graduates and their thesis titles

This appendix was compiled with information available until December 2002. Graduates are listed by date of admission to the degree by the ANU Council. The PhD degree was awarded except where otherwise indicated. In cases of a name change, the name on graduation has been listed.

1954

Antoni Przybylski	The maximum effect of convection in stellar atmospheres on the observed properties of stellar spectra

1955

Colin Stanley Gum	A study of diffuse southern H–α nebulae

1958

Alexander William Rodgers	A photometric study of the Southern Coalsack

1962

Roger Alistair Bell	White dwarf stars
John Bartlett Whiteoak	The structure of the southern Milky Way

1963

Donald Jack Faulkner	A study of the 30 Doradus and η Carinae nebulae

1964

Ivan John Danziger	A high dispersion spectral study of one carbon star and two barium stars
Robert Reginald Shobbrook	Photographic, spectrographic and photometric studies of southern galaxies
John Archibald Graham	The application of hydrogen line photometry to Milky Way research
David Sher	Distances of some open clusters near η Carinae

1965

Rawi Bhavilai	The structure and dynamics of the solar chromosphere
James Alfred Koehler	Intergalactic atomic neutral hydrogen

1966

Natarajan Visvanathan	Polarization in the Galaxy and the Large Magellanic Cloud
Richard Marcus Price	Cosmic background radiation at 408 Mc/s
Lindsey Fairfield Smith	Narrow band photometry of Wolf-Rayet stars
Betty Louise Webster	Southern planetary nebulae

1967

Gerard Henri de Vaucouleurs	'Manuel de photographie scientifique' (DSc)
Ronald David Ekers	The structure of southern radio sources
Michael Stanley Bessell	A study of some southern short period variables
Ardon Robin Hyland	A study of southern galactic clusters and the peculiar stars associated with them
John William Menzies	Three-colour photometry of southern globular clusters

1969

Anthony Lockington Brooke	Infrared observations of globular cluster giants
Donna Dee Hain	The Mira variable S Carinae
Beverley June Harris	Radio spectra of QSOs and galaxies
Edward Barry Newell	Hot extreme Population II stars
Brian Murray Lewis	The masses of some southern galaxies

1970

Edward George Schmidt	An astrophysical investigation of four classical Cepheid variable stars
Krzysztof Serkowski	On polarization of starlight (DSc)
John Edward Norris	Neutral-helium line strengths in the early-type stars
Jasper Vivian Wall	A deep sky survey at 2700 MHz

1971

Ian Andrew Lockhart	High resolution observations at 80 MHz of galactic and extragalactic radio sources
Robert Dudley Watson	A study of the Beta Cephei stars
Norman Roy Stokes	The variability of M-stars (MSc)

1972

John Bernard Hearnshaw	The abundances of the elements in the oldest disk stars
John William Robertson	The evolution of massive stars

1973

Peter Robert Wood	Stellar evolution and planetary nebula ejection

1974

Garth Darrel Illingworth	Masses of globular clusters
Harvey Raymond Butcher	Observational aspects of nucleosynthesis
Gary Royce Hovey	Software correction of astronomical telescope pointing errors

1976

Robert Arthur Gingold	The helium burning evolution of low mass stars
Jeremy Richard Mould	The lower main sequence and the atmospheres of M dwarfs
Chun Mun-Suk	Observational evidence for radial inhomogeneities in globular clusters

1977

Lee See-Woo	Accurate C-M diagrams of six southern globular clusters

Martha Noeleen Cleary	Local galactic structure
Gary Stewart Da Costa	The structure and content of globular clusters

1978

Don Allan VandenBerg	Globular cluster gas flow and stellar evolution studies
James Frederick Burrell	A study of several R Coronae Borealis variables
Peter Ledsam Cottrell	Spectrum synthesis of evolved stars

1979

Mark Philip Schwarz	Gas response to a rotating stellar bar

1980

David Griersmith	The CM effect for early-type spiral galaxies
Alan Leslie Duus	A Study of the Fornax I cluster of galaxies
Susan L. Barrell	Beat Cepheid studies

1981

Peter J. McGregor	Late-type supergiant stars in the Magellanic Clouds

1982

Warrick John Couch	The colour evolution of galaxies in clusters
Brian J. Jarvis	The dynamics of galactic bulges

1983

Luc Aurelle Binette	General spectral synthesis of nebular emission line regions
Peter Joseph Quinn	The dynamics of disk galaxy mergers
Graeme Hayward Smith	The nature of cyanogen inhomogeneities within star clusters
Elaine Margaret Sadler	Radio and optical observations of a complete sample of E and S0 galaxies
Ian Robert George Wilson	The evolution of H II regions in the Large Magellanic Cloud
Claude Carignan	Photometry and kinematics of pure disk galaxies
Dominic Michael Zarro	Stellar chromospheres: H-alpha and Ca II K profiles

1984

Desmond John Hillier	The extended atmosphere of the WN Star HD 50896
Andrew John Pickles	Differential population synthesis of 17 early-type galaxy nuclei in the Fornax Cluster
Colin S. Coleman	Axisymmetric models of extragalactic jets
Kavan Upajiva Ratnatunga	Outer regions of the galactic halo
Matthew W. Fox	The pulsation of long-period variables

1985

Neil E.B. Killeen	A radio to X-ray study of IC 4296 (PKS 1333–33) and NGC 1399 (PKS 0336–35)

1987

Ian Nigel Evans	Calibration of the H II region abundance sequence and abundances in Seyfert galaxies
James Robert Lewis	Kinematics and chemical properties of the old disk
Gregg Rowley	The structure of box-shaped galaxies
Rodney M. Smith	Observations of a sample of southern radio galaxies
Richard James Wainscoat	An infrared and optical study of edge-on galaxies
Catherine Margaret Lance	Young, high-velocity A stars
R. Alex Ruelas-Mayorga	Infrared studies of the structure of the Galaxy

1988

John Marshall Brett	Spectrum synthesis of cool stars
Peter Fletcher Teague	The dynamics and structure of rich clusters of galaxies
Stephen John Meatheringham	Dynamics and evolution of planetary nebulae

1989

Steven Maxwell Straw	The global morphologies of active star formation regions
Shaun Michael Grahame Hughes	The nature and kinematics of long-period variables in the Large Magellanic Cloud
Heather Lorraine Morrison	Metal weak giants of the inner halo and the disk

Michael Charles Brewster Ashley	The type-I planetary nebulae NGC 6302 and NGC 6537
George Paltoglou	Populations in old stellar systems

1990

Matthew Bailes	The origin of pulsar velocities
David A.H. Buckley	Optical and X-ray studies of HEAO-1 X-ray sources
Stephen Charles Russell	Abundances of the heavy elements in the Magellanic Clouds
Glen Howard Mackie	The stellar content of central dominant galaxies
Roger James Verge Brissenden	A multifrequency study of X-ray selected active galactic nuclei from the HEAO-1 survey
Lilia Ferrario	Accretion processes in AM Herculis systems
Chris M.L. Flynn	The galactic force law and the thick disk
Iain R. Hercus	A multiphase model of galactic collapse
Hans-Gerhardt R. Meurer	NGC 1705 and the evolution of blue compact dwarf galaxies
Robert Andrew Cameron	A radio, optical and X-ray study of PKS 2104–25
Sean Gerard Ryan	Abundances and kinematics of high proper motion stars
Simon Richard Wayte	The structure, dynamics and evolution of the Magellanic System

1991

Gregory Albert Wilson	Kinematic and observational studies of stellar streams
Ruan Kui	Photospheric studies of M dwarf stars
Mark L. Winsall	Kinematics of giant elliptical galaxies and a study of nucleated dwarf elliptical galaxies

1992

Nick Achilleos	Atmospheres and accretion halos of magnetic white dwarfs
Saul Caganoff	Optical emission lines in radio sources of intermediate power

1993

Emanuel Vassiliadis	Evolution from AGB star to planetary nebula
Markus Buchhorn	Dark matter in spiral galaxies
Yong-Ik Byun	Dust opacity and structure of spiral galaxies
Carl J. Grillmair	Dynamics of globular cluster systems
Angela E. Samuel	The nature of star formation in the Trapezium Cluster
Boudewijn Francois Roukema	Studies of the faint galaxy population
Ralph S. Sutherland	Interstellar shock fluorescence
Stuart D. Ryder	Massive star formation in galactic disks

1995

Christopher Erik Lidman	An optical search for distant galaxy clusters
Jayanne English	Gas dynamics and globular cluster formation in interacting galaxies
Stephanie M.O.H. Côté	Dwarf galaxies in nearby southern groups

1996

Jonathon Falder Bell	Binary and millisecond radio pulsars
Anton M. Koekemoer	Dynamics and excitation of extended emission-line regions in radio galaxies
Haida Liang	Distribution of matter in clusters of galaxies
Steven J. Tingay	Parsec and sub-parsec-scale structure and evolution in nearby compact radio sources and relationships to emission at other wavelengths
Sylvie F. Beaulieu	Dynamics of the galactic bulge using planetary nebulae

1997

Brett S. Wells	The far infrared to radio continuum correlation for nearby star-forming galaxies
Kim Michael Sebo	Cepheids and long-period variables near Magellanic Cloud clusters
Claudia Faye McCain	Morphology and kinematics of interacting galaxies: the spheroidal + irregular pairs

1998

Benjamin Willem Stappers	Observations of eclipsing and binary pulsars
Alister William McKenzie Graham	Elliptical galaxies: structure, dynamics and applications
Frank de Ruyter	IRAS galaxies behind the Milky Way (MSc)

1999

Martin Bureau	Bars in edge-on spiral galaxies
Sung-Eun Kim	An HI aperture synthesis mosaic and H-alpha survey of the Large Magellanic Cloud
Mark George Allen	Excitation of the narrow-line regions of active galaxies

2000

Roberto Soria	Accretion processes in black-hole binaries
David William Pfitzner	The structure of halos in CDM universes
David Heath Jones	Tunable filter surveys of star-forming galaxies

2001

Andrew John Drake	The detection of microlensing with difference image analysis
Stefan Claude Keller	Young populous clusters in the Magellanic Clouds
Mary Elaine Putman	Mapping the Galaxy's neutral hydrogen halo
Lisa Maree Germany	The Mount Stromlo Abell cluster supernova search

2002

Lisa Jennifer Kewley	The AGN–starburst connection in warm infrared galaxies
Michelle M. Buxton	Optical and infrared observations of low-mass X-ray binaries with black hole candidates

Bibliography

Note: Academic papers on astronomical topics and themes cited in the narrative are not included.

Primary Sources

Mount Stromlo Files

Australian Archives, Canberra

A1/1	Item 1918/6038, 'Minutes of a meeting in Melbourne regarding the proposed Solar Observatory with the Prime Minister'.
A1/15	Item 1925/10368, CSO Report.
A1/15	Item 1926/403, Creation of Positions, CSO Canberra.
A1/15	Item 1926/21338, First Aid Equipment, Solar Observatory.
A1/15	Item 1927/12579, Maintenance of Garden, Solar Observatory.
A1/15	Item 1927/21863, Solar Observatory Visiting Fellows.
A1/15	Item 1937/1943, Pines Tennis Club, Mount Stromlo.
A192	Item 1921/1905, Afforestation planting areas, 1913–1924.
A199	Item FC 1926/206, CSO Mount Stromlo, Erection, Works & Railways.
A202	Item 1914/3272, Solar Observatory, 1911–14.
A220	Item S39/698, Staff file, Richard van der Reit Woolley.
A361	Item DSG 24/1019, Kite House, Mount Stromlo.

A431 Item 1951/744, Solar Observatory, Building Program, Part 2.
A431 Item 1952/184, Establishment of Solar Observatory, Canberra.
A431 Item 1955/123, Solar Observatory, Restoration of Fire Damage.
A431/1 Item 1955/124, Solar Observatory, Administrative Buildings.
A432/86 Item 1929/3187, Rental of Quarters, CSO.
A432/86 Item 1929/4287, Solar Observatory, Canberra.
A461/1 Item B372/1/2, Reynolds Telescope.
A461/8 Item B372/1/2, Meteorological Solar Observatory.
A659/1 Item 1943/1/2792, CSO, Optical Munitions Branch, Estimates 1943–44.
A2502 Architectural Plans and Drawings with AB Prefix, 1925–1930.
A2503 Architectural Plans and Drawings with AC Prefix, 1925–1931.
A2506 Architectural Plans and Drawings with AG Prefix, 1930.
A3560 Mildenhall photographic collection of early Canberra, 1921–1935.
A9663 Aperture Cards for ACT Architectural Plans and Drawings, 1921–1959.

Commonwealth Departmental Files

Australian Archives, Canberra

CA 12 Prime Minister, 1 Jan 1957–13 Dec 1966
CA 15 Home & Territories, 1 Jan 1924–10 Dec 1928
CA 24 Home Affairs (II), 10 Dec 1928–12 Apr 1932
CA 27 Interior (I), 12 Apr 1932–26 Apr 1939
CA 31 Interior (II), 26 Apr 1939–1 Jan 1957
CA 1196 Education and Science, 13 Dec 1966–19 Dec 1972
CA 1478 Education (I), 19 Dec 1972–11 March 1983
CA 1486 Science (I), 19 Dec 1972–8 Feb 1975
CA 1888 Science & Consumer Affairs, 8 Feb 1975–22 Dec 1975
CA 1962 Science (II), 22 Dec 1975–5 Dec 1978
CA 2749 Science and the Environment, 5 Dec 1978–3 Nov 1980
CA 3069 Science & Technology, 3 Nov 1980–13 Dec 1984
CA 3494 Education & Youth Affairs, 11 Mar 1983–13 Dec 1984
CA 4134 Education (II), 13 Dec 1984–24 July 1987
CA 4136 Science (III), 13 Dec 1984–31 Dec 1986
CA 5986 Employment, Education & Training, 24 July 1987–1 Mar 1994
CP 698/30 Files relating to the construction and maintenance of CSO Mount Stromlo, 1925–1929

Australian Archives (Melbourne)

MP392/9	Box 2, 'Optical—Mount Stromlo'.
MP927, A177	Minutes of the Meetings of the Army Inventions Directorate.
MP1164/12	Register of Inventions, Army Inventions Directorate.
MP1472/4	Minutes of the Meetings of the Optical Munitions Panel.

ANU Archives

103/1/23	'Research School of Physical Sciences—Aims, Organisation and Finance', April 1948.
8/7/1/0	Part 1, Correspondence.
1871C/1985	'Mount Stromlo and Siding Spring Observatories.' A paper detailing proposed arrangements following the separation of MSSSO from RSPhysS.
2220/1986	Report of the Review Committee on the Mount Stromlo and Siding Spring Observatories, 1986 (ANU paper).
2610/1986	Minutes of the meeting of the Committee of the Observatories, 6 August 1986, Item 5.
2618/1986	Report of the meeting of Advisers on Astronomy Policy, 9 September 1986.

Basser Library, Australian Academy of Science

Box 37/1	'Board of Visitors of the Commonwealth Observatory—Meetings', and related correspondence.

CSIRO Archives

Series 372	Item 1/H, Draft resolutions of Committee on 'Faculty of Physical Science, University of Canberra'.

Printed Sources

Memorandum upon the Proposed Solar Observatory in Australia, Solar Physics Committee of the Australasian Association for the Advancement of Science, Sherring, Adelaide, 1909.

Commonwealth Observatory Annual Reports, 1930– (originally known as *Memoires*).

Supplement to the Report of the Commonwealth Astronomer to the Board of Visitors, 1950 (prepared by Dr Gwen Simonov).

'ANU: Report of the Council for the Period 1 January 1955 to 31 December 1955', *Commonwealth Parliamentary Paper* (CPP) No. 32.

Anglo-Australian Telescope Project, Report 1973–74.

Anglo-Australian Telescope, 1974/75.

Anglo-Australian Telescope, 1975/76.

A Space Policy for Australia, Report of the Madigan Committee, Australian Academy of Technological Sciences, 1985.

The Future of Australian Astronomy, A Report to the Prime Minister by the Australian Science and Technology Council, AGPS, Canberra, September 1989.

Australian Astronomy: Beyond 2000, National Board of Employment, Education and Training, Australian Research Council, Discipline Research Strategies, AGPS, Canberra, June 1995.

Manuscripts

Duffield, W.G., Transcribed diary for 1925 [copy held by authors].

Basser Library, Australian Academy of Science

MS 95, Walter Geoffrey Duffield—Records.

National Library of Australia

MS 7360, Clabon Walter Allen, Diaries.

Australian Archives (Melbourne)

MP730/11 Box 3, 'History of the Scientific Instruments and Optical Panel, Initially The Optical Munitions Panel', Ministry of Munitions, Ordnance Production Directorate, by J.S. Rogers.

Secondary Sources

Allen, C.W., 'Relation Between Magnetic Storms and Solar Activity', *MN*, vol. 104, 1944, pp. 13–21.

Allen, C.W., 'A sunspot cycle model', *Observatory*, vol. 80, 1960, pp. 94–8.

Allen, C.W., *Astrophysical Quantities*, 3rd edn, Athlone Press, London, 1973.

Allen, C.W., 'The beginnings of the Commonwealth Solar Observatory', *Records of the Australian Academy of Science*, vol. 4, no. 1, 1978, pp. 27–49.

Barsdell, Len, 'Sweeping the Skies from Stromlo', Commonwealth Department of Information, May 1946.

Beatty, J.K. and Goldman, S.J., 'The great crash of 1994: a first report', *S&T*, vol. 86, no. 4, October 1994, pp. 18–23.

Bhathal, R. and White, G., *Under the Southern Cross: A Brief History of Astronomy in Australia*, Kangaroo Press, Sydney, 1991.

Bok, B.J., 'Unlocking the secrets of the stars—an astronomer writes about Mount Stromlo Observatory', *Hemisphere*, vol. 2, no. 8, August 1958, pp. 2–7.

Bok, B.J., 'Proposed British Commonwealth Southern Observatory', *Australian Journal of Science*, vol. 26, 1963, p. 6.

Bok, B.J., 'The future of astronomy in Australia' (Presidential Address, ANZAAS Section A), *AJSci*, vol. 25, 1963, pp. 281–8.

Bok, B.J. and Bok, P.F., *The Milky Way*, 5th edn, Harvard University Press, Cambridge, Massachusetts, 1981.

Bolton, H.C., 'Some themes in the development of optics in Australia', *Australian Physicist*, vol. 21, December 1984, pp. 285–8.

Bolton, H.C., 'Optical Instruments in Australia in the 1939–45 war: successes and lost opportunities', *Australian Physicist*, vol. 27, no. 3, March 1990, pp. 31–43.

Bowen, E.G., 'The origins of radio astronomy in Australia', in W.T. Sullivan III (ed.), *The Early Years of Radio Astronomy*, Cambridge University Press, Cambridge, 1984, pp. 84–111.

Bray, R.J. and Loughhead, R.E., *Sunspots*, The International Astrophysics Series, vol. 7, Chapman and Hall, London, 1964.

Buscombe, W., 'Research at Mount Stromlo Observatory', *Journal of the Royal Astronomical Society of Canada*, vol. 52, 1958, pp. 68–81.

Christiansen, W.N., 'The first decade of solar radio astronomy in Australia', in W.T. Sullivan III (ed.), *The Early Years of Radio Astronomy*, Cambridge University Press, Cambridge, 1984, pp. 112–31.

Cockburn, S. and Ellyand, D., *Oliphant: The Life and Times of Sir Mark Oliphant*, Axiom Books, Adelaide, 1981.

Coombs, H.C., *Trial Balance*, Macmillan, Melbourne, 1981.

Craig, W.L. and Smith, Q., *Theism, Atheism and Big Bang Cosmology*, Oxford University Press, Oxford, 1993.

Crisp, S., 'Obituary—Sydney Ernest Williams', *Australian Physicist*, vol. 16, 1979, p. 80.

Davies, P., *God and the New Physics*, Simon & Schuster, New York, 1983.

Davies, P., *The Edge of Infinity: Beyond the Black Hole*, 2nd edn, Penguin, London, 1994.

Davies, S., 'R. v.d. R. Woolley in Australia', *Historical Records of Australian Science*, vol. 6, no. 1, 1982, pp. 57–69.

Dawkins, Richard, *The Blind Watchmaker*, Longman, London, 1986, preface.

Doobov, M. and Gascoigne, S.C.B., 'Refugee engineer had precision in his sights', Obituary, *The Australian*, 4 September 1995.

Duffield, Joan, 'Fond observations from the past', *Canberra Times*, 26 September 1996.

Edgeloe, V.A., 'Four nineteenth century graduates', *Journal of the Historical Society of South Australia*, vol. 14, 1986, pp. 137–48.

Eggen, O.J., 'The Australian Commonwealth Observatory', *Sky and Telescope*, vol. 15, June 1956, p. 339.

Eggen, O.J., 'Notes from a life in the dark', *ARAA*, vol. 31, 1993, pp. 1–11.

Ellison, M.A., *The Sun and its Influence*, Routledge and Kegan Paul, London, 1955.

Evans, D.S., *Under Capricorn: A History of Southern Hemisphere Astronomy*, Hilger, Bristol, 1984.

Foster, S.G. and Varghese, M.M., *The Making of the Australian National University, 1946–1996*, Allen & Unwin, Sydney, 1996.

Freeman, K., 'Olin J. Eggen, 1919–1998 (obituary)', *Astronomy & Geophysics*, vol. 41, no. 1, 2000, p. 36.

Gascoigne, S.C.B., 'Towards a Southern Commonwealth Observatory', *Nature*, vol. 197, 1963, pp. 1240–1.

Gascoigne, S.C.B., 'Arthur Robert Hogg', *Records of the Australian Academy of Science*, vol. 1, no. 3, November 1968, pp. 58–70.

Gascoigne, S.C.B., 'Astrophysics at Mount Stromlo: the Woolley era', *Proceedings of the Astronomical Society of Australia*, vol. 5, no. 4, 1984, pp. 597–605.

Gascoigne, S.C.B., Proust, K.M. and Robins, M.O., *The Creation of the Anglo-Australian Observatory*, Cambridge University Press, Cambridge, 1990.

Gascoigne, S.C.B., 'Bok, Woolley and Australian astronomy', *Historical Records of Australian Science*, vol. 9, no. 2, 1992, pp. 119–26.

Gascoigne, S.C.B., 'Mount Stromlo Observatory: the early years', *Sky & Space*, vol. ? June 1994, pp. 18–21.

Gascoigne, S.C.B., 'Mount Stromlo Observatory: the modern years', *Sky & Space*, vol. 8, no. 1, February 1995, pp. 24–8.

Gascoigne, S.C.B., 'The Great Melbourne Telescope and other 19th century reflectors', *Historical Records of Australian Science*, vol. 10., no. 3, June 1995, pp. 223–45.

Gillespie, L., *Canberra: 1820–1913*, AGPS, Canberra, 1991.

Gray, J., 'Charles Weston: pioneer of afforestation and conservation in the Federal Capital', *Canberra Historical Journal*, September 1999, pp. 9–18.

Gribbin, J., *In Search of the Big Bang: The Life and Death of the Universe*, rev. edn, Penguin, London, 1998.

Hartnett, L.J., *Big Wheels and Little Wheels*, 2nd edn, Gold Star, Hawthorn, Victoria, 1973.

Hartnett, L.J., 'Recollections of the Optical Munitions Panel in Australia', *Australian Physicist*, vol. 22, May 1985, pp. 158–60.

Hartung, E.J., 'Production of optical glass in Australia', *Nature*, vol. 149, 1942, pp. 518–19.

Hawking, S., *A Brief History of Time*, Bantam, New York, 1988.

Haynes, R., Haynes, R., Malin, D. and McGee, R., *Explorers of the Southern Sky: A History of Australian Astronomy*, Cambridge University Press, Cambridge, 1996.

Hogg, A.R., 'The Great Melbourne Telescope', *Australian Journal of Science*, vol. 21, 1958, pp. 2–5.

Hyde, W.I., 'The calamity of the Great Melbourne Telescope', *Proceedings of the Astronomical Society of Australia*, vol. 7, 1987, pp. 227–31.

Hyland, A.R. and Faulkner, D.J., 'From the Sun to the Universe—the Woolley and Bok directorships at Mount Stromlo', *Proceedings of the Astronomical Society of Australia*, vol. 8, 1989, pp. 216–28.

La Nauze, J.A., '"Other like services": physics and the Australian Constitution', *Records of the Australian Academy of Science*, vol. 1, no. 3, November 1968, pp. 36–44.

Ledger, E., *The Sun: Its Planets and Their Satellites*, Edward Stanford, London, 1982.

Leverington, D., *A History of Astronomy from 1890 to the Present*, Springer, London, 1995.

Levy, D.H., *The Man Who Sold the Milky Way: A Biography of Bart Bok*, University of Arizona Press, Tucson & London, 1993.

Lewis, M., 'The idea of a National University: the origins and establishment of the Australian National University', *Journal of the Australian and New Zealand History of Education Society*, vol. 8, no. 1, 1979, pp. 40–55.

Love, R., 'Science and government in Australia 1905–14: Geoffrey Duffield and the foundation of the Commonwealth Solar Observatory', *Historical Records of Australian Science*, vol. 6, no. 2, December 1985, pp. 171–88.

Lovell, B., 'The origin of the British enclave and the 48-inch Schmidt Telescope at Siding Spring', *Quarterly Journal of the Royal Astronomical Society*, vol. 26, 1985, pp. 456–62.

Martynov, D.I., *Observatories in Australia: Travel Impressions*, ANU, Canberra, 1965.

Massey, H.S.W., 'Obituary—T.H. Laby', *Nature*, vol. 158, 1946, p. 157.

Mathewson, D.S., 'Space astronomy in the 1980s: Australian role?', *ANU Reporter*, vol. 11, no. 2, 21 March 1980.

McCarthy, G. and Sankey, H. (eds), *The Papers of Walter Geoffrey Duffield (1879–1929)*, Australian Science Archives Project, Melbourne, 1987.

McCrea, W., 'Richard van der Reit Woolley, 1906–1986', *Historical Records of Australian Science*, vol. 7, no. 3, 1988, pp. 315–45.

McNally, D., 'A symposium on solar physics to mark the retirement of Professor C.W. Allen as Perren Professor of Astronomy, University College London, 1951–1972', *Observatory*, vol. 93, no. 992, 1973, p. 260.

McNally, D., 'Obituaries—C.W. Allen, 1904–1987', *Quarterly Journal of the Royal Astronomical Society*, vol. 31, no. 2, 1990, pp. 259–66.

Mellor, D.P., *The Role of Science and Industry*, Official History of Australia in the War of 1939–1945, Australian War Memorial, Canberra, 1958.

Menzel, D.H., *Our Sun*, Harvard University Press, Cambridge, Massachusetts, 1959.

Miller, M.J., *Mount Stromlo Observatory: A History and Description of the Observatory, its Work and its Equipment*, Research School of Physical Sciences, ANU, 1957.

Nelmes, M., 'Mr Oddie's Legacy', *Southern Sky*, November 1993, pp. 22–7.

Norris, R., 'The Australia Telescope', *S&T*, vol. 76, no. 6, 1988, pp. 615–20.

Orchiston, W., 'The emergence of professional astronomy in nineteenth century Australia', *Search*, vol. 19, March–April 1988, p. 76.

Pagels, H., *Perfect Symmetry: The Search for the Beginning of Time*, Michael Joseph, London, 1985.

Piddington, J.H., 'Ronald Gordon Giovanelli, 1915–1984', *Historical Records of Australian Science*, vol. 6, no. 2, December 1985, pp. 223–33.

Piddington, J.H. and Oliphant, M.L., 'David Forbes Martyn, 1906–70', *Records of the Australian Academy of Science*, vol. 2, no. 2, 1971, p. 47.

Pockley, P., 'Votes nudge Australia closer to joining telescope project', *Nature*, vol. 379, 1996, p. 668.

Pockley, P., 'Science loses cabinet seat after Australian election. Telescope prospects looking dimmer', *Nature*, vol. 380, 1996, p. 192.

Pockley, P., 'Australian astronomers are rebuffed in move to apply for ESO membership', *Nature*, vol. 381, 1996, pp. 100–1.

Pockley, P., 'Savings drive unsettles Australian science', *Nature*, vol. 381, 1996, p. 5.

Pockley, P., 'Australia eyes role in Gemini telescopes', *Nature*, vol. 389, 1997, p. 220.

Pockley, P., 'Seventh heaven for Australia's optical astronomers', *Nature*, vol. 391, 1998, p. 724.

Robertson, J.G., 'Two hundred years of astronomy in Australia', *Journal of the British Interplanetary Society*, vol. 41, 1988, pp. 567–80.

Robertson, P., *Beyond Southern Skies: Radio Astronomy and the Parkes Telescope*, Cambridge University Press, Cambridge, 1992.

Simmonds, D., 'Classical Gascoigne', *Bulletin*, 26 May 1998.

Sinnott, R.W. and Nyren, K., 'The world's largest telescopes', *S&T*, vol. 86, no. 1, July 1983, pp. 27–32.

Spaull, A., *John Dedman: A Most Unexpected Labor Man*, Hyland House, Melbourne, 1998.

Stetson, H.T., *Sunspots and Their Effects*, rev. edn, McGraw-Hill, New York, 1947.

Stibbs, D.W.N., 'A tribute to C.W. Allen', *Quarterly Journal of the Royal Astronomical Society*, vol. 14, no. 3, 1973, pp. 311–17.

Sullivan, W.T., 'The early days of Australian radio astronomy', in R.W. Home (ed.), *Australian Science in the Making, 1788–1988*, Cambridge University Press, Melbourne, 1984.

Turner, J., 'The Australian Ionospheric Prediction System', *Ionospheric Predictions Series X11*, Department of Science and Technology, December 1981, pp. 1–2.

White, F., 'Robert Gordon Menzies (1894–1978)', *Historical Records of Australian Science*, vol. 5, no. 1, 1980, pp. 68–102.

Wild, J.P., 'The beginnings of radio astronomy in Australia', *ProcASA*, vol. 7, no. 1, 1987, pp. 95–102.

Woolley, R., *Navigation by the Stars*, Victorian Railways Printing Works, Melbourne, 1942.

Woolley, R., 'Address to Session 2—Physical Sciences', in *Science in Australia: Proceedings of a Seminar Organized by the ANU on the Occasion of the Jubilee of the Commonwealth of Australia*, 24–27 July 1951, Melbourne, 1952.
Woolley, R., 'Mount Stromlo Observatory', in Mark Oliphant (ed.), *Records of the Australian Academy of Science*, vol. 1, no. 3, November 1968, pp. 53–77.

Notes

Preface

1. Jeremy Mould, *ANU Reporter*, vol. 30, no. 11, 4 August 1999.

Introduction

1. John Gribbin, *In Search of the Big Bang: The Life and Death of the Universe*, rev. edn, Penguin, London, 1998, pp. 184–5.
2. William Carroll, 'Thomas Aquinas and Big Bang cosmology', Jacques Maritain Centre: Thomistic Institute website http://www.nd.edu/Departments/Maritain/ti/carroll.htm, p. 3[August 2002]
3. Stephen Hawking, *A Brief History of Time*, Bantam, New York, 1988, p. 141.
4. Paul Davies, *God and the New Physics*, Simon & Schuster, New York, 1983, p. viii.
5. Paul Davies, *The Edge of Infinity: Beyond the Black Hole* [first published 1981], 2nd edn, Penguin, London, 1994, p. 169.
6. Heinz Pagels, *Perfect Symmetry: The Search for the Beginning of Time*, Michael Joseph, London, 1985, p. 17.
7. William Lane Craig and Quentin Smith, *Theism, Atheism and Big Bang Cosmology*, Oxford University Press, Oxford, 1993, p. 217.
8. Edmund Ledger, *The Sun: Its Planets and Their Satellites*, Edward Stanford, London, 1982.

9. David Leverington, *A History of Astronomy from 1890 to the Present*, Springer, London, 1995, p. ii.

CHAPTER 1

1. Rosaleen Love, 'Science and government in Australia 1905–14: Geoffrey Duffield and the foundation of the Commonwealth Solar Observatory', *Historical Records of Australian Science*, vol. 6, no. 2, December 1985, pp. 171–2.
2. For a very useful guide to the considerable body of papers left by Duffield, see Gavan McCarthy and Howard Sankey (eds), *The Papers of Walter Geoffrey Duffield (1879–1929)*, Australian Science Archives Project, Melbourne, 1987. The Adolph Basser Library, Australian Academy of Science, holds the largest single body of papers: 'MS 95—Walter Geoffrey Duffield—Records'. This body of records (1.16m shelf space) consists of material dating from 1880 to 1987 and includes items donated by his family after his death. The two most significant descriptions of his life and scientific achievements are V.A. Edgeloe, 'Four nineteenth century graduates', *Journal of the Historical Society of South Australia*, vol. 14, pp. 137–48; and Rosaleen Love, 'Science and government in Australia', pp. 171–88.
3. See also J.A. La Nauze, ' "Other like services": Physics and the Australian Constitution', *Records of the Australian Academy of Science*, vol. 1, no. 3, Canberra, November 1968, pp. 36–44.
4. For a detailed description of Canberra's selection as the site of the Federal Capital Territory see Lyall Gillespie, *Canberra: 1820–1913*, AGPS, Canberra, 1991, Ch. 10.
5. *Memorandum upon the Proposed Solar Observatory in Australia*, Solar Physics Committee of the Australasian Association for the Advancement of Science, Sherring, Adelaide, 1909, p. 12.
6. See *Who Was Who*, 1897–1915.
7. Quoted in Michael Nelmes, 'Mr Oddie's legacy', *Southern Sky*, November 1993, p. 22.
8. *Memorandum upon the Proposed Solar Observatory in Australia*, Solar Physics Committee of the Australasian Association for the Advancement of Science, Sherring, Adelaide, 1909.
9. ibid.

10. ibid., p. 12.
11. Secretary of the Royal Society, J. Larmor, to the Under Secretary of State for the Colonial Office, dated 10 February 1908. A copy of the statement was forwarded to Duffield, Duffield papers.
12. Love, 'Science and government in Australia', p. 176.
13. Quoted in Nelmes, 'Mr Oddie's legacy', p. 22.
14. Olin J. Eggen, 'The Australian Commonwealth Observatory', *Sky and Telescope*, vol. XV, no. 8, 1956, p. 5.
15. Nelmes, 'Mr Oddie's legacy', p. 22.
16. ibid., p. 23.
17. Australian Archives (hereafter AA) file A431 1965/1589.
18. 'Minutes of a meeting in Melbourne regarding the proposed Solar Observatory with the Prime Minister', AA A1/1 file 1918/6038.
19. Weston's contribution to Canberra is described in John Gray, 'Charles Weston: pioneer of afforestation and conservation in the Federal Capital', *Canberra Historical Journal*, September 1999, pp. 9–18.
20. ibid., p. 15.
21. 'Afforestation planting areas and projects, 1913–1924', AA, A192, Item FCL1921/1905.
22. Gray, 'Charles Weston', p. 16; AA, CP209/18, item B13, report by Charles Weston, 11 January 1924.
23. 'Meteorological Solar Observatory, Mount Stromlo, FCT', AA Series A461/8, Item B372/2.
24. Len Barsdell, *Sweeping the Skies from Stromlo*, Department of Information, May 1946, p. 3.
25. Pamphlet, 'Establishment of Solar Observatory Canberra', AA Series A431, Item 1952/184.
26. Commonwealth Department of Works and Railways, 'Commonwealth Observatory Mount Stromlo Erection—Notes on Projected Commonwealth Observatory, Stromlo, Federal Territory', 19 April 1923, AA series A199, Item FC 1926/206.
27. Eggen, 'The Australian Commonwealth Observatory', p. 5.
28. Joan D. Duffield, 'Recollections of an astronomer's daughter: a brief outline of life in the early days of Canberra and the beginning of Mount Stromlo Observatory', unpublished ms, dated July 1996, p. 2. (Later published in the *Canberra Times*.)
29. Love, 'Science and government in Australia', p. 186.

CHAPTER 2

1. See *Who Was Who*, 1916–1928.
2. 'Reynolds Telescope', AA Series A461/1 file B372/1/2.
3. Joan Duffield, 'Recollections of an astronomer's daughter', p. 2.
4. *Sydney Morning Herald*, 12 January 1925.
5. Walter Geoffrey Duffield, transcribed diary for 1925, copy held by authors.
6. Duffield diary, 25 July 1925.
7. Duffield diary, 1 August 1925.
8. 'Creation of positions at Commonwealth Solar Observatory Canberra', AA Series A1/15, Item 1926/403.
9. D. McNally, 'A symposium on solar physics to mark the retirement of Professor C.W. Allen as Perren Professor of Astronomy, University College London, 1951–1972', *Observatory*, vol. 93, no. 992, 1973, pp. 1–2.
10. S.C.B. Gascoigne, 'Arthur Robert Hogg', Proceedings of the *RAAS*, vol. 1, no. 3, November 1966.
11. National Library of Australia (hereafter NLA), MS 7360, 'Clabon Walter Allen, Diaries', see entries for 5 March 1926, 23 March 1926 and 17 June 1927.
12. 'Commonwealth Solar Observatory Report', AA Series A1/15, Item 1925/10368.
13. *CSO Annual Report 1928–29*, p. 4.
14. ibid., p. 5
15. Allen diary, 5 August 1927.
16. Dr Lewis Radford, Bishop of Goulburn, to Minister for Home & Territories (Perkins), 24 August 1933, Canberra & Goulburn diocesan archives, file P14/8, 'Canberra, Duffield Grave'.
17. Commonwealth Solar Observatory, *Memoire Number 1*, 'The luminosity of the night sky', Government Printer, 1935, p. i.
18. W.B. Rimmer, *Memoires of the Commonwealth Solar Observatory*, vol. 1, Commonwealth Government Printer, January 1935, p. 5.
19. ibid.

CHAPTER 3

1. His obituary was written by Severin Crisp and appeared in the *Australian Physicist*, vol. 16, 1979, p. 80.

2. J.H. Piddington, 'Ronald Gordon Giovanelli, 1915–1984', *Historical Records of Australian Science*, vol. 6, no. 2, December 1985, p. 225.
3. R.G. Giovanelli, *ApJ*, vol. 88, 1938, p. 204; *ApJ*, vol. 91, 1939, p. 334.
4. R.G. Giovanelli, *ApJ*, vol. 89, 1939, p. 555.
5. R.G. Giovanelli and A.J. Higgs, *Terr. Mag Atm. Electr*, vol. 44, 1939, p. 181.
6. R.G. Giovanelli, *ApJ*, vol. 91, 1940, p. 334.
7. Martyn, Munro, Higgs and Williams, *Nature*, vol. 140, 1937, p. 163.
8. C.W. Allen, *MN*, vol. 100, 1940, p. 635.
9. AA series A199/1 file FC 26/206.
10. AA series A6270/1 file E2/28/2959.
11. AA series A461/1 file B372/1/2.
12. *CSO Annual Report 1928–29*, p. 4.
13. Minnaert, Mulders and Houtgast, *Photometric Atlas of the Solar Spectrum*, 1940.
14. AA Series CP698/2/1 file 25/823.
15. AA Series A6270/1 file E2/28/2959.
16. ibid.
17. *CSO Annual Report 1929–30*.
18. *CSO Annual Report 1930–31*, p. 9.
19. See C.W. Allen, *Gerl. Beitr. Z. Geophys*, vol. 46, 1935, p. 32.
20. C.W. Allen, *Observatory*, vol. 70, 1950, p. 154.
21. See also C.W. Allen, *Quarterly Journal of the Royal Meteorological Society*, vol. 84, 1958, p. 307.
22. Gordon Newkirk Jnr, 'Variations in solar luminosity', *ARAA*, vol. 21, 1983, pp. 429–67.
23. H.S. Hudson, 'Observed variability of the solar luminosity', *ARAA*, vol. 26, 1988, pp. 473–507.
24. C.W. Allen, *Observatory*, vol. 70, 1950, p. 154.
25. C.W. Allen, *Astrophysical Quantities*, 3rd edn, Athlone Press, London, 1973, p. 169.
26. W.B. Rimmer, *Gerl. Beitr. Z. Geophys*, vol. 50, 1937, p. 388.
27. W.B. Rimmer and C.W. Allen, 'Solar radiation observatories', *CSO Memoires*, no. 11, 1950, pp. 1–44.
28. Allen (1973), *op. cit.*, p. 131.
29. A.R. Hogg, *Gerl. Beitr. Z. Geophys*, vol. 50, 1937, p. 394.
30. A.J. Higgs, 'Measurements of atmospheric ozone', *CSO Memoirs*, no. 3, 1934.
31. J.H. Dellinger, *Phys. Rev.*, vol. 50, 1937, p. 1189.
32. D.F. Martyn et al., *Nature*, vol. 140, 1937, p. 163.

33. C.W. Allen, *CSO Memoires*, no. 4, 1934; C.W. Allen, 'Potential gradient observations', *CSO Memoires*, no. 8, 1939.
34. S.J. Mauchley, *Terr. Mag. Atm. Electr.*, vol. 28, 1923, p. 61.
35. A.R. Hogg, *Gerl. Beitr. Z. Geophys*, vol. 41, 1934.
36. ibid., pp. 1–31, 32–57.
37. A.R. Hogg, *Gerl. Beitr. Z. Geophys*, vol. 47, 1936, pp. 31–59.
38. A.R. Hogg, *Proc. Phys Soc.*, vol. 51, 1939, pp. 10–14.
39. A.R. Hogg, 'Conduction of electricity at the lowest levels of the atmosphere', *CSO Memoire*, no. 7, 1939.
40. Gascoigne (1968), 'Arthur Robert Hogg', pp. 60–1.
41. A.R. Hogg, 'Measurements of the intensity of cosmic rays', *CSO Memoire*, no. 10, 1949.
42. Some of the results were published later than this date.
43. The testimonials supplied in support of Woolley's candidature are contained in his departmental staff file—AA CRS A220, S39/698.
44. Sir William McCrea, 'Richard van der Reit Woolley, 1906–86', *Historical Records of Australian Science*, vol. 7, no. 3, Canberra, 1988, pp. 315–45, p. 319 quoted.
45. Details of Woolley's early life are drawn from McCrea, 'Richard van der Reit Woolley'; Susan Davies, 'R. v.d. R. Woolley in Australia', *Historical Records of Australian Science*, vol. 6, no. 1, 1982; and Woolley's final entry (1987) in *Who's Who*.

Chapter 4

1. Sir Laurence Hartnett, *Big Wheels and Little Wheels*, 2nd edn, Gold Star, Melbourne, 1973, p. 129.
2. For a general account of the Panel see J.S. Rogers, 'The history of the Optical Munitions Panel: July 1940–December 1946', AA MP 730/11, Box 3. Correspondence relating to the production of this report can be consulted in AA MP 730/11, Box 4, particularly item 16 file 256.S.4.
3. 'Scientific Instruments and Optical Panel', Report of Meeting 16 February 1945, AA MP 1472/4, Box 2, vol. 7 (Meetings 28–29).
4. R. Woolley, 'Mount Stromlo Observatory', in Mark Oliphant (ed.), *Records of the Australian Academy of Science*, vol. 1, no. 3, 1968, p. 54.
5. D.P. Mellor, *The Role of Science and Industry, Official History of Australia in the War of 1939–45*, AWM, Canberra, 1958, p. 248.
6. ibid., p. 262.

7. 'Progress report from Commonwealth Solar Observatory', 24 January 1941, AA MP 1472/4, Box 1, vol. 1 (Meetings 1–8).
8. Woolley to F.S. Daley (Controller, OPD), 6 August 1941, AA MP 392/9, Box 2, 'Optical—Mount Stromlo'.
9. Woolley to Hartnett, 16 July 1941, AA MP 392/9, Box 2, 'Optical—Mount Stromlo'.
10. 'Commonwealth Solar Observatory—Optical Munitions Branch, Estimates 1943–44', AA Series A659/1, Item 1943/1/2792.
11. Mellor, *The Role of Science and Industry*, p. 271.
12. Woolley to Secretary, Department of the Interior, 25 August 1941, AA A431/1, 52/201.
13. *Canberra Times*, 24 February 1941.
14. Olin J. Eggen, 'The Australian Commonwealth Observatory', *Sky and Telescope*, vol. 15, no. 8, June 1956, p. 6.
15. Mervyn Doobov and Ben Gascoigne, 'Refugee engineer had precision in his sights', Obituary, *Australian*, 4 September 1995.
16. Interview with Mrs Gascoigne conducted by Louise Taylor, 2 July 1996. Transcript kindly made available to the authors by Mrs Taylor. See also her published interview with Diana Simmonds, 'Classical Gascoigne', *Bulletin*, 26 May 1998, pp. 72–3.
17. W.G. Jowett quoted in Gascoigne (1968), 'Robert Arthur Hogg', pp. 62–3.
18. A.J. Higgs, 'Ionospheric measurements made during the total solar eclipse of 1940 October 1', *MN*, vol. 102, London, 1942, pp. 24–34.
19. Higgs to Woolley, 28 July 1940, AA Series A9103, Item 4.
20. Woolley to Higgs, 17 August 1940, AA Series A9103, Item 4.
21. Tim Sherratt and Anne-Marie Conde, 'A wartime observatory observed', see the website www.discontents.com.au/words/stromlo.php
22. Hartnett, *Big Wheels*, p. 166.
23. Woolley, 'Mount Stromlo Observatory', p. 55.
24. Richard Woolley, *Navigation by the Stars*, Victorian Railways Printing Works, Melbourne, 1942, p. 1.
25. Memo from J.T. Malcolmson to T. Hedberg, 21 January 1943, AA MP 392/9, Box 12, 'Optical Machine Tools & Mount Stromlo'.
26. This estimate was made in the 'Report on Cost and Production of Optical Munitions' to the Director-General of Munitions (Essington Lewis) by the OMP Chairman.
27. Allen to Woolley, 29 August 1940, AA A9103, Item 4.
28. Allen, diary entry for 20 January 1941, vol. 17, Allen papers, NLA MS 7360.

29. Professor Walter Stibbs to the authors dated 9 January 2001.
30. Harland True Stetson, *Sunspots and Their Effects*, McGraw-Hill, New York; revised by the Ronald Press Company of New York in 1947 with a foreword by Sir Edward Appleton, pp. 46–59.
31. Mellor, *The Role of Science and Industry*, p. 508.
32. C.W. Allen, 'Relation between magnetic storms and solar activity', *MN*, vol. 104, 1944, pp. 13–21.
33. Donald H. Menzel, *Our Sun*, Harvard University Press, Cambridge, Massachusetts, 1959, p. 326.
34. R.J. Bray and R.E. Loughhead, *Sunspots*, The International Astrophysics Series, vol. 7, Chapman and Hall, London, 1964, pp. 277–8.
35. Menzel, *Our Sun*, p. 202.
36. Jack Turner, 'The Australian ionospheric prediction system', *Ionospheric Predictions Series XII*, Department of Science and Technology, December 1981, pp. 1–2.
37. H.S.W. Massey, 'Obituary—T.H. Laby', *Nature*, vol. 158, 1946, p. 157.
38. R. Woolley, 'Mount Stromlo Observatory', p. 54.
39. H.C. Bolton, 'Optical instruments in Australia in the 1939–45 war: successes and lost opportunities', *Australian Physicist*, vol. 27, no. 3, March 1990.
40. Obituary notice, *Canberra Times*, 3 July 1945.

Chapter 5

1. Bart J. Bok, *MN*, vol. 118, no. 4, 1958, pp. 363–4.
2. Woolley's own research gathered momentum in 1945. In December of that year he completed 'The mechanism of ionization', *Proceedings of the Royal Society*, vol. A 187, 1946, pp. 403–15.
3. D.W.N. Stibbs, 'A study of the spectrum and magnetic variable star HD 125248', *MN*, vol. 110, 1950, pp. 395–404.
4. AA series A431/1 file 46/1629.
5. One product of this activity was Gascoigne's paper 'Relative gradients for 166 southern stars', *MN*, vol. 110, no. 1, 1950, pp. 15–36.
6. AA CRS A431, Item 47/728.
7. Gascoigne (1968), 'Robert Arthur Hogg', p. 63.
8. A.R. Hogg, 'The eclipsing system Zeta Phoenicis', *MN*, vol. 111, 1951, pp. 315–24.
9. K.G. Jansky, 'Radio waves from outside the solar system', *Nature*, vol. 132, 1933, p. 66.

10. Grote Reber, 'Cosmic static', *ApJ*, vol. 91, 1940, pp. 621–4.
11. A comprehensive account has been given by Peter Robertson, *Beyond Southern Skies: Radio Astronomy and the Parkes Telescope*, Cambridge University Press, Cambridge, 1992, Ch. 2.
12. ibid., p. 31n; 'Microwave astronomy' was another early contender as a name for the new field.
13. An excellent account has been given by Robertson, *Beyond Southern Skies*, Chs 3 and 4.
14. J.P. Wild, 'The beginnings of radio astronomy in Australia', *ProcASA*, vol. 7, no. 1, 1987, pp. 95–102.
15. J.L. Pawsey, R. Payne-Scott and L.L. McCready, 'Radio-frequency energy from the Sun', *Nature*, vol. 157, 1946, pp. 158–9; J.L. Pawsey, 'Observation of million degree thermal radiation from the Sun at a wavelength of 1.5 metres', *Nature*, vol. 158, 1946, pp. 633–4.
16. L.L. McCready, J.L. Pawsey and Ruby Payne-Scott, 'Solar radiation at radio frequencies and its relation to sunspots', *Proceedings of the Royal Society*, London, vol. A 190, 1947, pp. 357–75.
17. J.G. Bolton and G.J. Stanley, 'Variable source of radio frequency radiation in the constellation of Cygnus', *Nature*, vol. 161, 1948, pp. 312–13.
18. J.G. Bolton, G.J. Stanley and O.B. Slee, 'Positions of three discrete sources of galactic radio-frequency radiation', *Nature*, vol. 164, 1949, pp. 101–2.
19. W. Baade and R. Minkowski, 'Identification of the radio sources in Cassiopeia, Cygnus A, and Puppis A', *ApJ*, vol. 119, 1954, pp. 206–14.
20. J.H. Piddington and H.C. Minnett, 'Microwave thermal radiation from the Moon', *Aust J Sci Res*, vol. A 2, 1949, pp. 63–77.
21. J.P. Wild and L.L. McCready, 'Observations of the spectrum of high-intensity solar radiation at metre wavelengths. I. The apparatus and spectral types of solar burst observed', *Aust J Sci Res*, vol. A 3, 1950, pp. 387–98.
22. W.N. Christiansen and J.V. Hindman, 'A preliminary survey of 1420 Mc/s line emission from Galactic hydrogen', *Aust J Sci Res*, vol. A 5, 1952, pp. 437–55.
23. W.N. Christiansen, 'A high-resolution aerial for radio astronomy', *Nature*, vol. 171, 1953, pp. 831ff; W.N. Christiansen, 'The first decade of solar radio astronomy in Australia', in W.T. Sullivan III (ed.), *The Early Years of Radio Astronomy*, Cambridge University Press, Cambridge, 1984, pp. 112–31.
24. The first complete 21-centimetre map of the Galaxy was published by J.H. Oort, F.J. Kerr and G. Westerhaut, 'The Galactic system as a spiral nebula', *MN*, vol. 118, 1958, p. 379.

25. B.Y. Mills and A.G. Little, 'A high-resolution aerial system of a new type', *AJPhys*, vol. 6, 1953, pp. 272–8.
26. B.Y. Mills, O.B. Slee and E.R. Hill, 'A catalogue of radio sources between declinations +10° and –20°', *AJPhys*, vol. 11, 1958, pp. 360–87; B.Y. Mills, 'Radio sources and the log N–log S controversy', in Sullivan, *The Early Years*, pp. 146–65.
27. R.X. McGee and J.G. Bolton, 'Probable observation of the Galactic nucleus at 400 Mc/s', *Nature*, vol. 173, 1954, pp. 985–7.
28. E.G. Bowen, 'The origins of radio astronomy in Australia', in Sullivan, *The Early Years*, p. 85.
29. J.H. Piddington and M.L. Oliphant, 'David Forbes Martyn 1906–70', *RAAS*, vol. 2, no. 2, 1971, p. 53.
30. C.W. Allen, 'Solar radio-noise of 200 Mc/s and its relation to solar observations', *MN*, vol. 107, 1947, p. 386; C.W. Allen and C.S. Gum, 'Survey of galactic radio-noise at 200 Mc/s', *Aust J Sci Res*, vol. A 3, 1950, p. 244.
31. R.v.d.R. Woolley, 'Galactic noise', *MN*, vol. 107, 1947, pp. 308–15.
32. R.v.d.R. Woolley, 'Opportunities for astrophysical work in Australia', *Report of the 26th Meeting of ANZAAS, Perth, August 1947* (Section A Presidential Address), pp. 64–81.
33. R.v.d.R. Woolley, Address to Session 2—Physical Sciences, in *Science in Australia: Proceedings of a Seminar organised by the Australian National University on the Occasion of the Jubilee of the Commonwealth of Australia, Canberra, July 24–27, 1951* (Melbourne, 1952), p. 51.
34. Recollection of Alex Rodgers who was in Woolley's office at the time.
35. Gerard de Vaucouleurs, 'Evidence for a local supergalaxy', *AJ*, vol. 58, 1953, pp. 30–2.
36. E.G. Bowen in Sullivan, *The Early Years*, p. 93; J.P. Wild, 'The beginnings of radio astronomy in Australia', *ProcASA*, vol. 7, no. 1, 1987, pp. 95–102, p. 100 footnote; Robertson, *Beyond Southern Skies*, pp. 109–10.
37. Wild, ibid., Appendix 1.
38. McCrea, 'Richard van der Reit Woolley', p. 321.
39. A.R. Hyland and D.J. Faulkner, 'From the Sun to the Universe: the Woolley and Bok Directorships at Mount Stromlo', *ProcASA*, vol. 8, 1989, pp. 216–28, note 6. Those present at this meeting were Professor Oscar Vonwiller, Emeritus Professor of Physics, University of Sydney (Chairman); Professor K.E. Bullen, Professor of Mathematics, University of Sydney; Professor Sir Kerr Grant, Emeritus Professor of Physics, University of Adelaide; Professor Ernst Hartung, Professor of Chemistry,

University of Melbourne; Sir John Madsen, Chairman, Radio Research Board, Sydney; Mr W.A. McLaren, Secretary, Department of the Interior; Father Daniel O'Connell, Director, Riverview College Observatory; Professor Mark Oliphant, Director, Research School of Physical Sciences, ANU; and Dr Richard Woolley, the Commonwealth Astronomer. Dr Arthur Hogg was Secretary.

40. Minutes of the 10th meeting of the Board of Visitors, 31 January 1951, Resolution 1.
41. For an example see Robertson, *Beyond Southern Skies*, pp. 109–113.
42. ibid., Ch. 5, quotation at note 10.
43. *Radiophysics Laboratory, Sydney (Annual Report)*, *QJRAS*, vol. 3, 1962, pp. 328–37, p. 328 cited.
44. Walter Stibbs, 'A tribute to C.W. Allen', *QJRAS*, vol. 14, no. 3, 1973, p. 392.
45. M.A. Ellison, *The Sun and its Influence*, Routledge and Kegan Paul, London, 1955.
46. D. McNally, 'Obituaries—C.W. Allen, 1904–1987', *QJRAS*, vol. 31, no. 2, 1990, p. 260.

Chapter 6

1. The failure of the Great Melbourne Telescope to produce any results is described in great detail by S.C.B. Gascoigne, 'The Great Melbourne Telescope and other 19th century reflectors', *Historical Records of Australian Science*, vol. 10, no. 3, June 1995, pp. 223–45.
2. ibid., p. 223.
3. S.C.B. Gascoigne, 'Astrophysics at Mount Stromlo: the Woolley era', *ProcASA*, vol. 5, 1984, pp. 597–605.
4. *MSOAR*, 1947, p. 3.
5. *MSOAR*, 1955, p. 2.
6. S.C.B. Gascoigne, 'Bok, Woolley and Australian astronomy', *Historical Records of Australian Science*, vol. 9, 1992, pp. 119–26.
7. G. Simonow, 'A history of the Commonwealth Observatory covering 1905–50', Supplement to the Report of the Commonwealth Astronomer to the Board of Visitors, 1950, p. 11.
8. *MSOAR*, 1951, p. 3.
9. W. Baade, 'The resolution of Messier 32, NGC 205, and the central region of the Andromeda Nebula', *ApJ*, vol. 100, 1944, pp. 137–46.
10. S.C.B. Gascoigne and G.E. Kron, 'Photoelectric observations of Magellanic Cloud Cepheids', *MN*, vol. 130, 1965, pp. 333–60; S.C.B.

Gascoigne, 'Further observations of Magellanic Cloud Cepheids', *MN*, vol. 146, 1969, pp. 1–36.

11. Kim M. Sebo, Daya Rawson et al., 'The Cepheid period-luminosity relation in the Large Magellanic Cloud', *ApJS*, vol. 142, 2002, pp. 71–8.
12. 'Part 2, Building Program, Solar Observatory', AA Series A431, Item 1951/744.
13. Woolley, 'Mount Stromlo Observatory', p. 55.
14. Doris Hogg, '1952 Mt Stromlo Fire Remembered: a recollection of a drama that turned out well for me', an unpublished reminiscence dated 1977, copy held by authors.
15. 'Solar Observatory, Mount Stromlo: Restoration of Fire Damage', AA Series A431, Item 1955/123.
16. Woolley, 'Mount Stromlo Observatory', p. 56.
17. Ragbir Bhathal and Graeme White, *Under the Southern Cross: A Brief History of Astronomy in Australia*, Kangaroo Press, Sydney, 1991, p. 56.
18. S.G. Foster and Margaret M. Varghese, *The Making of the Australian National University, 1946–1996*, Allen & Unwin, Sydney, 1996, pp. 2–9.
19. See H.C. Coombs, *Trial Balance*, Macmillan, Melbourne, 1981.
20. See Milton Lewis, 'The idea of a National University: the origins and establishment of the Australian National University', *Journal of the Australian and New Zealand History of Education Society*, vol. 8, no. 1, 1979, pp. 40–55.
21. ibid., pp. 10–15. See also Andrew Spaull, *John Dedman: A Most Unexpected Labor Man*, Hyland House, Melbourne, 1998, Ch. 6.
22. Foster and Varghese, *The Making of the ANU*, p. 16.
23. ANU Archives 103/1/23, M.L. Oliphant, 'Research School of Physical Sciences—aims, organisation and finance', April 1948.
24. CSIRO Archives, Series 372, Item 1/H, Draft resolutions of Committee on 'Faculty of Physical Science, University of Canberra'.
25. ANU Archives, 8.7.1.0. Part 1, Oliphant to Vice-Chancellor, 19 October 1949.
26. G. de Vaucouleurs, 'Survey of bright galaxies south of –35° declination', *Commonwealth Observatory Memoire*, no. 13, 1956.
27. G. de Vaucouleurs, 'Evidence for a local supergalaxy', *AJ*, vol. 58, 1953, pp. 30–2.
28. *QJRAS*, vol. 22, no. 352, 1981.
29. R.v.d.R. Woolley and D.W.N. Stibbs, *The Outer Layers of a Star*, The Clarendon Press, Oxford, 1953.

30. See Colin S. Gum, 'A survey of southern HII regions', *MN*, vol. 67, 1955, pp. 155–77.
31. Susan Davies, 'R.v.d.R. Woolley in Australia', *Historical Records of Australian Science*, vol. 6, no. 1, 1982, pp. 57–69, p. 59 quoted.
32. Richard Woolley, 'Address to Session 2—Physical Sciences', in *Science in Australia: Proceedings of a Seminar Organized by the ANU on the Occasion of the Jubilee of the Commonwealth of Australia*, 24–27 July 1951, Melbourne, 1952, p. 51.
33. ibid.
34. ANU Archives, 8.8.1.0. (A) Part 1, Woolley to Registrar, 7 September 1953.
35. Basser Library, Box 37/1, 'Minutes of the 13th meeting of the Board of Visitors of the Commonwealth Observatory', 4 February 1954.
36. Basser Library, Box 37/1, Woolley to the Secretary, Department of the Interior, 30 October 1953.
37. Woolley, 'Mount Stromlo Observatory', p. 57.
38. McCrea, 'Richard van der Reit Woolley', p. 321.
39. ibid., p. 328.
40. Davies, 'R.v.d.R. Woolley', p. 4.
41. Woolley, 'Mount Stromlo Observatory', p. 57.
42. ibid., pp. 98–9; Gascoigne, 'Astrophysics at Mount Stromlo: the Woolley era', pp. 597–605.
43. ANU Archives, 8.8.1.0. (A) Part 1, Prime Minister to Vice-Chancellor, 14 November 1955.
44. *Mount Stromlo Observatory Act* 1956 (No. 79).
45. McCrea, 'Richard van der Reit Woolley', p. 334.
46. Gascoigne, 'Bok, Woolley and Australian Astronomy', pp. 121–2.
47. Gascoigne, 'Astrophysics at Mount Stromlo: the Woolley era', pp. 597–605, p. 604 cited.
48. Gascoigne, 'Bok, Woolley and Australian astronomy', p. 121.
49. ANU file 8.2.5.34, folio 160A.
50. MSO Annual Report reprinted in *MN*, vol. 118, no. 4, 1957. See also 'ANU: Report of the Council for the Period 1 January 1955 to 31 December 1955', *Commonwealth Parliamentary Paper* (CPP) No. 32.
51. Gascoigne, 'Robert Arthur Hogg', p. 65.
52. Gascoigne, 'Bok, Woolley and Australian astronomy', p. 119.

53. See 'Report of the Council of the ANU for 1957', CPP No. 56 and *Commonwealth Yearbook* No. 44, 1958.

Chapter 7

1. O.J. Eggen, interview, April 1988; Sir Mark Oliphant, interview, April 1988.
2. O.J. Eggen, interview, April 1988.
3. Woolley, 'Mount Stromlo Observatory', pp. 53–7. In concluding this article, Woolley describes his departure from Stromlo in these words: 'I needed fresh contacts to get back into contemporary ideas, and Mount Stromlo needed a man who was soaked in them. It has got one now (1967).'
4. O.J. Eggen, interview, April 1988. Pawsey, in turn, was acting on a suggestion made to him by Fr Daniel O'Connell at the 1955 Dublin General Assembly of the IAU (Ben Gascoigne, private communication).
5. David H. Levy, *The Man Who Sold the Milky Way: A Biography of Bart Bok*, University of Arizona Press, Tuscon & London, 1993. Bok granted Levy over 50 interviews during the last year of his life, and access to his correspondence files.
6. ibid., Chs 2–3.
7. Harlow Shapley, 'Studies based on the colors and magnitudes in stellar clusters'. The three papers of Shapley's series most relevant to this discussion are: 'VI. On the determination of the distances of globular clusters', *ApJ*, vol. 48, 1918, pp. 89–124; 'VII. The distances, distribution in space, and dimensions of 69 globular clusters', *ApJ*, vol. 48, 1918, pp. 154–81; 'XII. Remarks on the arrangement of the siderial universe', *ApJ*, vol. 49, 1919, pp. 311–36.
8. Henrietta S. Leavitt, '1777 variables in the Magellanic Clouds', *Harvard Ann*, vol. 60, 1908, pp. 87–108; 'Periods of 25 variable stars in the Small Magellanic Cloud', *Harvard Circ*, no. 173, 1912, pp. 1–3.
9. Bart J. Bok and Priscilla F. Bok, *The Milky Way*, 5th edn, Harvard University Press, Cambridge, Massachusetts, 1981.
10. Bart J. Bok, *A Study of the η Carinae Region*, Hoitsema Brothers, Groningen, 1932. Bok's thesis also appeared as a *Harvard College Observatory Reprint*, no. 77, 1932. Levy, *The Man Who Sold the Milky Way*, Ch. 4.
11. ibid., Ch. 9.
12. Harvard College Observatory, *Annual Report*, 1953, p. 1.
13. Harvard College Observatory, *Annual Report*, 1956, p. 1.

14. Levy, *The Man Who Sold the Milky Way*, pp. 84–9.
15. H.C. van de Hulst, *Ned. Tijdschr. Natuurkunde*, vol. 11, 1945, p. 201; H.I. Ewen and E.M. Purcell, 'Radiation from Galactic hydrogen at 1420 Mc/sec', *Nature*, vol. 168, 1951, p. 356.
16. J.H. Oort, F.J. Kerr and G. Westerhaut, 'The Galactic system as a spiral nebula', *MN*, vol. 118, 1958, pp. 379–89.
17. Bart J. Bok, 'Unlocking the secrets of the stars—an astronomer writes about Mount Stromlo Observatory', *Hemisphere*, vol. 2, no. 8, August 1958, pp. 2–7.
18. F.J. Kerr and A.W. Rodgers (eds), *The Galaxy and the Magellanic Clouds*, IAU–URSI Symposium No. 20, Australian Academy of Science, 1964.
19. *MSOAR*, 1957, p. 19.
20. ibid.
21. *MSOAR*, 1964, p. 16.
22. The last of these regular Visitors' Nights was held on 30 March 1966, eleven days after Bok left. It was conducted by Arthur Hogg as Acting Director; he died later that same night. Hogg became Acting Director following the departure of both Woolley and Bok; and Ben Gascoigne assumed the role following Hogg's death.
23. Z. Kopal, 'Dr Harlow Shapley' [obituary], *Nature*, vol. 240, 1972, pp. 429–30.
24. *MSOAR*, 1957, p. 18.
25. *MSOAR*, 1958, p. 7.
26. *MSOAR*, 1957, pp. 15–16.
27. S.C.B. Gascoigne, K.M. Proust and M.O. Robins, *The Creation of the Anglo-Australian Observatory*, Cambridge University Press, Cambridge, 1990, p. 51.
28. *MSOAR*, 1957, p. 16.
29. The Yale–Columbia report became available in 1959, see *MSOAR*, 1959, p. 8.
30. *MSOAR*, 1958, pp. 8–11.
31. ibid.
32. *MSOAR*, 1959, pp. 6–8.
33. *MSOAR*, 1960, pp. 5–6; 1961, pp. 4–6.
34. *MSOAR*, 1962, pp. 1, 3–4; Ben Gascoigne, Alex Rodgers, private communications.
35. Gascoigne, 'Bok, Woolley and Australian astronomy', see p. 123.
36. *MSOAR*, 1957, p. 4.
37. *MSOAR*, 1961, p. 3.

38. S.C.B. Gascoigne, 'Astrophysics at Mount Stromlo—the Woolley era', *ProcASA*, vol. 5, 1984, pp. 597–605, p. 604 quoted.
39. *MSOAR*, 1957, p. 4.
40. *MSOAR*, 1961, p. 3.
41. Grubb Parsons alone made half a dozen 74-inch-class telescopes in the mid-twentieth century, for the David Dunlap Observatory at Toronto (1935), the Radcliffe Observatory at Pretoria (1948), Mount Stromlo Observatory (1955), Haute Provence in France (1958; a 76-inch), the Okayama Astrophysical Observatory in Japan (1960) and the Helwan (Kottamia) Observatory in Egypt (1963).
42. Very high-dispersion spectrographs are so large that they need to be positioned away from the telescope itself, the light beam reaching them via a series of reflections. Because the beam is bent through so many angles, this is called a '*coudé*' (meaning 'elbow') arrangement.
43. A.R. Hyland and D.J. Faulkner, 'From the Sun to the Universe—the Woolley and Bok directorships at Mount Stromlo', *ProcASA*, vol. 8, 1989, pp. 216–28. Material from this paper has been quoted, sometimes without further specific reference, by kind permission of the authors.
44. *MSOAR*, 1960, p. 3; 1965, p. 11.
45. *MSOAR*, 1962, p. 4.
46. *MSOAR*, 1960, p. 4.
47. *MSOAR*, 1964, p. 8; 1972, p. 5.
48. *MSOAR*, 1962, p. 6. N. Visvanathan, 'Polarisation in the Galaxy and the Large Magellanic Cloud', ANU PhD thesis, 1965.
49. *MSOAR*, 1963, p. 7.
50. *MSOAR*, 1966, p. 1.
51. *MSOAR*, 1960, p. 11.
52. *MSOAR*, 1964, pp. 1, 8–9.
53. Gascoigne and Kron, 'Photoelectric observations', pp. 333–60; Gascoigne, 'Further observations', pp. 1–36.
54. S.C.B. Gascoigne, 'Intermediate-age clusters in the Magellanic Clouds', *Obs*, vol. 83, 1963, pp. 71–4; S.C.B. Gascoigne, 'Colour-magnitude diagrams for nine globular-like clusters in the Magellanic Clouds', *MN*, vol. 134, 1966, pp. 59–82.
55. J.W. Menzies, 'Three-colour photometry of southern globular clusters', ANU PhD thesis, 1967.
56. Hugh M. Johnson, 'The spectrum and mass of the 30 Doradus Nebula', *PASP*, vol. 71, 1959, pp. 425–40.

57. L.H. Aller and D.J. Faulkner, 'The helium to hydrogen ratio in the Small Magellanic Cloud', *PASP*, vol. 74, 1962, pp. 219–22; H.R. Dickel, L.H. Aller and D.J. Faulkner, 'Photoelectric spectrophotometry of emission nebulosities in the Magellanic Clouds', in F.J. Kerr and A.W. Rodgers (eds), *The Galaxy and the Magellanic Clouds*, IAU Symposium no. 20, 1964, pp. 294–310; D.J. Faulkner and L.H. Aller, 'Photoelectric spectrophotometry of the η Carinae and 30 Doradus Nebulae', *MN*, vol. 130, 1965, pp. 393–409; D.J. Faulkner, 'Isophotometry of the 30 Doradus Nebula', *MN*, vol. 135, 1967, pp. 401–12.
58. A. Przybylski, 'Abundance analysis of a supergiant in the Large Magellanic Cloud', *Nature*, vol. 205, 1965, p. 163.
59. Bengt E. Westerlund and Lindsey F. Smith, 'Wolf-Rayet stars in the Large Magellanic Cloud', *MN*, vol. 128, 1964, pp. 311–25; B.L. Webster, 'Southern planetary nebulae', ANU PhD thesis, 1966.
60. J.B. Whiteoak, 'The structure of the southern Milky Way', ANU PhD thesis, 1961; D. Sher, 'Distances of some open clusters near η Carinae', ANU PhD thesis, 1963; J.A. Graham, 'The application of hydrogen line photometry to Milky Way research', ANU PhD thesis, 1964.
61. A.W. Rodgers, C.T. Campbell and J.B. Whiteoak, 'A catalogue of H-α-emission regions in the southern Milky Way', *MN*, vol. 121, 1960, pp. 103–10.
62. B.J. Bok, 'The arms of the Galaxy', *Scientific American*, vol. 201, no. 6, December 1959, pp. 92–104; Bart J. Bok, 'The spiral structure of our Galaxy', *Obs*, vol. 79, 1959, pp. 58–62.
63. J.A. Koehler, 'Intergalactic atomic neutral hydrogen', ANU PhD thesis, 1965; R.M. Price, 'Cosmic background radiation at 408 Mc/s', ANU PhD thesis, 1965; R.D. Ekers, 'The structure of southern radio sources', ANU PhD thesis, 1967.
64. R.R. Shobbrook, 'Photographic, spectrographic and photometric studies of southern galaxies', ANU PhD thesis, 1963; 'Southern groups and clusters of galaxies. I. spectra and radial velocities of nineteen southern galaxies', *MN*, vol. 131, 1966, pp. 293–7; 'II. UBV photometry of fifty-three southern galaxies', *MN*, vol. 131, 1966, pp. 351–63; 'III. distances of some southern galaxies', *MN*, vol. 131, 1966, pp. 365–9.
65. A.W. Rodgers and R.A. Bell, 'The atmosphere of β Doradus. I. differential curves of growth', *MN*, vol. 127, 1964, pp. 471–91; R.A. Bell and A.W. Rodgers, 'The atmosphere of β Doradus. II. line profiles, half-widths and central depths', *MN*, vol. 128, 1964, pp. 365–87; 'III. hydrogen line

profiles', *MN*, vol. 135, 1967, pp. 121–9; R.A. Bell, 'White dwarf stars', ANU PhD thesis, 1961; I.J. Danziger, 'A high dispersion spectral study of one carbon and two barium stars', ANU PhD thesis, 1963; A.R. Hyland, 'A study of southern galactic clusters and the peculiar stars associated with them', ANU PhD thesis, 1967; M.S. Bessell, 'A study of some southern short period variables', ANU PhD thesis, 1967.

66. A. Przybylski, 'HD 101065—a G0 star with high metal content', *Nature*, vol. 189, 1961, p. 739; A. Przybylski, 'Abundance analysis of the peculiar star HD 101065', *Nature*, vol. 210, 1966, pp. 20–2.
67. D.J. Bord, '*Ab initio* calculations of oscillator strengths and Landé factors for Nd III', *A&AS*, vol. 144, 2000, pp. 517–22.
68. A.W. Rodgers and Leonard Searle, 'Spectrophotometry of the object η Carinae', *MN*, vol. 135, 1967, pp. 99–119. See *S&T*, vol. 92, no. 4, October 1996, p. 13, for a reproduction of the HST image of η Carinae.
69. N. Visvanathan, 'Polarization in η Carinae', *MN*, vol. 135, 1967, pp. 275–86.
70. Ben Gascoigne, 'Mount Stromlo Observatory—the early years', *Sky & Space*, June 1994, pp. 18–21.
71. S. Cockburn and D. Ellyard, *Oliphant: The Life and Times of Sir Mark Oliphant*, Axiom Books, Adelaide, 1981, p. 178.
72. Frederick White, 'Robert Gordon Menzies (1894–1978)', *Historical Records of Australian Science*, vol. 5, no. 1, 1980, pp. 68–102.
73. Gascoigne, Proust and Robins, *The Creation of the Anglo-Australian Observatory*, p. 30.
74. ibid., pp. 30–1.
75. B.J. Bok, 'The future of astronomy in Australia' (Presidential Address, ANZAAS Section A), *AJSci*, vol. 25, 1963, pp. 281–8.
76. Bok's own account of the incident, as quoted in Gascoigne, Proust and Robins, *The Creation of the Anglo-Australian Observatory*, p. 31.
77. Gascoigne, 'Bok, Woolley and Australian astronomy', see p. 119.
78. Gascoigne, Proust and Robins, *The Creation of the Anglo-Australian Observatory*, p. 33.
79. ibid., p. 35.
80. *MSOAR*, 1965 (original version), pp. 1–3.
81. 'Steward 90-inch telescope dedicated', *S&T*, vol. 38, 1969, pp. 164–5.
82. *Mercury*, vol. 13, no. 2, Astronomical Society of the Pacific, March–April 1984.

CHAPTER 8

1. S.C.B. Gascoigne, 'Astrophysics at Mount Stromlo: The Woolley Era', *ProcASA*, vol. 5, no. 4, 1984, pp. 597–605.
2. Olin J. Eggen, 'Notes from a life in the dark', *ARAA*, vol. 31, 1993, pp. 1–11, p. 1 quoted.
3. Historical note by M.G. Bouquet, ANU file: 4469 8.2.5.34 12.1.7.1, folio 269.
4. O.J. Eggen, D. Lynden-Bell and A.R. Sandage, 'Evidence from the motions of old stars that the Galaxy collapsed', *ApJ*, vol. 136, 1962, pp. 748–66; *American Astronomical Society Centennial Issue*, ed. Helmut A. Abt, *ApJ*, vol. 525, no. 1C, Pt 3, 1999. This celebratory issue reprints and discusses 53 key twentieth century papers.
5. This title first appeared in Eggen's first annual report. See *QJRAS*, vol. 8, 1967, pp. 373–81.
6. Ken Freeman, 'Olin J. Eggen 1919–1998 (obituary)', *Astron & Geophys*, vol. 41, no. 1, 2000, p. 36.
7. It is uncertain whether Eggen's review was ever actually published. A search in several of the obvious places has proved fruitless. The text here is 'quoted' from a recollection.
8. 'Society business—the Inaugural Meeting of the Astronomical Society of Australia', *ProcASA*, vol. 1, no. 1, 1967, p. 36.
9. *Trans IAU*, vols XV A and XV B, 1974.
10. Although the IAU's *Information Bulletin* regularly reported a list of *deceased* members, there seems to have been no corresponding mechanism, at the time, for recording *resignations*. Eggen's departure is indicated by the disappearance of his name from the 1973 triennial membership list: see *Trans IAU*, XV B, 1974, p. 261.
11. Astronomers and biologists attach different meanings to the word 'evolve'. Biologists use it to describe the changes that occur in the properties of *a whole class* of entities over time—*species* are said to evolve. Astronomers use it to describe the changes that occur in the properties of an *individual entity* itself—a *star* is said to evolve.
12. The 'Doppler effect' is named for the Austrian physicist Christian Doppler (1803–53), who first enunciated the principle that the observed wavelength of light (or sound) is affected by the motion of its source towards, or away from, the observer. A shift to shorter wavelengths indicates approach, while a longward shift indicates recession. The size of the shift is proportional to the velocity of the source. Thus the 'radial velocity' of an astronomical object may be obtained from its spectrum.

13. The 'heavy elements' are also called 'metals' by some writers.
14. O.J. Eggen, K.C. Freeman and A.W. Rodgers, 'Structure of the Galaxy,' *Rep. Prog. Physics*, vol. 36, 1973, pp. 625–94.
15. A.W. Rodgers, 'Solar composition A-stars in the Galactic Halo', *ApJ*, vol. 165, 1971, pp. 581–97.
16. A.W. Rodgers, P. Harding and Elaine Sadler, 'The nature of the metal-rich stellar population in the Galactic Halo', *ApJ*, vol. 244, 1981, pp. 912–18.
17. Ben Gascoigne, 'Mount Stromlo Observatory: the modern years', *Sky & Space*, vol. 8, no. 1, February 1985, pp. 24–8.
18. D.F. Martyn was elected an FRS in 1950, for the work on atmospheric tides he carried out while on secondment at Mount Stromlo during 1944–56. R.v.d.R. Woolley also became (1953) an FRS while he was Director at Mount Stromlo.
19. K.C. Freeman, 'On the disks of spiral and S0 galaxies', *ApJ*, vol. 160, 1970, pp. 811–30; G. de Vaucouleurs, 'General physical properties of external galaxies', *Handbuch der Physik*, vol. 53, 1959, pp. 311–72.
20. K.C. Freeman, 'Stellar dynamics and the structure of galaxies', in Allan Sandage, Mary Sandage and Jerome Kristian (eds), *Galaxies and the Universe*, University of Chicago Press, Chicago, 1975, Ch. 11, pp. 409–507.
21. Garth D. Illingworth, 'Masses of globular clusters', ANU PhD thesis, 1973.
22. Susan M. Simkin, 'Measurement of velocity dispersions and Doppler shifts from digitized optical spectra', *A&A*, vol. 31, 1974, pp. 129–36.
23. Harvey R. Butcher, 'Observational aspects of nucleosynthesis', ANU PhD thesis, 1974; Jeremy Mould, 'The lower main sequence and the atmospheres of M dwarfs', ANU PhD thesis, 1976.
24. K.C. Freeman and A.W. Rodgers, 'The chemical inhomogeneity of ω Centauri', *ApJ*, vol. 201, 1975, L71–4; John Norris and M. Bessell, 'Abundance variations on the lower giant branch of ω Centauri', *ApJ*, vol. 201, 1975, L75–9; M.S. Bessell and John Norris, 'The giant branch of ω Centauri. I. Abundance variations due to mixing', *ApJ*, vol. 208, 1976, pp. 369–81; John Norris and M.S. Bessell, 'The giant branch of ω Centauri. II. Mixing versus primordial abundance variations', *ApJ*, vol. 211, 1977, L91–5.
25. Peter Robert Wood, 'Stellar evolution and planetary nebula ejection', ANU PhD thesis, 1973; Robert A. Gingold, 'The helium burning evolution of low mass stars', ANU PhD thesis, 1975.

26. John William Robertson, 'The evolution of massive stars', ANU PhD thesis, 1972; J.W. Robertson, 'Color-magnitude diagrams for eleven young clusters in the Magellanic Clouds', *A&AS*, vol. 15, 1974, pp. 261–309; J.W. Robertson, 'Core-helium-burning stars in young clusters in the Large Magellanic Cloud', *ApJ*, vol. 191, 1974, pp. 67–78.
27. S.C.B. Gascoigne, 'Further observations of Magellanic Cloud Cepheids', *MN*, vol. 146, 1969, pp. 1–36; 'The Magellanic Clouds and the distance scale', *QJ*, vol. 13, 1972, pp. 274–81; 'Metal abundance and the luminosities of Cepheids', *MN*, vol. 166, 1974, 25P–7P.
28. R.S. Stobie, 'Cepheid pulsation. I. Numerical techniques and test calculations', *MN*, vol. 144, 1969, pp. 461–84; 'II. Models fitted to evolutionary tracks', *MN*, vol. 144, 1969, pp. 485–510; 'III. Models fitted to a new mass-luminosity relation', *MN*, vol. 144, 1969, pp. 511–35; 'On the difference between the Oosterhoff Types I and II globular clusters', *ApJ*, vol. 168, 1971, pp. 381–91; A.W. Rodgers, 'The masses of pulsating stars', *MN*, vol. 151, 1970, pp. 133–40.
29. *MSOAR*, 1977, pp. 25–8.
30. C. Hazard, M.B. Mackey and A.J. Shimmins, 'Investigation of the radio source 3C 273 by the method of lunar occultations', *Nature*, vol. 197, 1963, pp. 1037–9; M. Schmidt, '3C 273: A star-like object with large red-shift', *Nature*, vol. 197, 1963, p. 1040; J.B. Oke, 'Absolute energy distribution in the optical spectrum of 3C 273', *Nature*, vol. 197, 1963, pp. 1040–1; Jesse L. Greenstein and Thomas A. Matthews, 'Red-shift of the unusual radio source: 3C 48', *Nature*, vol. 197, 1963, pp. 1041–2. The term 'redshift' derives from the fact that the spectrum of an object moving away from an observer has all of its lines shifted to longer wavelengths. (See note 12 of this chapter for a description of the 'Doppler effect'.) Each line's shift is proportional to the 'rest' wavelength of the line in question, and to the recession velocity of the source. Thus a single quantity—the redshift, usually denoted z—can be used to describe the whole effect. In the Hubble expansion of the Universe, the further away an object is, the greater is its apparent velocity of recession, and the greater its redshift.
31. Edwin Hubble, 'A relation between distance and radial velocity among extra-galactic nebulae', *Proc. Nat. Acad. Sci.*, vol. 15, 1929, pp. 168–73; = *Mount Wilson Communication* No. 105; Edwin Hubble and Milton L. Humason, 'The velocity–distance relation among extra-galactic nebulae', *ApJ*, vol. 74, 1931, pp. 43–80. Hubble's discovery of the expansion of the

Universe is one of the most fundamental ever made, and his 1931 follow-up paper with Humason was also judged to be one of the top 50 astronomical contributions of the twentieth century; see note 4 of this chapter.

32. Harlan J. Smith and Dorrit Hoffleit, 'Light variations in the superluminous radio galaxy 3C 273', *Nature*, vol. 198, 1963, pp. 650–1.
33. A good description of the discovery events has been given by Peter Robertson, *Beyond Southern Skies: Radio Astronomy and the Parkes Telescope*, Cambridge University Press, Cambridge, 1992, pp. 227–41.
34. *MSOAR*, 1970, p. 3.
35. A.J. Shimmins, J.G. Bolton, B.A. Peterson and J.V. Wall, 'Identification of southern quasi-stellar objects', *Astrophys. Lett.*, vol. 8, 1971, pp. 139–43; B.A. Peterson and J.G. Bolton, 'Identification of southern quasi-stellar objects. II', *Astrophys. Lett.*, vol. 10, 1972, pp. 105–8; 'Identification of southern quasi-stellar objects. III', *Astrophys. Lett.*, vol. 13, 1973, pp. 187–92; B.A. Peterson, J.G. Bolton and A.J. Shimmins, 'Identification of southern quasi-stellar objects. IV', *Astrophys. Lett.*, vol. 15, 1973, pp. 109–13.
36. Leonard Searle and J.G. Bolton, 'Redshifts of fifteen radio sources', *ApJ*, vol. 154, 1968, pp. L101–4; Bruce A. Peterson and J.G. Bolton, 'Redshifts of southern radio sources', *ApJ*, vol. 173, 1972, pp. L19–L21.
37. See *S&T*, vol. 41, January 1971, pp. 14–17; June 1971, p. 341.
38. O.J. Eggen, Kenneth C. Freeman and Allan Sandage, 'On the optical identification of the X-ray source Cen XR-2 as WX Cen', *ApJ*, vol. 154, 1968, L27–31; N.V. Vidal, D.T. Wickramasinghe, B.A. Peterson and M.S. Bessell, 'Photoelectric and spectroscopic observations of WRA 795', *ApJ*, vol. 188, 1974, pp. 163–5; 'Spectroscopic studies of a suggested candidate for Cen X-3', *ApJ*, vol. 191, 1974, pp. L23–L24.
39. D.T. Wickramasinghe and M.S. Bessell, 'Gas streaming in 2U 0900–40 and Cyg X-1', *Nature*, vol. 251, 1974 pp. 25–8; D.T. Wickramasinghe, 'Radiation pressure driven mass loss from accretion discs in X-ray binaries', *MN*, vol. 168, 1974, pp. 297–305.
40. D.T. Wickramasinghe, N.V. Vidal, M.S. Bessell, B.A. Peterson and M.E. Perry, '2U 0900–40: A black hole?', *ApJ*, vol. 188, 1974, pp. 167–72.
41. *MSOAR*, 1977, pp. 30–1.
42. Robert A. Cameron, 'NASA's 91-centimetre airborne telescope', *S&T*, vol. 52, 1976, pp. 327–31.
43. A.R. Hyland, R.A. Hirst, G. Robinson and J.A. Thomas, 'Infrared observations of some IR-OH sources', *Astrophys. Lett.*, vol. 11, 1972, pp. 7–11;

A.R. Hyland, E.E. Becklin, J.A. Frogel and G. Neugebauer, 'Infrared observations of 1612 MHz IR/OH sources', *A&A*, vol. 16, 1972, pp. 204–19.

44. D.S. Mathewson and V.L. Ford, 'Polarization observations of 1800 stars', *Memoirs R. Astron. Soc*, vol. 74, 1970, pp. 139–82. Vince Ford has been a Research Officer at Mount Stromlo for over 30 years.
45. D.S. Mathewson, 'The local Galactic magnetic field and the nature of the radio spurs', *ApJ*, vol. 153, 1968, pp. L47–53.
46. D.S. Mathewson and V.L. Ford, 'Polarization measurements of stars in the Magellanic Clouds', *AJ*, vol. 75, 1970, pp. 778–84; 'The magnetic-field structure of the Magellanic Clouds', *ApJ*, vol. 160, 1970, pp. L43–6.
47. D.S. Mathewson and J.N. Clarke, 'A supernova remnant in the Small Magellanic Cloud', *ApJ*, vol. 178, 1972, pp. L105–7; 'Supernova remnants in the Large Magellanic Cloud', *ApJ*, vol. 180, 1973, pp. 725–38; 'Supernova remnants in the Magellanic Clouds', *ApJ*, vol. 182, 1973, pp. 697–8.
48. R. Giacconi et al., 'The *Einstein* (HEAO2) X-ray Observatory', *ApJ*, vol. 230, 1979, pp. 540–50; Dennis Overbye, 'The X-ray eyes of *Einstein*', *S&T*, vol. 57, 1979, pp. 527–34.
49. D.S. Mathewson et al., 'A new oxygen-rich supernova remnant in the Large Magellanic Cloud', *ApJ*, vol. 242, 1980, pp. L73–6; M.A. Dopita, I.R. Tuohy and D.S. Mathewson, 'An oxygen-rich supernova remnant in the Small Magellanic Cloud', *ApJ*, vol. 248, 1981, pp. L105–8; I.R. Tuohy et al., 'Optical identification of Balmer-dominated supernova remnants in the Large Magellanic Cloud', *ApJ*, vol. 261, 1982, pp. 473–84. Ian Tuohy was a Senior Research Fellow at Mount Stromlo during most of the 1980s.
50. D.S. Mathewson et al., 'Supernova remnants in the Magellanic Clouds', *ApJS*, vol. 51, 1983, pp. 345–55; vol. 55, 1984, pp. 189–210.
51. M.A. Dopita, D.S. Mathewson and V.L. Ford, 'Optical emission from shock waves. III. Abundances in supernova remnants', *ApJ*, vol. 214, 1977, pp. 179–88, Plate 4.
52. *Canberra Times*, 12 April 1978.
53. S.C.B. Gascoigne, K.M. Proust and M.O. Robins, *The Creation of the Anglo-Australian Observatory*, Cambridge University Press, Cambridge, 1990.
54. ibid., p. 39. The final capital cost was A$15.9 million; see *Anglo-Australian Telescope 1974/75*, p. 22. Allowing for inflation, this was within the $11 million budget estimate made in 1967.
55. ibid., p. 45.
56. *Anglo-Australian Telescope Agreement Act* 1970 (No. 57); date of assent: 1 September 1970; date of commencement: 22 February 1971 (see *Gazette* 1971, p. 1358).

57. Gascoigne, Proust & Robins, *The Creation of the Anglo-Australian Observatory*, p. 41.
58. ibid., pp. 21–3.
59. ibid., p. 24.
60. *MSOAR*, 1970, p. 3; 1972, p. 6.
61. Gascoigne, Proust and Robins, *The Creation of the Anglo-Australian Observatory*, pp. 118–22. Not all of Mount Stromlo's technical contributions to the AAT had happy outcomes. The Observatory was awarded contracts to build a number of instruments in the early 1970s, which were notably unsuccessful; ibid., pp. 183–4; Raymond Haynes, Roslynn Haynes, David Malin and Richard McGee, *Explorers of the Southern Sky: A History of Australian Astronomy*, Cambridge University Press, Cambridge, 1996, p. 180.
62. Roger W. Sinnott and Kari Nyren, 'The world's largest telescopes', *S&T*, vol. 86, no. 1, July 1983, pp. 27–32.
63. 'Proposed administrative arrangements for the AAT submitted to the two Governments by the Royal Society of London and the Australian Academy of Science in March 1965'. See Gascoigne, Proust and Robins, *The Creation of the Anglo-Australian Observatory*, Appendix 2.
64. *The Anglo-Australian Telescope Agreement Act* 1970, The Schedule, Article 4, para. (3).
65. Gascoigne, Proust and Robins, *The Creation of the Anglo-Australian Observatory*, Ch. 8.
66. ibid., p. 139.
67. The minutes of the 9 February 1973 meeting of the ANU Council record that it resolved, *inter alia*: '(a) to authorize the Vice-Chancellor to approve an establishment and budget for the Mount Stromlo–Siding Spring Observatory Services Unit as a separate unit in the University effective immediately; and (b) that the Vice-Chancellor be responsible for the conduct and management of the Observatory Services Unit with appropriate delegations to the Director of the Observatory.'
68. *Anglo-Australian Telescope Project, Report 1973–74*, pp. 1, 2.
69. Gascoigne, Proust and Robins, *The Creation of the Anglo-Australian Observatory*, p. 25.
70. W.N. Christiansen, G.R.A. Ellis, S.C.B. Gascoigne, R.G. Giovanelli, B.Y. Mills and J.H. Piddington.
71. J.G. Bolton, E.G. Bowen, R. Hanbury Brown, J.P. Wild and R.v.d.R. Woolley.

72. Gascoigne, Proust and Robins, *The Creation of the Anglo-Australian Observatory*, p. 132.
73. Chris Christiansen was appointed in 1960, Hanbury Brown in 1964 and Bernard Mills in 1965.
74. *Anglo-Australian Telescope Project, Report 1973–74*, pp. 3–4. *Anglo-Australian Telescope, 1974/75*, p. 12.
75. *Anglo-Australian Telescope, 1975/76*, p. 7. Sir Harrie Massey played a conciliatory role, as a Board member, in the discussions resulting in the final Marsfield decision, and the AAO building there is named for him.
76. See time allocation statistics published in the annual reports of the Anglo-Australian Telescope/Observatory for the years 1980–81 until 1990–91.
77. The AAT's annual report for 1984–85 lists 107 scientific papers based on observations made in the total time allocation (both British and Australian). The names of Mount Stromlo staff or students appear as authors on 30 of them.
78. *MSOAR*, 1973, p. 4
79. *MSOAR*, 1973, p. 4; 1976, p. 3; 1977, p. 6.
80. *MSOAR*, 1972, p. 5.
81. *MSOAR*, 1977, p. 10.
82. *MSOAR*, 1967, p. 4.
83. *MSOAR*, 1967, p. 3; 1970, p. 3; 1972, pp. 4, 5; 1974, p. 19; 1975, p. 4; 1976, p. 4. Peter Rudge was the Observatory's Chief Engineer from 1964 until his tragic death in 1971.
84. Letter from O.J. Eggen to the Vice-Chancellor, D.A. Low, on 10 July 1975, shortly after the latter's appointment. Eggen was outraged that the University had agreed to the establishment of a four-person AAO–ANU liaison committee without any consultation with him, even though he was to be one of the four. In fact, he had learned of this development in a casual conversation with the Director of the Research School of Physical Sciences, and AATB member, Robert Street.
85. Olin J. Eggen, 'Notes from a life in the dark', *ARAA*, vol. 31, 1993, pp. 1–11, p. 1 quoted.

CHAPTER 9

1. Robert Street (Director of RSPhysS at the time) consulted Gascoigne, Mathewson and Rodgers before making his recommendation on the acting headship. (Street to the Vice-Chancellor, 16 September 1977, ANU File 4469 8.2.5.34 12.1.7.1, part 2, folio 164.)

2. ANU File 8.2.5.34, part 5, folio 160A. The text of Woolley's essay is reproduced as an appendix in Gascoigne, 'Bok, Woolley and Australian astronomy', pp. 119–26.
3. 'England's great radio telescope', *S&T*, vol. 16, no. 11, 1957, pp. 516–18; 'Jodrell Bank Observatory', in Paul Murdin (ed.), *Encyclopedia of Astronomy & Astrophysics*, IOP Publishing Ltd & Nature Publishing Group, Bristol & London, 2001, vol. 2, p. 1328; M.I. Large, D.S. Mathewson and C.G.T Haslam, 'A high-resolution survey of the Andromeda Nebula at 408 Mc/s', *Nature*, vol. 183, 1959, pp. 1250–1; 'A high-resolution survey of the Coma Cluster of galaxies at 408 Mc/s', *Nature*, vol. 183, 1959, pp. 1663–4; D.S. Mathewson, M.I. Large and C.G.T. Haslam, 'A spectral analysis of the radio sources in Cygnus X at 1390 Mc/s and 408 Mc/s', *MN*, vol. 120, 1960, pp. 242–7; 'Radio emission from the Cygnus Loop', *MN*, vol. 121, 1960, pp. 543–50.
4. 'Westerbork Synthesis Radio Telescope', in Paul Murdin (ed.), *Encyclopaedia of Astronomy & Astrophysics*, IOP Publishing Ltd & Nature Publishing Group, Bristol & London, 2001, vol. 4, p. 3517.
5. W.N. Christiansen, 'A high-resolution aerial for radio astronomy', *Nature*, vol. 171, 1953, pp. 831–3; 'The first decade of solar radio astronomy in Australia', in Sullivan (ed.), *The Early Years*, pp. 117–22.
6. 'Radio "photography" of the sky', *S&T*, vol. 42, no. 4, October 1971, pp. 205–6.
7. D.S. Mathewson, P.C. van der Kruit and W.N. Brouw, 'A high-resolution radio continuum survey of M51 and NGC 5195 at 1415 MHz', *A&A*, vol. 17, 1972, pp. 468–86; P.C. van der Kruit, J.H. Oort and D.S. Mathewson, 'The radio emission of NGC 4258 and the possible origin of spiral structure', *A&A*, vol. 21, 1972, pp. 169–84.
8. Electromagnetic radiation emitted by electrons travelling at high speed in a magnetic field is called 'synchrotron' emission; it was first discovered in synchrotron particle accelerators in nuclear laboratories.
9. Gerard de Vaucouleurs, 'The Magellanic Clouds and the Galaxy', *Obs*, vol. 74, 1954, pp. 23–31, 158–64.
10. J.V. Hindman, F.J. Kerr and R.X. McGee, 'A low-resolution hydrogen-line survey of the Magellanic System II. Interpretation of results', *AJPhys*, vol. 16, 1963, pp. 570–83; R.X. McGee and J.A. Milton, '21 centimetre hydrogen-line survey of the Large Magellanic Cloud II. Distribution and motions of neutral hydrogen', *AJPhys*, vol. 19, 1966, pp. 343–74; J.V. Hindman, 'A high-resolution study of the distribution and motions of neutral hydrogen in the Small Cloud of Magellan', *AJPhys*, vol. 20, 1967, pp. 147–71.

11. J. v. Kuilenburg, 'A systematic search for high-velocity hydrogen outside the Galactic plane II', *A&A*, vol. 16, 1972, pp. 276–81; P. Wannier and G.T. Wrixon, 'An unusual high-velocity hydrogen feature', *ApJ*, vol. 173, 1972, pp. L119–23.
12. D.S. Mathewson, M.N. Cleary and J.D. Murray, 'The Magellanic Stream', *ApJ*, vol. 190, 1974, pp. 291–6.
13. D.S. Mathewson, M.P. Schwarz and J.D. Murray, 'The Magellanic Stream: the turbulent wake of the Magellanic Clouds in the halo of the Galaxy', *ApJ*, vol. 217, 1977, pp. L5–8.
14. M.T. Brück and M.R.S. Hawkins, 'An investigation of faint stars in a region of the Magellanic Stream', *A&A*, vol. 124, 1983, pp. 216–22.
15. Gerhardt R. Meurer, G.V. Bicknell and R.A. Gingold, 'A drag dominated model of the Magellanic Stream', *ProcASA*, vol. 6, 1985, pp. 195–8.
16. Ben Moore and Marc Davis, 'The origin of the Magellanic Stream', *MN*, vol. 270, 1994, pp. 209–21; Lance T. Gardiner and Masafumi Noguchi, 'N-body simulations of the Small Magellanic Cloud and the Magellanic Stream', *MN*, vol. 278, 1996, pp. 191–208.
17. M.J. Irwin, W.E. Kunkel and S. Demers, 'A blue stellar population in the HI bridge between the two Magellanic Clouds', *Nature*, vol. 318, 1985, pp. 160–1.
18. This 'Doppler effect' is described in Chapter 8, note 12.
19. Hindman, 'A high-resolution study . . . in the Small Cloud of Magellan'.
20. D.S. Mathewson, V.L. Ford and N. Visvanathan, 'The structure of the Small Magellanic Cloud', *ApJ*, vol. 301, 1986, pp. 664–74; 'The structure of the Small Magellanic Cloud. II.', *ApJ*, vol. 333, 1988, pp. 617–45.
21. Don Mathewson, 'The Mini-Magellanic Cloud', *Mercury*, vol. 13, no. 2, 1984, pp. 57–9.
22. Tadayuki Murai and Mitsuaki Fujimoto, 'The Magellanic Stream and the Galaxy with a massive halo', *Publ. Astr. Soc. Japan*, vol. 32, 1980, pp. 581–603.
23. D.S. Mathewson, S.R. Wayte, V.L. Ford and K. Ruan, 'The "high velocity cloud" origin of the Magellanic System', *ProcASA*, vol. 7, 1987, pp. 19–25.
24. D.G. Barnes et al., 'The HI Parkes All Sky Survey: southern observations, calibration and robust imaging', *MN*, vol. 322, 2001, pp. 486–98.
25. M.E. Putman et al., 'Tidal disruption of the Magellanic Clouds by the Milky Way', *Nature*, vol. 394, 1998, pp. 752–4; M.E. Putman, 'The Magellanic System's interactive formations', *Publ. Astron. Soc. Aust.*, vol. 17, 2000, pp. 1–5.

26. Typescript agenda for a seminar in the Duffield Lecture Theatre on the afternoon of 17 November 1977, to discuss new developments at the Observatory.
27. *MSOAR*, 1977, pp. 3–4.
28. W.L. Peters, 'A barred spiral undergoing a collision: neutral hydrogen aperture synthesis observations of NGC 1313', *Bulletin of the American Astronomical Society*, vol. 17, 1985, p. 612; J.G. Ables et al., 'An HI study of the galaxy NGC 4945 with a two-element synthesis telescope', *MN*, vol. 226, 1987, pp. 157–71; J.R. Forster et al., 'Aperture synthesis observations of galactic atomic hydrogen in the direction l=333.5°, b=–0.3°', *MN*, vol. 226, 1987, pp. 173–85.
29. Haynes, Haynes, Malin and McGee, *Explorers of the Southern Sky*, p. 417.
30. Ray Norris, 'The Australia Telescope', *S&T*, vol. 76, no. 6, 1988, pp. 615–20.
31. *MSOAR*, 1978, pp. 8–15, 26–7.
32. The Standing Committee of the ANU Council approved the new arrangements in April 1978.
33. *MSOAR*, 1980, pp. 17–18.
34. Recollection of John Hart, mechanical engineer for the project.
35. *MSOAR*, 1978, pp. 7–8.
36. S.G. Foster and Margaret M. Varghese, *The Making of the Australian National University*, Allen & Unwin, Sydney, 1996, p. 286.
37. ibid., p. 291.
38. Another crucial meeting regarding the 2.3-metre Telescope took place on 18 March 1981, when the Vice-Chancellor, the Deputy Vice-Chancellor, the Bursar and the Director of the Research School of Physical Sciences received an update on the project. In the course of the afternoon, Tony Low indicated how appreciative the Chancelry was to have such progress reports, so as to avoid any nasty surprises as time went on.
39. Roger W. Sinnott and Kari Nyren, 'The world's largest telescopes', *S&T*, vol. 86, no. 1, 1993, pp. 27–32; 'Giant mirror blanks poured for Chile and Australia', *S&T*, vol. 28, no. 3, September 1969, pp. 140–3 and cover.
40. *MSOAR*, 1980, p. 39.
41. James Cornell, 'Six new eyes peer from Mount Hopkins', *S&T*, vol. 58, no. 1, July 1979, pp. 23–4.
42. All contributing companies are listed in *The 2.3m Telescope*, an eight-page colour brochure produced for the official opening by the Prime Minister on 16 May 1984.

43. Entry in the Mount Stromlo inventory dated 31 December 1985.
44. A.W. Rodgers, P. Conroy and G. Bloxham, 'A dual-beam Nasmyth spectrograph', *PASP*, vol. 100, 1988, pp. 627–34.
45. *MSOAR*, 1987, pp. 19–20.
46. *MSOAR*, 1991, p. 36; A.W. Rodgers, G. Bloxham and P. Conroy, 'A Nasmyth combined imager and low-resolution spectrograph', *PASP*, vol. 105, 1993, pp. 315–18.
47. Terry Jay Jones et al., 'A cooled grating spectrometer using cylindrical optics', *PASP*, vol. 94, 1982, pp. 207–12; *MSOAR*, 1986, p. 24. Terry Jones was a Research Fellow at Mount Stromlo during 1978–82.
48. Peter J. McGregor, *'CASPIR'*, *Southern Sky*, no. 7, Sept–Oct 1994, pp. 18–24.
49. *MSOAR*, 1994, p. 39.
50. J. Kelly Beatty and Stuart J. Goldman, 'The great crash of 1994: a first report', *S&T*, vol. 86, no. 4, October 1994, pp. 18–23.
51. The text of the Prime Minister's speech was reported by *The Coonabarabran Times*, 17 May 1984, pp. 1, 11.
52. Ellen Peterson, 'Hawke goes to see the stars but finds a problem back on Earth', *The Australian*, 17 May 1984, p. 2.
53. 'Power scheme, observatory win awards', *Engineers Australia*, 1 November 1985, pp. 26–7.
54. A. Boggess et al., 'The IUE spacecraft and instrumentation', *Nature*, vol. 275, 1978, pp. 372–7; Giacconi et al., 'The *Einstein* (HEA02) X-ray observatory'; Overbye, 'The X-ray eyes of *Einstein*'.
55. In the event, the launch of the HST was much delayed—especially following the 1986 *Challenger* tragedy—finally taking place on the Space Shuttle *Discovery* in 1990. It was also marred by the notorious optical design flaw which only then became apparent. In spite of this, the HST has proved an outstanding success, vastly increasing our understanding of the Universe at large.
56. Ian Tuohy, 'Space activities', *The Australian Encyclopaedia*, vol. 7, 6th edn, Australian Geographic, Sydney, 1996, pp. 2769–75.
57. Piero Benvenuti, Michael Dopita and Sandro D'Odorico, 'Far-ultraviolet spectrophotometry of supernova remnants: observations and astrophysical interpretation', *ApJ*, vol. 238, 1980, pp. 601–13; Ian Tuohy and Gordon Garmire, 'Discovery of a compact X-ray source at the center of the supernova remnant RCW 103', *ApJ*, vol. 239, 1980, L107–10.
58. Don Mathewson, 'Space astronomy in the 1980s: Australian role?', *ANU*

Reporter, vol. 11, no. 2, 21 March 1980. The quotation is from the opening paragraphs of an article introducing the STARLAB project.

59. *MSOAR*, 1980, pp. 12–17.
60. This initial costing proved far too conservative. By February 1984 the estimate for the instrument package alone had grown to $93 million, or $65 million for a cut-down version omitting the spectrograph. ANU file, 'STARLAB Project', 8.8.1.42, Part 3. Jane Ford, 'Funds shortage could eclipse STARLAB', *The Australian*, 14 February 1984, p. 3.
61. T.E. Stapinski, A.W. Rodgers and M.J. Ellis, 'A two-dimensional photon counting array', *PASP*, vol. 93, 1981, pp. 242–6; A.W. Rodgers, J. van Harmelen, D. King, P. Conroy and P. Harding, 'Large-format photon-counting arrays', *PASP*, vol. 100, 1988, pp. 841–52.
62. 'Astronomers are winners in Australian budget', *New Scientist*, 26 August 1982, p. 542. The grant was made via the Australian Industrial Research and Development Incentives Board, and the initial funding agreement with the ANU was signed on 15 October 1982. ANU file, 'STARLAB Project', 8.8.1.42, Part 1.
63. *MSOAR*, 1983, pp. 16–23.
64. Jane Ford, 'Sky's not the limit for Auspace Ltd', *The Australian*, 12 January 1983.
65. The name *Lyman* referred to the fact that the instrument's proposed wavelength range (0.0900–0.1200 microns) was chosen to include the highest energy feature of the Lyman series in the spectrum of neutral hydrogen.
66. *Hansard*, House of Representatives, 10 September 1981, pp. 1223–4.
67. *MSOAR*, 1984, pp. 21–5; Wayne Deeker, 'COSSA: vision and perseverance', *CSIRO Space Industry News*, issue 77, pp. 3–9, December 1997.
68. *A Space Policy for Australia*, Report of the Madigan Committee, Australian Academy of Technological Sciences, 1985.
69. Tuohy, 'Space activities'.
70. *MSOAR*, 1986, pp. 20–1. These transfers took place three months after the conclusion of Don Mathewson's term of office when Alex Rodgers had become Acting Director.
71. *MSOAR*, 1988, p. 19.
72. The FUSE project survived within NASA itself, although it underwent a good deal of cost-cutting redesign. It was finally launched into orbit on 24 June 1999. Jonathan McDowell, 'Far Ultraviolet Spectroscopic Explorer', *S&T*, vol. 88, no. 6, December 1994, p. 16; vol. 98, no. 3, September 1999, p. 30; 'The far ultraviolet unveiled', *S&T*, vol. 100, no. 6, June 2000, p. 26.

73. *MSOAR*, 1996, pp. 51–2.
74. The Australian Space Board was reconstituted as the Australian Space Council in 1993, but then abolished, together with the Australian Space Office, in the federal Budget cuts of August 1996. Some of its functions were absorbed by other government agencies. For instance, in January 1998 the Cooperative Research Centre program was used to fund a CRC for Satellite Systems (for seven years). For a press report of the 1996 Budget changes, see James Woodford and Leigh Dayton, 'Australia to develop and launch satellites', *SMH*, 21 August 1996, Budget '96, p. 14.
75. 'Satellite makes first contact', *The Canberra Times*, 16 December 2002, p. 3.
76. *MSOAR*, 1981, p. 11.
77. The *Annual Reports* record the publication of 43 papers during the year in which Mathewson became Acting Director (1977), and 92 during the last full year of his term of office (1985).
78. The star α Centauri is actually a *triple* system, but the third component, Proxima Centauri, is much smaller and fainter than the two principal companions, α Centauri A and B. Proxima Centauri, however, lies slightly on our side of the triple system, so it has the distinction of being 'the closest star to our Solar System'.
79. M.S. Bessell, 'α Centauri', *ProcASA*, vol. 4, 1981, pp. 212–14.
80. Bart J. Bok and Edith F. Riley, 'Small dark nebulae', *ApJ*, vol. 105, 1947, pp. 255–7; Bart J. Bok, 'Dark nebulae, globules and protostars', *PASP*, vol. 89, 1977, pp. 597–611: address on receiving the Bruce Medal of the Astronomical Society of the Pacific.
81. The Southern Coalsack is the name given to a prominent, irregular, dark cloud in the Milky Way, adjoining the Southern Cross and of a similar size. Globule 2 is a dense structure embedded in this much larger cloud.
82. Bart J. Bok, M. Elizabeth Sim and T.G. Hawarden, 'A conspicuous globule in the Southern Coalsack', *Nature*, vol. 266, 1977, pp. 145–7.
83. Terry J. Jones et al., 'Infrared observations of a Bok globule in the Southern Coalsack', *ApJ*, vol. 242, 1980, pp. 132–40; Terry J. Jones, A.R. Hyland and Jeremy Bailey, 'The inner core of a Bok globule', *ApJ*, vol. 282, 1984, pp. 675–82.
84. P.R. Wood, 'Pulsation and mass loss in Mira variables', *ApJ*, vol. 227, 1979, pp. 220–31; P.R. Wood and D.M. Zarro, 'Helium-shell flashing in low-mass stars and period changes in Mira variables', *ApJ*, vol. 247, 1981, pp. 247–56; P.R. Wood, 'Stellar evolution, mass loss, and nucleosynthesis on the asymptotic giant branch', *ProcASA*, vol. 4, 1981, pp. 145–8.

85. Michael A. Dopita, Holland C. Ford, Catherine J. Lawrence and B. Louise Webster, 'The kinematics and internal dynamics of planetary nebulae in the Small Magellanic Cloud', *ApJ*, vol. 296, 1985, pp. 390–8; Michael A. Dopita, Holland C. Ford and B. Louise Webster, 'Extremely energetic planetary nebulae in the Large Magellanic Cloud', *ApJ*, vol. 297, 1985, pp. 593–8.
86. M.S. Bessell and John Norris, 'The ultra-metal-deficient (Population III?) red giant CD –38°245', *ApJ*, vol. 285, 1984, pp. 622–36. See Ch. 8, note 13 for the significance of 'heavy elements' or 'metals'.
87. N. Christlieb et al., 'A stellar relic from the early Milky Way', *Nature*, vol. 419, 2002, pp. 904–6.
88. Bruce A. Peterson et al., 'PKS 2000–330: a quasi-stellar radio source with a redshift of 3.78', *ApJ*, vol. 260, 1982, pp. L27–L29.
89. See Ch. 8, note 30 for a description of the term 'redshift'. In the Hubble expansion of the Universe, the further away an object is, the greater is its apparent velocity of recession, and the greater its redshift. Thus finding an object with a redshift greater than any previously observed is equivalent to setting a new record for the most distant object yet discovered.
90. E.M. Hu et al., 'A redshift $\zeta = 6.56$ galaxy behind the cluster Abell 370', *ApJ*, vol. 568, 2002, pp. L75–9. Remoteness records are usually held by QSOs since they are more luminous than normal galaxies. In this case, however, the intrinsic brightness of the galaxy HCM 6A is enhanced because it is being 'lensed' by an intervening cluster of galaxies, Abell 370.
91. Peter J. Quinn, 'The dynamics of disk galaxy mergers', ANU PhD thesis, 1982.
92. David F. Malin and David Carter, 'Giant shells around normal elliptical galaxies', *Nature*, vol. 285, 1980, pp. 643–5.
93. See Ch. 8, notes 29–31.
94. G.V. Bicknell and R.A. Gingold, 'On tidal detonation of stars by massive black holes', *ApJ*, vol. 273, 1983, pp. 749–60.
95. See Ch. 8, note 31 and the more detailed explanation in the paragraphs following Ch. 11, note 4.
96. N. Visvanathan, 'A global value of the Hubble Constant', *ApJ*, vol. 275, 1983, pp. 430–4.
97. *MSOAR*, 1978, pp. 1–5; 1985, pp. 1–7.
98. *MSOAR*, 2001, pp. 5–9.
99. ANU 1871C/1985, 'Mount Stromlo and Siding Spring Observatories', a paper detailing proposed arrangements following the separation of MSSSO from RSPhysS, p. 2.

100. I.G. Ross, 'Speaking notes at meeting with MSSSO Staff, 14 November 1985', handwritten by the Deputy Vice-Chancellor at 12.30 a.m. on the day in question, and photocopied by his office for distribution.
101. Resolution of the ANU Council at its meeting of 11 October 1985.
102. Ross, 'Speaking notes'.
103. ibid.
104. *MSOAR*, 1986, pp. 21–2.
105. *MSOAR*, 1991, pp. 32–3; 1994, p. 37.

CHAPTER 10

1. See Ch. 9, note 102.
2. Report of the Review Committee on the Mount Stromlo and Siding Spring Observatories, 1986 (ANU paper 2220/1986).
3. Minutes of the meeting of the Committee of the Observatories, 6 August 1986, Item 5 (ANU paper 2610/1986); Report of the meeting of Advisers on Astronomy Policy, 9 September 1986 (ANU paper 2618/1986).
4. The Electoral Committee was established at the Council meeting of 14 November 1986. The Registrar's letter of invitation to senior staff was dated 12 December 1986.
5. K.C. Freeman to the Registrar, 22 December 1986. His interview took place on 12 May 1987.
6. King Baudouin of the Belgians and Queen Beatrix of the Netherlands both visited Mount Stromlo in the late 1980s.
7. *Mount Stromlo and Siding Spring Observatories Alternative Annual Report*, 1989–90, p. 4.
8. He also spent six months as a Fulbright scholar at the Royal Greenwich Observatory, Herstmonceux, where Woolley had by then taken up residence as Astronomer Royal.
9. *MSOAR*, 1982, pp. 17–18.
10. Commissioned in 1964; *MSOAR*, 1964, p. 8; 1965, p. 12.
11. Among them were: M.S. Bessell, 'A study of some southern short period variables', ANU PhD thesis, 1967; A.R. Hyland, 'A study of southern galactic clusters and the peculiar stars associated with them', ANU PhD thesis, 1967; A.L. Brooke, 'Infrared observations of globular cluster giants', ANU PhD thesis, 1969; E.G. Schmidt, 'An astrophysical investigation of four classical Cepheid variable stars', ANU PhD thesis, 1969; J.E. Norris, 'Neutral-helium line strengths in the early-type stars', ANU PhD thesis, 1970. R.D. Watson, 'The β Cephei stars', ANU PhD thesis, 1971.

12. *MSOAR*, 1969, p. 4.
13. T.E. Stapinski, A.W. Rodgers and M.J. Ellis, 'A two-dimensional photon counting array', *PASP*, vol. 93, 1981, pp. 242–6; A.W. Rodgers, J. van Harmelen, D. King, P. Conroy and P. Harding, 'Large-format photon counting arrays', *PASP*, vol. 100, 1988, pp. 841–52.
14. *MSOAR*, 1989, p. 25.
15. *MSOAR*, 1989, p. 3.
16. *MSOAR*, 1990, p. 22; 1991, pp. 3, 36; 1992, p. 42; 1993, pp. 7, 48; 1994, p. 40.
17. David O. Woodbury, *The Glass Giant of Palomar*, William Heinemann Ltd, London, 1940; Robert R. Coles, 'The 200-inch Telescope', *S&T*, vol. 7, 1948, pp. 267–9 and cover; Bagrat K. Ioannisiani, 'The Soviet 6-metre altazimuth reflector', *S&T*, vol. 54, 1977, pp. 356–62.
18. Roger W. Sinnott, 'The Keck Telescope's giant eye', *S&T*, vol. 80, 1990, pp. 15–22. The telescopes are named for their benefactor, the W.M. Keck Foundation, which contributed grants totalling more than US$140 million to the project.
19. Massimo Tarenghi, 'Eyewitness view: first sight for a glass giant', *S&T*, vol. 96, no. 5, 1998, pp. 46–51.
20. Gary Stix and Andy Ryan, 'A new glass eye opens on the cosmos', *Scientific American*, vol. 280, no. 4, 1999, pp. 104–11.
21. *The Future of Australian Astronomy*, A Report to the Prime Minister by the Australian Science and Technology Council, AGPS, Canberra, September 1989, p. 17, Recommendation 5.
22. ibid., p. 12, Recommendation 3.
23. *MSOAR*, 1991, front cover.
24. *MSOAR*, 1987, p. 20; 1988, p. 19; 1990, pp. 2–3; 1992, pp. 6–7. N.C. Coops, J.F. Le Marshall and A.W. Rodgers, 'A review of cloud cover—part of the site testing program for a large Australian optical/infrared telescope', unpublished Bureau of Meteorology–MSSSO report, 1991.
25. The altitude of Siding Spring Observatory is 1149 metres.
26. Peter Aldhous and David Lindley, 'Canada gets cold feet', *Nature*, vol. 351, 1991, p. 680; Jeffrey Mervis, 'Gemini telescope project shifts into high gear', *Nature*, vol. 357, 1992, p. 430.
27. David Dickson and Colin Macilwain, 'Gemini telescopes in suspended animation', *Nature*, vol. 364, 1993, p. 566. A second opportunity arose subsequently, however, and Australia became the seventh Gemini partner in 1998 (see next chapter).

28. *MSOAR*, 1993, pp. 6–7. Peter Wood, 'The quest for better seeing', *Sky & Space*, vol. 9, no. 3, 1996, pp. 16–18.
29. P.R. Wood, A.W. Rodgers and K.S. Russell, 'Seeing measurements at Freeling Heights and Siding Spring Observatory', *Publ. Astron. Soc. Aust.*, vol. 12, 1995, pp. 97–105.
30. Peter Quinn, 'Searching for the dark side: the story of the MACHO Project', *Southern Sky*, Sept–Oct 1993, pp. 16–19; Nov–Dec 1993, pp. 28–30.
31. The majority of astrophysicists accept this conclusion, although Moti Milgrom of the Weizmann Institute in Israel suggested in 1983 that the apparent mass discrepancy can be explained by a low-acceleration modification to Newtonian gravity or inertia. See M. Milgrom, 'A modification of the Newtonian dynamics as a possible alternative to the hidden mass hypothesis', *ApJ*, vol. 270, 1983, pp. 365–70. His 'Modified Newtonian Dynamics' (MOND) theory has proved surprisingly resilient. The whole dark matter puzzle may yet turn out to be due not to *an excess* of mass in the Universe, but to *a deficiency* in our understanding of its fundamental physics. Nevertheless, in this book we will adopt the conventional view of the dark matter conundrum. A recent account of MOND has been given by Robert H. Sanders and Stacy S. McGaugh, 'Modified Newtonian Dynamics as an alternative to dark matter', *ARAA*, vol. 40, 2002, pp. 263–317.
32. F. Zwicky, 'On the masses of nebulae and of clusters of nebulae', *ApJ*, vol. 86, 1937, pp. 217–46. This paper too was judged to be one for the top 50 astronomical contributions of the twentieth century: see Ch. 8, note 4.
33. See Ch. 8, note 19.
34. M.S. Roberts, 'The rotation curves of galaxies', IAU Symposium No. 69, in Avram Hayli (ed.), *Dynamics of Stellar Systems*, D. Reidel Publishing Company, Dordrecht, 1975, pp. 331–40.
35. IAU Symposium No. 100, *Internal Kinematics and Dynamics of Galaxies*, E. Athanassoula (ed.), D. Reidel Publishing Company, Dordrecht, 1983, pp. 87–8.
36. 'Missing *mass*' is an unfortunate nomenclature. It is not the *mass* that is missing, but any associated *light* (or radio emission, X-rays, and so on). That the *mass* is there is quite evident from its observed gravitational pull.
37. John E. Norris and M.R.S. Hawkins, 'Population studies. X. constraints on the mass and extent of the Galaxy's dark corona', *ApJ*, vol. 380, 1991, pp. 104–15.

38. See review by Michel Fich and Scott Tremaine, 'The mass of the Galaxy', *ARAA*, vol. 29, 1992, pp. 409–45, § 5.1.
39. See reviews by Herbert Gursky and Daniel A. Schwartz, 'Extragalactic X-ray sources', *ARAA*, vol. 15, 1977, pp. 541–68; and W. Forman and C. Jones, 'X-ray-imaging observations of clusters of galaxies', *ARAA*, vol. 20, 1982, pp. 547–85.
40. For example, D. Fabricant, M. Lecar and P. Gorenstein, 'X-ray measurements of the mass of M87', *ApJ*, vol. 241, 1980, pp. 552–60.
41. For example, A.C. Fabian, P.E.J. Nulsen and C.R. Canizares, 'Cooling flows in clusters of galaxies', *Nature*, vol. 310, 1984, pp. 733–40; P.E.J. Nulsen, 'Thermal instability in cooling flows', *MN*, vol. 221, 1986, pp. 377–92; P.E.J. Nulsen, R.M. Johnstone and A.C. Fabian, 'Star formation in cooling flows', *ProcASA*, vol. 7, 1987, pp. 132–5.
42. The concept of inflation was introduced into cosmology by Alan H. Guth, 'Inflationary Universe: a possible solution to the horizon and flatness problems', *Phys. Rev. D*, vol. 23, 1981, pp. 347–56. Two reviews, at different levels, are: Alan H. Guth and Paul J. Steinhardt, 'The inflationary Universe', *Scientific American*, vol. 250, no. 5, 1984, pp. 90–102; and J.V. Narlikar and T. Padmanabhan, 'Inflation for astronomers', *ARAA*, vol. 29, 1991, pp. 325–62.
43. IAU Symposium No. 117, *Dark Matter in the Universe*, J. Kormendy and G.R. Knapp (eds), D. Reidel Publishing Company, Dordrecht, 1987.
44. C. Carignan and K.C. Freeman, 'Basic parameters of dark halos in late-type spirals', *ApJ*, vol. 294, 1985, pp. 494–501.
45. Freeman, 'Parameters for dark halos', pp. 119–30.
46. Arno Penzias and Robert Wilson received the Nobel Prize for their serendipitous discovery, in 1965, of the CMB radiation: R.H. Dicke et al., 'Cosmic black-body radiation', *ApJ*, vol. 142, 1965, pp. 414–19; A.A. Penzias and R.W. Wilson, 'A measurement of excess antenna temperature at 4080 Mc/s', *ApJ*, vol. 142, 1965, pp. 419–21; P.J.E. Peebles and David T. Wilkinson, 'The primeval fireball', *Scientific American*, vol. 216, no. 6, 1967, pp. 28–37; R.B. Partridge, 'The primeval fireball today', *American Scientist*, vol. 57, 1969, pp. 37–74. Exquisite observations of the CMB's properties have been made by NASA's Cosmic Background Explorer (COBE) satellite, launched in November 1989: David H. Smith and Alan MacRobert, 'Looking into the Big Bang', *S&T*, vol. 77, 1989, p. 593; Joseph Silk, 'Probing the primeval fireball', *S&T*, vol. 79, 1990, pp. 600–3; Richard Tresch Fienberg, 'COBE confronts the Big Bang', *S&T*, vol. 84, no. 1, 1992, pp. 34–5.

47. The acronym WIMP was coined by Mike Turner, of the University of Chicago: Gary Steigman and Michael S. Turner, 'Cosmological constraints on the properties of weakly interacting massive particles', *Nuclear Physics*, vol. B 253, 1985, pp. 375–86.
48. The acronym MACHO was coined by a member of the team that conducted the Stromlo experiment: Kim Griest, 'Galactic microlensing as a method of detecting massive compact halo objects', *ApJ*, vol. 366, 1991, pp. 412–21. Its use quickly became widespread.
49. Bohdan Paczynski, 'Gravitational microlensing by the Galactic Halo', *ApJ*, vol. 304, 1986, pp. 1–5.
50. F.W. Dyson, A.S. Eddington and C. Davidson, 'A determination of the deflection of light by the Sun's gravitational field, from observations made at the total eclipse of May 29, 1919', *Memoirs of the Royal Astronomical Society*, vol. 32, 1919, pp. 291–333.
51. Charles Alcock to Ken Freeman, 20 July 1989; Dave Bennett to Ken Freeman, 2 August 1989.
52. 'Memorandum of Understanding—Microlensing Survey', signed for the ANU by Alan D. Barton, Treasurer, on 22 November 1990.
53. J. Hart et al., 'The telescope system of the MACHO Program', *PASP*, vol. 108, 1996, pp. 220–2; *MSOAR*, 1990, pp. 1–2; 1992, p. 7.
54. Mario Mateo, 'Searching for dark matter', *S&T*, vol. 87, no. 1, 1994, pp. 20–4.
55. C. Alcock et al., 'Possible gravitational microlensing of a star in the Large Magellanic Cloud', *Nature*, vol. 365, 1993, pp. 621–3 and cover.
56. E. Aubourg et al., 'Evidence for gravitational microlensing by dark objects in the Galactic halo', *Nature*, vol. 365, 1993, pp. 623–5; A. Udalski et al., 'The optical gravitational lensing experiment. Discovery of the first candidate microlensing event in the direction of the Galactic bulge,' *Acta Astron.*, vol. 43, 1993, pp. 289–94.
57. Events seen towards the Galactic centre are probably caused by normal stars in the Galaxy; they have no bearing on the dark matter problem.
58. C. Afonso et al., 'Combined analysis of the binary lens caustic-crossing event MACHO 98–SMC-1', *ApJ*, vol. 532, 2000, pp. 340–52.
59. C. Alcock et al., 'The MACHO Project: microlensing results from 5.7 years of Large Magellanic Cloud observations', *ApJ*, vol. 542, 2000, pp. 281–307.
60. *MSOAR*, 1999, p. 32; 2000, p. 31.
61. *MSOAR*, 2001, pp. 29–30.

62. *MSOAR*, 1989, p. 2.
63. The Observatory was advised of this by a letter (15 January 1993) from Scott & Furphy Pty Ltd, acting for the ACT Planning Authority.
64. The high level of public support it received pleasantly surprised the Observatory. *The Canberra Times*, for instance, said in an editorial (13 August 1993, p. 10) that 'the importance of Stromlo as a centre of intellectual excellence for Canberra should take precedence over a housing estate'—even though the newspaper's usual stance was strongly in favour of commercial development in the national capital.
65. Media Release (18 February 1994) by the ACT Minister for the Environment, Land and Planning, Mr Bill Wood MLA; *MSOAR*, 1994, p. 8.
66. Stan Woosley and Tom Weaver, 'The great supernova of 1987', *Scientific American*, vol. 261, no. 2, 1989, pp. 24–32.
67. David H. Clark and F. Richard Stephenson, *The Historical Supernovae*, Pergamon Press, Oxford, 1977. Karl Kamper and Sidney van den Bergh, 'Cassiopeia A—an unseen supernova', *S&T*, vol. 51, 1976, pp. 236–9; William B. Ashworth Jr, 'A probable Flamsteed observation of the Cassiopeia A supernova', *J. Hist. Astron.*, vol. 11, 1980, pp. 1–9.
68. Ronald A. Schorn, 'Neutrinos from hell', *S&T*, vol. 73, 1987, pp. 477–9; James M. Lattimer and Adam S. Burrows, 'Neutrinos from Supernova 1987A', *S&T*, vol. 76, 1988, pp. 348–50.
69. *MSOAR*, 1987, p. 19.
70. M.A. Dopita et al., '250 days of SN 1987A: *UBVRI* photometry at MSSSO', *AJ*, vol. 95, 1988, pp. 1717–23.
71. Peter J. McGregor, 'Observations of SN 1987A in the near-infrared', *ProcASA*, vol. 7, 1988, pp. 450–61.
72. P.R. Wood, 'The progenitor star and its evolution', *ProcASA*, vol. 7, 1988, pp. 386–9.
73. P.R. Wood et al., 'Speckle interferometry of SN 1987A up to one year after explosion', *ApJ*, vol. 339, 1989, pp. 1073–7; P.E.J. Nulsen et al., 'Speckle interferometry of SN 1987A: final measurements', *ApJ*, vol. 358, 1990, pp. 266–7.
74. The Elizabeth and Frederick White Research Conference on Supernova 1987A, K.M. Proust and W.J. Couch (eds), *ProcASA*, vol. 7, 1988, pp. 343–562.
75. Michael A. Dopita, 'SN 1987A: an overview', *ProcASA*, vol. 7, 1988, pp. 344–51.
76. For the significance of the term 'heavy elements', see Ch. 8, note 13.

77. This work led to the publication of over a dozen papers, all cited in the conference review: M.A. Dopita, 'The evolution of the planetary nebulae in the Magellanic Clouds and the Galactic Bulge', IAU Symposium No. 155, *Planetary Nebulae*, R. Weinberger and A. Acker (eds), Kluwer Academic Press, Dordrecht, 1993, pp. 433–41.
78. M.A. Dopita and S.J. Meatheringham, 'Photoionization modelling of Magellanic Cloud planetary nebulae', *ApJ*, vol. 367, 1991, pp. 115–25; vol. 377, 1991, pp. 480–91. E. Vassiliadis and P.R. Wood, 'Evolution of low- and intermediate-mass stars to the end of the asymptotic giant branch with mass loss', *ApJ*, vol. 413, 1993, pp. 641–57; 'Post-asymptotic giant branch evolution of low- to intermediate-mass stars', *ApJS*, vol. 92, 1994, pp. 125–44.
79. Michael A. Dopita et al., '*Hubble Space Telescope* observations of planetary nebulae in the Magellanic Clouds. I. the extreme type I SMP 83/WS 35', *ApJ*, vol. 418, 1993, pp. 804–11; Michael A. Dopita, et al., '*Hubble Space Telescope* observations of planetary nebulae in the Magellanic Clouds. II. SMP 85, a young planetary', *ApJ*, vol. 426, 1994, pp. 150–8.
80. *MSOAR*, 1993, p. 9.
81. G.S. Da Costa et al., 'The dwarf spheroidal companions to M31: WFPC2 observations of Andromera I', *AJ*, vol. 112, 1996, pp. 2276–95; 'WFPC2 observations of Andromera II', *AJ*, vol. 119, 2000, pp. 705–26.
82. John Bally, Ralph S. Sutherland et al., 'Externally illuminated young stellar environments in the Orion Nebula: *Hubble Space Telescope* Planetary Camera and ultraviolet observations', *AJ*, vol. 116, 1998, pp. 293–321.
83. Kris Davidson et al., 'An unusual brightening of η Carinae', *AJ*, vol. 118, 1999, pp. 1777–83.
84. Mark B. Vincent et al., 'Mapping Jupiter's latitudinal bands and Great Red Spot using HST/WFPC2 far-ultraviolet imaging', *Icarus*, vol. 143, 2000, pp. 189–204; Mark B. Vincent et al., 'Jupiter's polar regions in the ultraviolet as imaged by HST/WFPC2: auroral-aligned features and zonal motions', *Icarus*, vol. 143, 2000, pp. 205–22; Nahum Arav, Martijn de Kool et al., '*HST* STIS observations of PG 0946+301: the highest quality UV spectrum of a BALQSO', *ApJ*, vol. 561, 2001, pp. 118–30.
85. See Ch. 8, note 29.
86. See Ch. 5, note 18.
87. Matthew Malkan, 'Active galaxies: overview', in Paul Murdin (ed.), *Encyclopedia of Astronomy and Astrophysics*, Nature Publishing Group, London, 2001, pp. 26–31.
88. Bradley M. Peterson and Belinda J. Wilkes, 'Active galaxies: unified model', pp. 31–5.

89. Two such were F.J. Kerr and A.W. Rodgers (eds), *The Galaxy and the Magellanic Clouds*, IAU Symposium No. 20, Australian Academy of Science, Canberra, 1964; P. Ledoux, A. Noels and A.W. Rodgers (eds), *Stellar Instability and Evolution*, IAU Symposium No. 59, D. Reidel Publishing Company, Dordrecht, 1974.
90. *MSOAR*, 1993, p. 6.
91. G.V. Bicknell, M.A. Dopita and P.J. Quinn (eds), *The First Stromlo Symposium: The Physics of Active Galaxies*, ASP Conference Series, vol. 54, Astronomical Society of the Pacific, San Francisco, 1994.
92. M. Arnaboldi, G.S. Da Costa and P. Saha (eds), *The Second Stromlo Symposium: The Nature of Elliptical Galaxies*, ASP Conference Series, vol. 116, Astronomical Society of the Pacific, San Francisco, 1997; B.K. Gibson, T.S. Axelrod and M.E. Putman (eds), *The Third Stromlo Symposium: The Galactic Halo*, ASP Conference Series, vol. 165, Astronomical Society of the Pacific, San Francisco, 1999. This conference was held as a memorial to Alex Rodgers. *The Fourth Stromlo Symposium: Planetary Nebulae: Their Role in the Universe*, IAU Symposium No. 209, in press.
93. *MSOAR*, 1999, pp. 21–4; 2000, pp. 18–19.
94. No Mount Stromlo Director has ever *retired* in office. Duffield died while Director; Woolley, Bok, Eggen and Mould resigned to take up appointments overseas. Mathewson and Rodgers resumed research work on the Stromlo staff at the conclusion of their terms.
95. C. Alcock et al., 'The MACHO Project LMC variable star inventory. II. LMC RR Lyrae stars—pulsational characteristics and indications of a global youth of the LMC', *AJ*, vol. 111, 1996, pp. 1146–55.
96. P.R. Wood et al., 'MACHO observations of LMC red giants: Mira and semi-regular pulsators, and contact and semi-detached binaries', in T. Le Bertre, A. Lèbre and C. Waelkens (eds), IAU Symposium No. 191, *Asymptotic Giant Branch Stars*, 1999, pp. 151–8.
97. As Minister for Post-War Reconstruction, Dedman had the carriage of the 1946 Act of Parliament that established the ANU (see Ch. 6, notes 20 and 21).
98. Letter from Alex Rodgers to Peter Baume, 28 August 1997. Reply, 23 September 1997.
99. Bruce Juddery, 'Obituary: a MACHO man who looked to the stars', *The Canberra Times*, 22 October 1997.
100. *MSOAR*, 1998, pp. 35–6.

CHAPTER 11

1. The Extragalactic Distance Scale was designated a Key Project in 1984. The other two were: (i) that on Quasar Absorption Lines; and (ii) the HST Medium-Deep Survey.
2. Marc Aaronson and Jeremy Mould, 'The distance scale: present status and future prospects', *ApJ*, vol. 303, 1986, pp. 1–9; Pierce Prize Lecture, American Astronomical Society.
3. The other two co-Principal Investigators were Wendy Freedman (of the Observatories, Carnegie Institution of Washington, Pasadena) and Robert Kennicutt (of the University of Arizona). An invited talk given by them to the American Astronomical Society in 1995 has been published as Robert C. Kennicutt Jr, Wendy L. Freedman and Jeremy R. Mould, 'Measuring the Hubble Constant with the Hubble Space Telescope', *AJ*, vol. 110, 1995, pp. 1476–91.
4. See Ch. 8, note 31.
5. Strictly speaking, what we are measuring is the value of Hubble's Constant *at the present epoch* in the history of the Universe. The subscript zero in the H_0 symbol is there to remind us of that. Naming it Hubble's *Constant* was somewhat unfortunate, since it can vary with time. Some writers call it Hubble's *Parameter* instead.
6. The *parsec* is the unit of distance used routinely by astronomers, although in this book we have generally adopted the more easily envisaged *light year*. We introduce the parsec in this section because, by convention, Hubble's Constant is always quoted in units of kilometres/second/megaparsec. One parsec is 3.2616 light years (or 30,857 thousand million kilometres). A megaparsec is a million parsecs.
7. When Hubble first discovered his relationship, he obtained a value of about 500 for H_0—seven times larger than the most recent (HST Key Project) value.
8. The expressions 'flat', 'open' and 'closed' refer to the *geometry* of space–time. In a Universe with flat geometry a two-dimensional slice resembles a sheet of paper (extended so as to be infinite and unbounded). In one with closed geometry the slice resembles the surface of an orange (*finite*, but still unbounded). In one with open geometry the slice resembles the surface of a saddle (infinite and unbounded). The statements made in this paragraph relating the *ultimate fate* of the Universe to its *geometry* are true for the class of cosmological models that were in vogue

when the HST Key Project commenced. By the time it concluded, however, cosmologists felt compelled to adopt another, more complex approach, as will soon be described.

9. *MSOAR*, 1994, pp. 2, 36.
10. Wendy L. Freedman et al., 'Distance to the Virgo Cluster galaxy M100 from Hubble Space Telescope observations of Cepheids', *Nature*, vol. 371, 1994, pp. 757–62; Laura Ferrarese et al., 'The extragalactic distance scale key project. IV. The discovery of Cepheids and a new distance to M100 using the Hubble Space Telescope', *ApJ*, vol. 464, 1996, pp. 568–99.
11. Jeremy Mould et al., 'Interim report on calibration of the Tully–Fisher relation in the HST Key Project to measure H_0', in M. Livio, M. Donahue and N. Panagia (eds), *The Extragalactic Distance Scale*, Cambridge University Press, Cambridge, 1997, pp. 158–70.
12. Don A. VandenBerg, Michael Bolte and Peter B. Stetson, 'The age of the Galactic globular cluster system', *ARAA*, vol. 34, 1996, pp. 461–510.
13. Wendy L. Freedman et al., 'Final results from the *Hubble Space Telescope* Key Project to measure the Hubble Constant', *ApJ*, vol. 553, 2001, pp. 47–72.
14. S.C.B. Gascoigne and G.E. Kron, 'Photoelectric observations of Magellanic Cloud Cepheids', *MN*, vol. 130, 1965, pp. 333–60; S.C.B. Gascoigne, 'Further observations of Magellanic Cloud Cepheids', *MN*, vol. 146, 1969, pp. 1–36.
15. Kim M. Sebo, Daya Rawson et al., 'The Cepheid period–luminosity relation in the Large Magellanic Cloud', *ApJS*, vol. 142, 2002, pp. 71–8.
16. Brian Chaboyer et al., 'The age of globular clusters in light of *Hipparcos*: resolving the age problem?', *ApJ*, vol. 494, 1998, pp. 96–110.
17. Jeremy Mould, 'The age of the globular clusters', *Nature*, vol. 395, 1998, pp. A20–2.
18. Astronomers use the symbol '*z*' to denote the redshift; thus 'high-*z*' indicates large redshift. For a popular account of the High-*z* Supernova Search see Craig J. Hogan, Robert P. Kirshner and Nicholas B. Suntzeff, 'Surveying space–time with supernovae', *Scientific American*, vol. 280, no. 1, 1999, pp. 46–51.
19. S. Perlmutter et al., 'A supernova at $z = 0.458$ and implications for measuring the cosmological deceleration', *ApJ*, vol. 440, 1995, pp. L41–4; S. Perlmutter et al., 'Measurements of the cosmological parameters Ω and Λ from the first seven supernovae at $z \geqslant 0.35$', *ApJ*, vol. 483, 1997, pp. 565–81.

20. *MSOAR*, 1996, p. 47; 'Beyond the Hubble Constant', ESO Press Release 09/95, 17 August 1995.
21. The precise relationship between distance and redshift depends on the cosmological model used to describe the real Universe. Thus statements purporting to quote *distances in light years* beg the question somewhat; they should be viewed with caution, especially for more remote objects. On the other hand, it is perfectly safe to use an observed redshift to make statements about *the size of the Universe when the light from the galaxy concerned was emitted, relative to its present size*. We adopt this approach for SN 1997ff in the last sentence of this paragraph, and thereafter.
22. This inflationary episode lasted from when the Universe was about 10^{-35} seconds old until about 10^{-32} seconds.
23. Guth and Steinhardt, 'The inflationary Universe'.
24. Adam G. Riess et al., 'Observational evidence from supernovae for an accelerating Universe and a Cosmological Constant', *AJ*, vol. 116, 1998, pp. 1009–38.
25. S. Perlmutter et al., 'Measurements of Ω and Λ from 42 high-redshift supernovae', *ApJ*, vol. 517, 1999, pp. 565–86.
26. James Glanz, 'Breakthrough of the year: cosmic motion revealed', *Science*, vol. 282, 1998, pp. 2156–7.
27. Two excellent descriptions of the implications of a non-zero Cosmological Constant are Lawrence M. Krauss, 'Cosmological antigravity', *Scientific American*, vol. 280, no. 1, 1999, pp. 52–9; Sean M. Carroll, 'Cosmological Constant', in Paul Murdin (ed.), *Encyclopedia of Astronomy and Astrophysics*, Nature Publishing Group, London, 2001, pp. 547–50.
28. The cosmological term could also introduce an *attractive* rather than a *repulsive* force—the Cosmological Constant Λ can be *negative* rather than *positive*. In the context of the SN Ia results, however, it is the *repulsive* alternative that is indicated.
29. Even this is an over-simplification. What we have called 'matter-energy' must be further subdivided into a component related to 'baryonic matter', and another to 'cold dark matter'. Baryonic matter comprises all the ordinary material we see about us in the Universe—in the planets, stars, nebulae and so on, including objects which are so faint that they are hard to detect. Cold dark matter is a name given to some mysterious *additional* material known to exist from several lines of evidence, but which currently remains unidentified. It may consist of one or more types of unknown particle. (See the discussion of dark matter in Chapter 10.)

30. Adam G. Riess et al., 'The farthest known supernova: support for an accelerating Universe and a glimpse of the epoch of deceleration', *ApJ*, vol. 560, 2001, pp. 49–71.
31. Guth and Steinhardt, 'The inflationary Universe'.
32. See Ch. 10, note 46.
33. G.F. Smoot et al., 'Structure in the COBE differential microwave radiometer first-year maps', *ApJ*, vol. 396, 1992, pp. L1–5; Richard Tresch Fienberg, 'COBE confronts the Big Bang', *S&T*, vol. 84, no. 1, 1992, pp. 34–5.
34. *MSOAR*, 1997, p. 40; 1998, p. 30; 1999, pp. 42–3; 2000, pp. 12–15; 2001, pp. 36–7. Govert Schilling, 'Cosmology's treasure map: the 2dF Galaxy Redshift Survey', *S&T*, vol. 105, no. 2, 2003, pp. 32–9. Matthew Colless et al., 'The 2dF Galaxy Redshift Survey: spectra and redshifts', *MN*, vol. 328, 2001, pp. 1039–63.
35. I.J. Lewis et al., 'The Anglo-Australian Observatory 2dF facility', *MN*, vol. 333, 2002, pp. 279–98.
36. Colless, 'The 2dF Galaxy Redshift Survey'.
37. G. Efstathiou et al., 'Evidence for a non-zero Λ and a low matter density from a combined analysis of the 2dF Galaxy Redshift Survey and cosmic microwave background anisotropies', *MN*, vol. 330, 2002, L29–35.
38. Yuzuru Yoshii and Bruce A. Peterson, 'Interpretation of faint galaxy number counts in the K band', *ApJ*, vol. 444, 1995, pp. 15–20.
39. Efstathiou, 'Evidence for a non-zero Λ', p. L30.
40. Peter J. McGregor, Philip D. Nicholson and Marek G. Allen, 'CASPIR observations of the collision of Comet Shoemaker-Levy 9 with Jupiter', *Icarus*, vol. 21, 1996, pp. 361–88.
41. Marvin Kleine et al., 'The carbon isotope abundance ratio in comet Halley', *ApJ*, vol. 439, 1995, pp. 1021–33.
42. P.R. Wood, 'Final stages of AGB evolution', IAU Symposium No. 180, *Planetary Nebulae*, H.J. Habing and G.L.M. Lamers (eds), Kluwer Academic Publishers, Dordrecht, 1997, pp. 297–302; P.R. Wood, 'Properties of Cepheids and long-period variables', IAU Symposium No. 189, *Fundamental Stellar Properties: The Interaction between Observation and Theory*, T.R. Bedding, A.J. Booth and J. Davis (eds), Kluwer Academic Publishers, Dordrecht, 1997, pp. 299–304; P.R. Wood, 'Convection-induced oscillatory thermal modes in red giants: a new type of stellar oscillation', IAU Colloquium No. 176, *The Impact of Large-scale Surveys on Pulsating Star Research*, L. Szabados and D. Kurtz (eds), Astronomical Society of the Pacific, San Francisco, 2000, p. 379–80.

43. J.F. Bell, 'Binary and millisecond radio pulsars', ANU PhD thesis, 1995; B.W. Stappers, 'Observations of eclipsing and binary pulsars', ANU PhD thesis, 1997.
44. *MSOAR*, 2000, pp. 25–6. For the significance of the term 'metal', see Ch. 8, note 13.
45. N. Christlieb, M.S. Bessell et al., 'A stellar relic from the early Milky Way', *Nature*, vol. 419, 2002, pp. 904–6.
46. See Ch. 8, note 24.
47. John E. Norris and G.S. Da Costa, 'The giant branch of ω Centauri. IV. Abundance patterns based on echelle spectra of 40 red giants', *ApJ*, vol. 447, 1995, pp. 680–705; John E. Norris, K.C. Freeman and K.J. Mighell, 'The giant branch of ω Centauri. V. The calcium abundance distribution', *ApJ*, vol. 462, 1996, pp. 241–54.
48. F. van Leeuwen et al., 'A proper motion study of the globular cluster ω Centauri', *A&A*, vol. 360, 2000, pp. 472–98.
49. Sylvie F. Beaulieu et al., 'Dynamics of the Galactic bulge using planetary nebulae', *AJ*, vol. 120, 2000, pp. 855–71; Maartje N. Sevenster and Agris J. Kalnajs, 'An inner ring and the microlensing toward the bulge', *AJ*, vol. 122, 2001, pp. 885–94; R. Fux, 'Gas dynamics in the Galactic bar region from n-body and SPH simulations', *ApSS*, vol. 276, 2001, pp. 367–74; R. Fux, 'Order and chaos in the local disc stellar kinematics induced by the Galactic bar', *A&A*, vol. 373, 2001, pp. 511–35.
50. H. Jerjen, A. Kalnajs and B. Binggeli, 'IC 3328: a "dwarf elliptical galaxy" with spiral structure', *A&A*, vol. 358, 2000, pp. 845–9.
51. R.H. Méndez et al., 'Detection, photometry, and slitless radial velocities of 535 planetary nebulae in the flattened elliptical galaxy NGC 4697', *ApJ*, vol. 563, 2001, pp. 135–50.
52. *MSOAR*, 1998, p. 32; 2001, p. 41.
53. Sean G. Ryan, John E. Norris and Timothy C. Beers, 'The Spite lithium plateau: ultrathin but postprimordial', *ApJ*, vol. 523, 1999, pp. 654–77; Sean G. Ryan et al., 'Primordial lithium and Big Bang nucleosynthesis', *ApJ*, vol. 530, 2000, pp. L57–L60.
54. The ARC had assumed a major science policy role at the time, and was actively encouraging and facilitating such discipline reviews. The review report appeared as *Australian Astronomy: Beyond 2000*, National Board of Employment, Education and Training, Australian Research Council, Discipline Research Strategies, AGPS, Canberra, June 1995.
55. *Australian Astronomy: Beyond 2000*, pp. 71–2.

56. ibid., Executive Summary, pp. x–xi.
57. For a description of the European Southern Observatory see Ch. 10, note 19.
58. Department of Industry, Science and Technology, *Annual Report 1994–95*, Australia, Parliamentary Papers, 1995, vol. 18, Paper 232, Ch. 8, 'Science and Technology', pp. 134–5, Major National Research Facilities.
59. Gavin Gilchrist, 'How Australia blew the best job in the universe', *Sydney Morning Herald*, 8 December 1995, p. 1; David Swinbanks, 'Australia backs innovation, shuns telescope', *Nature*, vol. 378, 1995, pp. 653.
60. Peter Pockley, 'Votes nudge Australia closer to joining telescope project', *Nature*, vol. 379, 1996, p. 668.
61. Julian Cribb, 'Europe deal a quantum leap for astronomy', *The Australian*, 15 February 1996, p. 2.
62. The new government's arrangements also placed direct responsibility for science matters in the hands of a minister without cabinet rank. Peter Pockley, 'Science loses cabinet seat after Australian election. Telescope prospects looking dimmer', *Nature*, vol. 380, 1996, p. 192; 'Savings drive unsettles Australian science', *Nature*, vol. 381, 1996, p. 5.
63. Peter McGauran to Director General, ESO, 23 May 1996.
64. Peter Pockley, 'Australian astronomers are rebuffed in move to apply for ESO membership', *Nature*, vol. 381, 1996, pp. 100–1.
65. European Southern Observatory, *Annual Report*, 1997, p. 65.
66. For a description of the Gemini Project see Ch. 10, note 20; Peter Pockley, 'Australia eyes role in Gemini telescopes', *Nature*, vol. 389, 1997, p. 220.
67. 'Life in outer space?', Media Release K8/98, Minister for Education, Training and Youth Affairs, 18 February 1998; Peter Pockley, 'Seventh heaven for Australia's optical astronomers', *Nature*, vol. 391, 1998, p. 724. The total Australian commitment to Gemini (capital and recurrent) was $17.7 million over the first five years, about 70 per cent of the corresponding cost of joining ESO.
68. *MSOAR*, 1998, p. 13.
69. *MSOAR*, 1999, pp. 55–6; 2000, pp. 15–17; Peter J. McGregor et al., 'Near-infrared Integral-Field Spectrograph (NIFS): an instrument proposed for Gemini', *Publ. Astron. Soc. Aust.*, vol. 16, 1999, pp. 273–87.
70. *MSOAR*, 1995, p. 31; 1996, p. 53; 1999, p. 54.
71. 'ANU wins competition to build $6.3 M sharp-eyed imager', ANU media release, 2 December 2002; Stacey Lucas, 'ANU's $6.3m contract out of this world', *Canberra Times*, 3 December 2002, p. 1.

72. *MSOAR*, 1996, p. 11; 1997, p. 42; 1998, p. 31; 1999, p. 53; 2000, pp. 45–6.
73. *MSOAR*, 1995, p. 33; 1996, p. 50.
74. *MSOAR*, 1996, p. 17; 1997, p. 17; 1998, p. 17.
75. *MSOAR*, 1994, p. 3.
76. *MSOAR*, 1996, pp. 15, 98–9; 1997, pp. 98–100.
77. *MSOAR*, 1995, pp. 4–5; 1996, pp. 9, 16–17; 1997, pp. 10, 42–4.
78. *MSOAR*, 1999, p. 15; 2000, p. 18.
79. Michael Burton (ed.), 'The scientific potential for astronomy from the Antarctic Plateau', *ProcASA*, vol. 11, 1994, pp. 127–50.
80. *MSOAR*, 1994, pp. 5–6; 1995, p. 31; 1997, pp. 11–12; 1998, pp. 32–3; Michael G. Burton et al., 'JACARA's plans', *Publ. Astron. Soc. Aust.*, vol. 13, 1996, pp. 33–4.
81. Michael A. Dopita, Peter R. Wood and Gary R. Hovey, 'An automated DIMM telescope for Antarctica', *Publ. Astron. Soc. Aust.*, vol. 13, 1996, pp. 39–43.
82. *MSOAR*, 2001, p. 42.
83. *MSOAR*, 1997, p. 18.
84. Simon Grose, 'British students get to see southern skies', *Canberra Times*, 23 January 2002, p. 4.
85. *MSOAR*, 2000, pp. 21–2.
86. *MSOAR*, 2001, p. 21.
87. ibid.
88. *MSOAR*, 2001, pp. 12–13.
89. *MSOAR*, 2001, pp. 21–2.
90. A tale recounted by Professor Mike Seaton, of University College London, in the early 1960s.
91. *MSOAR*, 2001, pp. 11–12 and front cover.
92. Research School of Astronomy and Astrophysics data: The staff list in the 2001 *Annual Report* contains the names of 24 astronomers. The annual budget for 2001 was $8.9 million. Australia-wide statistics are from *Australian Science and Technology at a Glance 2002*, Department of Education, Science and Training, Science and Technology Policy Branch. They are for 1998–99, the most recent year with data available. Page 2: Government and higher education R&D funding was $4,610 million (in constant 2001–02 dollars); p. 10: there were 46,998 research scientists and engineers employed in the government and higher education sectors (excluding postgraduate and research students).

93. Other graduate schools which were awarded five or more Hubble Fellowships during 1990–2002 were: UC Berkeley (12), Harvard (11), Caltech (10), Arizona (9), Cambridge (6), Chicago (6), UC Santa Cruz (5), Leiden (5), Texas (5) and Yale (5).
94. *MSOAR*, 1999, p. 19.

Epilogue

1. Megan Doherty, 'Stanhope helps save pilot', *Canberra Times*, 14 January 2003, p. 1.
2. *Canberra Times*, 18 January 2003, p. 1.
3. See Ch. 11, note 69.
4. The Honourable Peter McGauran, MP, Minister for Science, 'Mount Stromlo Tribute from PM', media release, MIN 122/03, 4 February 2003.

Index